深入浅出智能物联网 OpenWrt 操作系统

跟 hoowa 学 AIoT

孙冰 郑淇文 李兴仁 著

人民邮电出版社

北京

图书在版编目（CIP）数据

深入浅出智能物联网OpenWrt操作系统：跟hoowa学AIoT / 孙冰，郑淇文，李兴仁著. -- 北京：人民邮电出版社，2022.8(2023.12重印)

（i创客）

ISBN 978-7-115-58623-0

Ⅰ．①深… Ⅱ．①孙… ②郑… ③李… Ⅲ．①物联网－操作系统－程序设计 Ⅳ．①TP393.4②TP18

中国版本图书馆CIP数据核字(2022)第023156号

内 容 提 要

互联网的下一步是物联网，又称为"万物互联"。人和人、人和物、物和物之间的连接在信息革命中不断演进，其中长盛不衰的技术就是网络和路由。MIPS指令集在过去的30多年中持续创新，基于MIPS指令集芯片的出货速度持续增长，然而直到2018年4月，来自上海的SF16A18芯片出现，基于MIPS的路由器芯片才实现了国产化。OpenWrt则是路由产品必备的操作系统。

自主、可控、开源正是未来掌握信息技术产业的密钥，本书即围绕自主的SF16A18芯片、开源的OpenWrt操作系统展开，让你能够把握万物互联时代的脉络和先机。你将学习到的知识图谱涉及目前行业中流行的网络设备使用的操作系统，涵盖路由器、交换机、物联网控制器、私有云服务器、边缘计算等多个领域。本书从最易开始的使用环节循序渐进地涵盖应用开发、系统交叉编译、软件包制作，直指系统与硬件搭配的核心环节——启动流程、ubus、Netifd与Hotplug，最终涉及ZigBee物联网、工业物联网开发技术。

本书既能满足OpenWrt开发爱好者学习与高校教学需求，又适合作为路由产品开发人员的参考资料。

♦ 著　　孙　冰　郑淇文　李兴仁

责任编辑　周　明

责任印制　马振武

♦ 人民邮电出版社出版发行　北京市丰台区成寿寺路11号

邮编　100164　电子邮件　315@ptpress.com.cn

网址　https://www.ptpress.com.cn

北京盛通印刷股份有限公司印刷

♦ 开本：787×1092　1/16

印张：28.25　　　　　　　　　2022年8月第1版

字数：758千字　　　　　　　　2023年12月北京第3次印刷

定价：119.80元

读者服务热线：(010)81055493　印装质量热线：(010)81055316
反盗版热线：(010)81055315
广告经营许可证：京东市监广登字20170147号

序

2014年，我还在工业和信息化部软件与集成电路促进中心（CSIP）从事开源项目扶持和产业生态工作时，认识了活跃在MIPS领域的孙冰，也就是hoowa，他当时刚刚写完一本名叫《OpenWrt智能路由系统开发——跟hoowa学智能路由》的技术书籍。当年也是"智能硬件"概念爆发的一年，随着大厂争夺移动互联网入口，小米、百度、360等互联网公司，用低价甚至免费的硬件获取用户，再以硬件为基础搭建自己的生态系统，智能路由器一度被解读为智能客厅的入口。

再后来，我到鹏城实验室负责新一代人工智能产业技术战略联盟OpenI开源社区的筹划和具体实施，与华为、小米、浪潮等硬件公司及百度、旷视、商汤、腾讯、讯飞等偏AI的网络公司或软件公司都保持着频繁的技术交流，同时与Linaro、Google、Facebook、Linux基金会、伯克利RISC-V、MIT、CMU等国外公司和组织也有不少的技术交流。在这个过程中，我持续地感受到中国IT基础技术能力在这些年间得到迅速的提升，同时也能清楚地看到我们与美国的信息技术发展水平依然有着很大的差距。

自主创新无论在信息安全还是产业的可持续发展上，都起着无法替代的关键作用。当我们从一个比较低的起点开始加速奔跑，去追赶产业上游的一个个巨人时，开源也许是最有力的手段之一。互联网发展至今已有50多年，开源也发展了50多年，从1969年贝尔实验室UNIX脱胎出来的Linux到今年鹏城实验室带头发起OpenI和华为开源鸿蒙，我们的技术人员在开源社区学习、实践、积累并发展出了比较成体系的、自主可控的IT基础能力。当然，我们在这个领域还有一个明显的短板，就是通用芯片设计能力。龙芯、寒武纪、华为等公司在芯片技术上投入了大量研发力量，我们也欣喜地看到这些公司的芯片工艺、性能和稳定性都在快速接近美国同行。OpenI在2019年也推出了一个全开源的芯片项目——海藻。

很多数据和实例的支持，让我们对中国在不远的未来在通用芯片这个产业链上游的高精尖领域有机会跻身顶端有信心。然而对RISC-V、MIPS、ARM等这些比较开放的硬件体系结构来说，我们能够提供的底层软件的支持能力还是很薄弱的，如果我们不单单是做课题研究，而是要从产业角度全面追赶上国际一流水平，就必须借助开源的力量构造完整的生态系统，如IP核、芯片、驱动、

IDE、开发板、文档资料、社区、应用样例等。因此我觉得 hoowa 的这本书具有很高的产业生态价值，对于知识产权分享更友好的 MIPS 和 RISC-V 硬件体系结构来说，基础软件层面的支持能力，特别是社区支持能力往往决定了产品在市场上的成败，hoowa 和矽昌公司以教程加开源社区的方式建立开发者生态是一次非常有益的尝试，我非常期待 hoowa 和矽昌公司的物联网社区生态能以此为开端建立起来，同时我们也在计划借鉴这次经验，将鸿蒙操作系统、翼辉实时操作系统等国内其他优秀的自主硬核底层开源软件和开源硬件项目引入进来，写出国产底层技术开源项目的系列丛书，建立一个中国自主硬核技术根目录。

刘明

2021 年 12 月 25 日

前言

本书全部内容都基于矽昌团队的首颗国产边缘计算网络芯片 SF16A18 来完成，内容分为三篇。

第一篇：主要介绍关于系统、芯片、指令集的历史。

第二篇：介绍 OpenWrt 系统的特点以及如何使用 SF16A18 这颗芯片完成基本的网络和操作系统功能。

第三篇：介绍 OpenWrt 系统的深入内容，包括基于该芯片的各种硬件接口的实验，以及操作系统的启动原理和定制技术。

FreeIRIS 团队基于本书要求开发了一款开源电路板，该电路板的设计完全开源，而 OpenWrt 操作系统部分则由矽昌团队提供。

参与本书编写的人员有（排名不分先后）：孙冰（hoowa）、郑淇文、李兴仁博士（SF16A18 总设计师）、何铮、王明海、王胜、陈进东。没有大家一起合作，我们很难完成本书。

参与本书修正及对本书编写进行指导的公司有（排名不分先后）：青岛矽昌通信技术有限公司、杭州精云智能科技有限公司、北京我爱帮科技有限公司、成都极企科技有限公司、杭州天进科技有限公司，专家有杨凯、谢晓秋。

由于读者基础知识各异，本书从 OpenWrt 的系统命令使用入手开始，高手请耐心查阅。我们采用了循序渐进的方法进行描述，逐渐增加知识点深度。读者必须具备一定的 Linux 系统操作知识，如果同时具备 C 语言编程、TCP/IP 通信知识则更佳。

本书能让希望了解和学习国产路由芯片方案的朋友顺利地拿到整套的方案资料；能帮助读者朋友快速学习，更好地掌握相关知识；能提供硬件和软件环境以实际操作练习，为国产路由芯片产业链培养更多优秀的人才。

学生朋友通过本书能了解到 SF16A18 芯片及方案，通过配套的开发板和 SDK 可以开发一些新功能或新产品，将其作为毕业设计也是不错的选择，相信这对毕业后找工作也是非常有价值的。

我们希望更多的潜在合作伙伴通过本书了解 SF16A18 芯片，同时也希望更多志同道合的合作伙伴一起为国产路由器芯片产业的发展做贡献，早日实现路由器芯片的国产替代，为普通消费者和国家的网络安全贡献力量。

读者对象

- 计算机、电子、信息、通信相关专业学生
- 智能设备、智能网关、物联网从业人员
- OpenWrt 爱好者
- 硬件设计厂商人才

孙冰

2021 年 11 月

目录

第一篇 让我们开始吧 1

1 从芯片开始 2
1.1 OpenWrt 系统介绍 3
1.2 MIPS 处理器体系结构 4
1.3 SF16A18 芯片 11
1.4 本书背景介绍 12
1.5 AIoT 的技术应用 13
1.6 表达约束 17

第二篇 SF16A18 芯片的 OpenWrt 系统 18

2 环境与工具准备 19
2.1 SF16A18 芯片的规格 19
2.2 DF1A 开发板介绍 21
2.3 U-Boot 网页刷机 23
2.4 TTL 串口调试 25
2.5 SSH 远程登录 30
2.6 SCP 文件传输 33

3 分区与软件包 37
3.1 SPI Flash 分区原理 37
3.2 文件系统与透明挂载 43
3.3 OPKG 软件包管理 55

4 UCI 统一配置 62
4.1 UCI 介绍 62
4.2 UCI 的配置文件 62
4.3 UCI 配置文件语法 63
4.4 UCI 命令行接口 65

· I ·

CONTENTS

    4.5 UCI 的 Lua 接口 71

5 网络配置 82

    5.1 配置文件 82

    5.2 WAN 口配置 88

    5.3 LAN 口配置 90

    5.4 配置无线网络 92

    5.5 DHCP 服务 100

    5.6 如何连接外网 103

6 服务功能 104

    6.1 防火墙 104

    6.2 UPnP 与 NATPMP 116

    6.3 dropbear 远程登录 119

    6.4 系统、时钟、日志 121

    6.5 用命令刷固件 123

    6.6 域名劫持 129

    6.7 服务与常用命令 131

7 存储器扩展 137

    7.1 存储器的准备 137

    7.2 存储器的使用 141

    7.3 Windows 文件共享 146

    7.4 FTP 文件共享 149

    7.5 BT 远程下载 151

    7.6 PPTP 客户端 156

8 SF16A18 的 LuCI 界面 159

    8.1 SF16A18-LuCI 目录结构 159

    8.2 界面的简易定制 171

9 工具与命令 177

    9.1 iPerf 177

目录

  9.2 网络测试工具 190

  9.3 Wi-Fi 命令 197

第三篇 深入浅出 OpenWrt 系统 213

10 交叉编译 OpenWrt 214

  10.1 安装 VirtualBox 虚拟机 214

  10.2 准备 Ubuntu 16.04 环境 218

  10.3 编译 OpenWrt 固件 228

  10.4 U-Boot 固件编译 234

11 软件包开发 238

  11.1 软件包构建基础 238

  11.2 创建常规软件包 269

  11.3 内核软件包创建 282

12 硬件定制 294

  12.1 源代码结构 294

  12.2 定制案例 299

13 总线原理分析 317

  13.1 系统启动原理 317

  13.2 ubus 总线原理 340

  13.3 Netifd 原理 365

  13.4 Hotplug 原理 391

14 扩展与实战 405

  14.1 PHP/Python 开发环境 405

  14.2 GPIO 灯与按键控制 413

  14.3 UART-TTL 串口 420

  14.4 ZigBee 物联网通信 423

  14.5 工业物联网网关 428

第一篇 让我们开始吧

1 从芯片开始

传统路由器： 全世界近80%的路由器是中国生产的，传统的路由器厂商使用几大路由器芯片厂商提供的设计方案进行产品设计。这些芯片厂商会给路由器厂商提供资料、电路板设计原理图、软件SDK。

路由器CPU： 据统计，仅2014年，中国制造的路由器数量就达到1.8亿台，而路由器产品的平均迭代周期为3年。从2017年开始，中国制造的智能音箱、物联网网关也大量采用路由器CPU作为控制中心。如果从ADSL宽带网络开始普及的时间来算，那么路由器进入寻常家庭市场已经有18个年头，而这一重要的处理器产品在2018年之前竟然不能国产化。本书所描述的内容就是基于国产化CPU的路由器的操作系统开发方法。

智能路由器： 智能路由器是从2012年开始出现在国内市场的一种新形式的路由器产品，它具备安装软件的能力，具有主频更高的CPU与容量更大的内存，可实现类似智能手机的产品。目前小米等互联网厂商出品的智能路由器进入了大量年轻人的家中。虽然智能路由器发展过程中有很多知名的产品消失，但是整个产业的发展向全行业应用边缘计算与人工智能的目标迈出了重要的一步。

OpenWrt系统： OpenWrt系统几乎是现在路由器产品必备的操作系统，即使是芯片厂出品的SDK系统，也均基于OpenWrt优化完成。如果说几年前OpenWrt系统是智能路由器的事实性标准系统，那么如今OpenWrt就是边缘计算的事实性标准系统。

国外OpenWrt社区对OpenWrt做了一个简短而精准的说明："OpenWrt就是为嵌入式设备研发的Linux发行版"。当年Linksys开放了一款路由器的源代码，然后就有不同的黑客对这个源代码打补丁来实现不同的功能，最后就出现了针对不同市场的杂乱无章的路由器固件。

OpenWrt选择了另外一条路，从开始的那一刻起，它就采用了一种非常灵活并且开放式的方法一点点把各种软件加到系统中，这样就令全世界所有的厂商和爱好者都能加强OpenWrt的功能，而开放式体系结构也令OpenWrt支持数量繁多、体系结构不同的芯片，从x86到ARM、MIPS等。而开发者使用OpenWrt只需要通过简单的编译方法，就可以将一套操作系统编译出不同芯片的版本；把OpenWrt烧写到自己的路由器上，就拥有了一台Linux服务器。OpenWrt还支持一种称作OPKG（OpenWrt Package Management）的增强型安装技术。OPKG是OpenWrt下的一款轻量级软件包管理工具，使用起来就像CentOS下的YUM一样，用一个命令就可以将已预编译好的软件安装到系统中。

OpenWrt的优点可以总结如下。

- 与Android被特定厂商控制不同，OpenWrt完全由开源社区支持发展，全球几十万个工程师参与研发。

- OpenWrt是嵌入式设备的Linux发行版，尤其适用于各类网络产品。
- OpenWrt对硬件的需求是可伸缩的，既可以运行在20年前的低性能设备中，又可运行在如今的PC服务器中。
- OpenWrt具备几千种核心服务器软件，这些核心软件的一个特点就是稳定，这与常见死机、卡顿的Android系统有着明显的区别。

边缘计算与人工智能：这两个名词进入大众视野已有2年多时间，相信现在各年龄层的人都或多或少地了解它们的概念。越来越多的家庭电器产品具备智能化的特性或物联网的功能，而传统的路由器又具备了新的使命：达到智能化和边缘计算的要求。早期的智能产品由于技术的局限性，都需要通过互联网连接使用。我们设想一下，如果你家开门、开灯都需要连接互联网，一旦断网怎么办？也因此，市场一直向远程可控（云计算）+本地可控（边缘计算）的方向发展，这就需要这种可控的设备具备以下几个特点。

- 可以支持各种联网方式；
- 具有CPU和操作系统，可以运行常见软件；
- 具备相比单片机较强的运算能力；
- 价格要远低于手机和计算机；
- 能耗小；
- 噪声低。

而路由器作为家庭必备电器（试想你可以租没有电视机、洗衣机、冰箱的房子，你会能忍受你的手机、平板电脑都不能上网吗？）是最适合完成边缘计算这件事情的设备。未来3~5年的智能设备中心如图1-1所示。

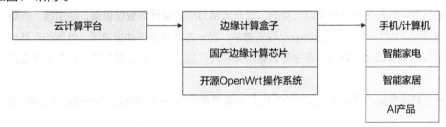

图1-1 未来3~5年的智能设备中心

1.1 OpenWrt系统介绍

OpenWrt是现在很多网络通信设备（尤其是路由器）的操作系统，这也是由维护一个完善或功能丰富的操作系统的难度决定的。这与我们目前所熟知的一些操作系统的情况有所不同。

- **Windows**：美国微软公司1985年发布的一款商业、收费的闭源操作系统，广泛应用于个人计算机产品中。
- **UNIX**：最早由Kenneth Thompson于1969年开发的服务器操作系统，目前有大量分支版本分布于大型计算机厂商，例如AIX、Solaris、HP-UX等商业系统。
- **Linux**：真正的开源操作系统，风格类似UNIX，由分布在全球的几十万工程师研发，没有任何一家公司能直接控制其发展，各公司均基于Linux衍生发行版，如Ubuntu、RedHat、Arch等。

● **Android**：基于Linux核心的操作系统，目前影响力最大的移动设备操作系统，虽然部分软件是完全开源的，但受控于一些厂商。

● **OpenWrt**：基于Linux设计的嵌入式网络设备操作系统，设计完全开源，目前没有任何一家公司直接控制其发展，各硬件厂商只是基于该系统进行功能修订来推出产品。

OpenWrt社区对OpenWrt系统做了一个介绍，原文翻译如下："OpenWrt是一个为嵌入式设备设计的Linux操作系统。它创建了整合的、固件形态的、提供完整可写的文件系统和软件包管理技术。它允许你通过软件选择配置菜单自由地自定义设备使用的软件包。对于开发者，OpenWrt系统是一个软件编译框架（可以在不重新编译固件的情况下单独编译软件包）。对于用户，它可以让你的设备使用在任何场景中。"

当年Linksys开放了一款路由器的源代码，然后就有不同的黑客对这个源代码打补丁来实现不同的功能，最后就出现了针对不同市场的杂乱无章的路由器固件。从2005年2月至今，OpenWrt已经支持了一千多种软件，并且自己将软件移植到OpenWrt中相比以前将软件移植到嵌入式Linux中更容易。OpenWrt系统高度模块化，有人说我们的产品是深度定制的OpenWrt，现在跟随我们的步伐把本书学完，人人都可以定制OpenWrt系统，人人都可以开发智能产品。

边缘计算智能设备是从2014年开始出现在国内市场的一种新形态的产品，与路由器既有相同之处，也有不同之处：相同之处在于不论智能设备如何管理，家庭采购的最便宜、最实用的24小时开机的硬件只有路由器一种，并且该设备的CPU运算能力与内存情况完全满足家庭的边缘计算需要；不同之处在于，相比路由器，边缘计算智能设备对软件和功能要求更多。

OpenWrt是如何支持各类CPU芯片的：这种支持要么是水平极高的并且是芯片厂的合作伙伴实现的，要么就是芯片厂的内部人员实现的。想要支持某种CPU说起来并不难，只需要两方面的支持。一方面就是在汇编层面让OpenWrt编译时支持所属CPU的指令集，方便编译出可以在硬件上运行的软件。这部分大都是可以支持的，现在大部分的边缘计算智能设备、路由器、网络通信产品采用的是MIPS指令集，这是Linux已经支持的。另一方面就是驱动程序了，外围的厂商也不容易写出驱动程序，所以大部分驱动程序要么直接是芯片厂提供的，要么是和芯片厂有更亲密关系的厂商写的。

一个好消息是，SF16A18这款CPU的厂商非常重视OpenWrt系统，从一开始就支持该操作系统。

1.2 MIPS处理器体系结构

1.2.1 MIPS指令集体系结构

在介绍MIPS之前，我们先了解一下指令集，通俗地讲，指令集就是CPU在执行任务时需要遵循的规范和语言，是计算机硬件提供给软件的编程语言。

指令集体系结构（Instruction Set Architecture，ISA）又称指令集，是计算机体系结构中与程序设计有关的部分，包含了基本数据类型、指令集、寄存器、寻址模式、存储体系、中断、异常处理以及外部I/O。指令集体系结构包含一系列的opcode即操作码（机器语言），以及由特定处理器执行的基本命令。

常见的指令集有以下几种。

- 复杂指令集计算机（Complex Instruction Set Computer，CISC）：代表有Intel和AMD公司的x86、Intel的x86-64、AMD的AMD64等。
- 精简指令集计算机（Reduced Instruction Set Computer，RISC）：代表有HP的PA-RISC、Alpha，IBM的PowerPC，MIPS公司的MIPS，SUN公司的SPARC，ARM公司的ARM、Cortex等。另外，RISC-V也是一个基于RISC的开源体系结构。
- 显式并行指令运算（Explicitly Parallel Instruction Computing，EPIC）：这是一套全新指令集，代表只有Intel的IA-64体系结构。
- 超长指令字（Very Long Instruction Word，VLIW）：它通过将多条指令放入一个指令字，有效地提高了CPU各个计算功能部件的利用效率，提高了程序的性能。超长指令字是指令级并行的，是美国Multiflow和Cydrome公司于20世纪80年代设计的体系结构，EPIC体系结构就是从VLIW中衍生出来的。

MIPS指令集体系结构即MIPS体系结构，从MIPS I、MIPS II、MIPS III、MIPS IV和MIPS V体系结构，发展到目前的MIPS32、MIPS64和microMIPS体系结构。从1985年发布第一个版本开始，MIPS指令集经过30多年的不断演进和创新，目前已经发展成为业界最高效的RISC指令集体系结构，具有绝佳的性能和最低的功耗。截止到2014年，MIPS指令集一共发布了6个Release版本，其中最新版本为2014年发布的Release 6，如表1-1所示。

表1-1 MIPS指令集的6个Release版本

版本	发布时间
Release 1	1985年
Release 2	2002年
Release 3	2010年
Release 4	2012年
Release 5	2013年
Release 6	2014年

MIPS指令集在发展过程中基本保持了新版本向下兼容旧版本的特点。例如Release 5可以向下兼容之前的所有版本。但是在最新的Release 6版本中，这种情况发生了变化。这个版本中增加了一些新的指令并对指令集进行了简化，删除了一些不常用的指令，重新排布了指令的编码，预留了大量的指令槽用于将来的扩展。因此，我们可以把MIPS Release 6看成一个几乎全新的版本，它并不与之前的Release版本兼容。

RISC-V的快速发展使MIPS感到危机，2018年12月17日，MIPS的当前所有者——Wave Computing公司，宣布开放其MIPS指令集体系结构，而这个MIPS指令集体系结构包含的是MIPS Release 6版本。

MIPS指令集体系结构演进历程如图1-2所示。

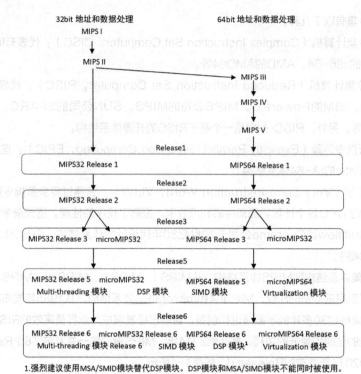

图1-2 MIPS指令集体系结构演进历程

从图1-2中,我们可以看到MIPS32和MIPS64指令集体系结构从Release 1到Release 6版本一直在并行演进。microMIPS32和microMIPS64指令集体系结构直到Release 3版本开始才迅速发展。2010年(Release 3版本)以后,MIPS指令集体系结构飞速发展,在2010年到2014年的短短5年时间里,更新了4个版本,在传统的整数浮点指令基础上,逐步增加了多线程(Multi-threading)模块、DSP模块、SIMD模块以及虚拟化(Virtualization)模块。

访问MIPS官网可以查看到MIPS目前支持的指令集体系结构和支持的模块,如图1-3所示。

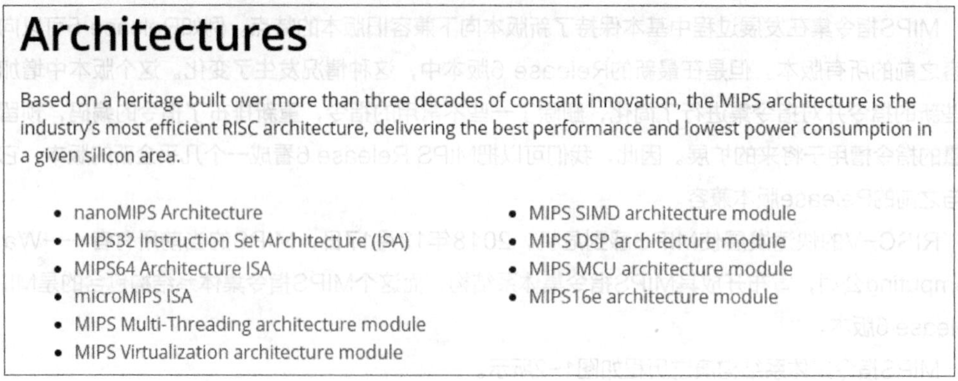

图1-3 MIPS指令集体系结构和支持的模块

随着市场需求不断变化，IoT（物联网）和移动互联网等迅速发展，MIPS指令集体系结构也在同步演进，我们也将看到更多新的MIPS指令集体系结构和越来越多的功能模块。

1.2.2 MIPS的发展历史

基于MIPS的产品覆盖范围非常广泛，并且种类繁多。MIPS广泛应用于消费娱乐、家庭网络和基础设施设备、LTE调制解调器和嵌入式应用等产品，越来越多应用于物联网设备、汽车高级驾驶员辅助系统（ADAS）和自动驾驶汽车。基于MIPS设计的芯片已在全球出货数十亿颗，并且以每年10亿颗的速度继续增长。MIPS的发展与MIPS公司密不可分，MIPS公司有着悠久的发展历史，被多次并购。

1981年，美国斯坦福大学的约翰·亨尼西（John Hennessy）教授发布了第一款MIPS芯片。

1984年，斯坦福大学的一组科研人员成立了MIPS计算机系统公司（MIPS Computer Systems Inc.）。

1989年，MIPS计算机系统公司上市。

1991年，MIPS计算机系统公司发布了世界上第一款64位芯片R4000。

1992年，Silicon Graphics Inc.（SGI）并购了MIPS计算机系统公司。

1998年，被从SGI中分拆后，MIPS技术公司（MIPS Technologies Inc.，MTI）再次上市。

2013年，英国Imagination公司以6000万美元现金收购MIPS技术公司。

2018年6月，MIPS技术公司被来自硅谷的AI创新公司Wave Computing收购。

Wave Computing是致力于开发AI和深度学习数据流芯片的美国硅谷公司。Wave Computing于2018年12月17日宣布开放其部分MIPS指令集体系结构，以便半导体公司、开发者以及大学快速采用MIPS体系结构开发下一代SoC芯片。开放MIPS指令集是该公司"All in AI"战略的重要组成部分。

1.2.3 MIPS的授权方式

MIPS是一个开放体系结构，MIPS公司本身不生产和销售芯片，而是把提供解决方案、服务和授权（License）作为经营范围。世界上很多大公司，如AMD、Cisco、SONY、Broadcom、NEC等都购买过MIPS授权。

MIPS授权分为处理器核授权（Core License）和体系结构授权（Architecture License）两类。

处理器核授权也称为IP Core License。IP（Intelligent Property）Core是具有知识产权的集成电路芯核总称。IP Core授权是指通过购买MIPS公司设计的MIPS处理器核而取得授权，可分为软核和硬核两类。软核包含电路的硬件描述语言（HDL）描述等内容，其优点是灵活性高、可移植性强，允许用户自配置；缺点是对模块的预测性较低，在后续设计中存在发生错误的可能性，有一定的设计风险。硬核是基于半导体工艺的物理设计，经过反复验证，可靠性高；缺点是在保护知识产权的要求下，不允许设计人员对其有任何改动。

体系结构授权也就是指令集体系结构授权，核心是指令集兼容，购买授权主要是为了使用"MIPS兼容"品牌以及通过加入MIPS兼容联名共享知识产权。购买MIPS兼容授权可以缩短产品进入市场的时间。

目前,MIPS官网列出的IP Core如图1-4所示。

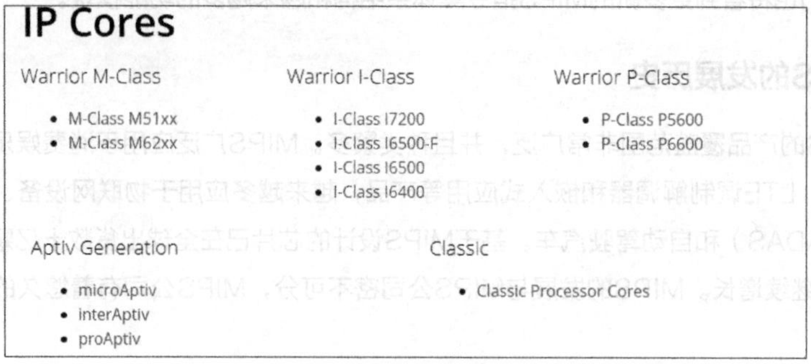

```
IP Cores

Warrior M-Class              Warrior I-Class              Warrior P-Class
  • M-Class M51xx              • I-Class I7200              • P-Class P5600
  • M-Class M62xx              • I-Class I6500-F            • P-Class P6600
                               • I-Class I6500
                               • I-Class I6400

Aptiv Generation                                          Classic
  • microAptiv                                              • Classic Processor Cores
  • interAptiv
  • proAptiv
```

图1-4 MIPS官网列出的IP Core

MIPS的IP Core产品主要被应用在人工智能、汽车电子、消费电子、IoT以及网络等5个市场。其中的人工智能领域是目前MIPS的母公司Wave Computing的主营方向。

1.2.4 MIPS IP Core

在2014年发布MIPS Release 6版本后,MIPS公司的IP Core产品研发全面转向MIPS Release 6版本,形成覆盖高、中、低不同性能和应用需求的Warrior产品系列,而把MIPS Release 6之前的产品归为Aptiv Generation系列和Classic产品系列,如表1-2所示。

表1-2 MIPS IP Core、指令集体系结构及对应的产品

IP Core	指令集体系结构	系列	产品
Warrior	Release 6	P-Class	P-Class P5600 P-Class P6600
		M-Class	M-Class M51xx M-Class M62xx
		I-Class	I-Class I7200 I-Class I6500-F I-Class I6500 I-Class I6400
Aptiv Generation	Release 3		proAptiv interAptiv microAptiv
Classic	Release 1 Release 2 Release 4 Release 5		Class Processor Cores

值得一提的是,自2014年MIPS Release 6发布以来,除MIPS公司自己的Warrior产品系列之外,尚未有其他厂商产品应用MIPS Release 6。相信MIPS开源MIPS Release 6后,会有越来越多基于MIPS Release 6的产品出现。

1.2.5 MIPS Aptiv Generation系列

Aptiv Generation系列是MIPS于2012年发布的,基于MIPS Release 3指令集体系结构,

包含proAptiv、interAptiv和microAptiv三个产品家族（见图1-5），其中proAptiv和interAptiv可选择多核配置，proAptiv最大核心数量为6个，interAptiv最大核心数量为4个，而最低端的microAptiv均为单核心的。它们分别面向高、中、低端不同市场，其特点是高性能、小面积和低功耗（见表1-3）。此系列产品推出后，频获大奖，如Linley Group"2012年最佳处理器IP"奖、嵌入式计算设计中的"2012年最创新产品"奖等。

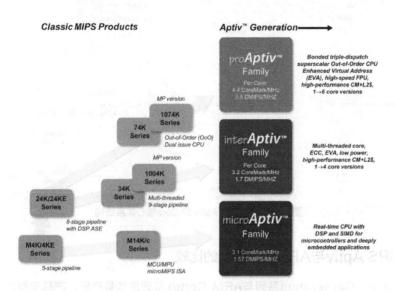

图1-5　MIPS Aptiv Generation产品家族

表1-3　MIPS Aptiv Generation 系列的核心数、性能跑分和用途

Aptiv 系列	核心	DMPS/MHz	CoreMark/MHz	用途
proAptiv	1~6	3.5	4.4	核心针对高性能应用，如智能手机、平板电脑、高清机顶盒、汽车信息娱乐和网关
interAptiv	1~4	1.7	3.2	针对中端应用，如无线基带以及汽车安全/传动系统控制等
microAptiv	1	1.57	3.1	针对低功耗成本敏感型应用：microAptiv MCU针对微控制器SoC开发设计，具有特定于应用的功能和实时性能；microAptiv MPU包括缓存控制器和MMU，可运行嵌入式系统、管理虚拟内存的操作系统（如Linux和Android）

值得一提的是，本书采用的矽昌SF16A18 SoC芯片就是基于MIPS interAptiv IP Core的。MIPS interAptiv IP Core是第3代MIPS IP Core，4核多线程32位处理器，电路规模和功耗比proAptiv更小。其最大的特点是支持超线程，3个interAptiv核心的性能表现与2个ARM Cortex-A9核心相当（基于DMIPS/MHz），核心面积也差不多。它主要应用于网络、存储以及图像和音频处理等，它可利用多线程并提供更好的整体吞吐量、QoS和功耗/性能效率。

MIPS给出的MIPS interAptiv IP Core核心体系结构如图1-6所示。

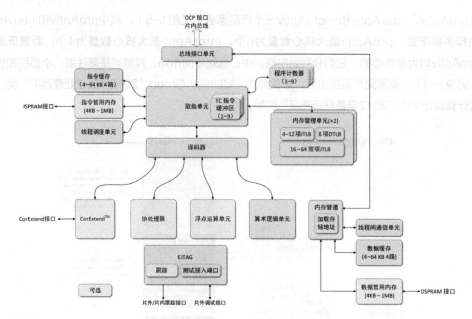

图1-6 MIPS interAptiv IP Core核心体系结构

1.2.6 MIPS Aptiv与ARM Cortex的比较

MIPS Aptiv Gerneration系列与ARM Cortex系列是竞争产品,产品定位也非常相近。MIPS proAptiv对标ARM Cortex-A系列,MIPS interAptiv对标ARM Cortex-R系列,MIPS microAptiv对标ARM Contex-M系列。图1-7是第三方网站给出的性能基准测试对比图。

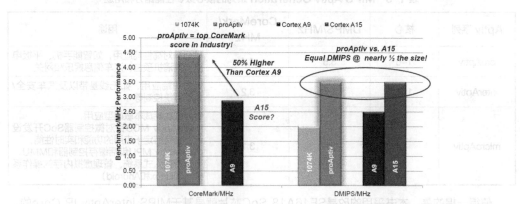

图1-7 MIPS proAptiv与ARM Cortex-A9、Cortex-A15性能基准测试对比

从第三方给出的数据中,我们可以看到MIPS proAptiv获得4.4 CoreMark/MHz分数,性能比ARM Cortex-A9高出50%(2.88CoreMark/MHz)。在DMIPS/MHz测试中,它还达到了3.5 DMIPS/MHz,与ARM Cortex-A15性能相当,但是每个核心的面积却为标准Cortex-A15的一半。

高端市场中,MIPS proAptiv系列的竞争对手是ARM Cortex-A系列;中端市场中,MIPS interAptiv系列的竞争对手是ARM Cortex-R系列;在微控制器核心领域,MIPS microAptiv系列的竞争对手是ARM Cortex-M系列。表1-4~表1-6所示是MIPS和ARM两者全系列产品的性能详细对比，采用DMIPS/MHz.Core数据进行比较。

表1-4 MIPS proAptiv 与 ARM Cortex A15/Cortex A9 等 DMIPS 评分比较

MIPS	DMIPS/MHz·Core	ARM
proAptiv	3.5	Cortex-A15
	2.5	Cortex-A9
1074K	2.03	
74K	2.0	Cortex-A8
	1.9	Cortex-A7
	1.57	Cortex-A5
M24K	1.46	

表1-5 MIPS interAptiv 与 ARM Cortex R7/R5/R6 等 DMIPS 评分比较

MIPS	DMIPS/MHz·Core	ARM
	2.5	Cortex-R7
interAptiv	1.7	
	1.66	Cortex-R5
34K	1.62	Cortex-R4
1004K	1.5	
24K	1.46	

表1-6 MIPS microAptiv 与 ARM Cortex M3/M4/M0/M1 等 DMIPS 评分比较

MIPS	DMIPS/MHz·Core	ARM
microAptiv	1.56	
M14K	1.5	
M4K	1.3	
	1.25	Cortex-M3/M4
	0.9	Cortex-M0
	0.8	Cortex-M1

除了第三方给出的数据，读者也可以访问Coremark官网，注册下载测试代码，自己动手测试。

1.3 SF16A18芯片

是的，没错，本书全部内容都基于SF16A18这颗国产芯片完成，这是一颗基于MIPS体系结构的边缘计算处理器。

设计这样一颗CPU芯片的难度超过我们的想象。SF16A18芯片实现了4核心12线程，提供了Wi-Fi芯片（2.4GHz+5GHz）、USB控制器、内存控制器、几十个I/O、各类串口和外部接口。在物联网方面，芯片的接口丰富程度达到甚至超越了市面上的单片机产品的水平；在边缘计算方面，CPU的性能是单片机的几十倍；而在Wi-Fi方面，它也是截至2019年，业界发布的唯一全集成芯片方案。除此之外，市面上还有很多传统厂商在提供这个级别的芯片。

Atheros：全球路由芯片顶级公司，其创始人便是MIPS体系结构的发明者约翰·亨尼西。该公司也是Wi-Fi标准的制定者之一，目前该公司已被高通收购，其芯片主要用于高端及企业级产品线。

Broadcom：美国Broadcom（博通）公司的无线路由芯片产品以稳定可靠著称。小米第一代路由器便采用了其BCM470X系列处理器，该处理器基于ARM体系结构，不过Broadcom其他路由芯片大都基于MIPS体系结构。

Ralink：Ralink（雷凌科技）也是一家常年耕耘在路由器领域的芯片厂商，而且是目前国内出货量较大的家庭路由器芯片厂商，其芯片价格也相对较低，因此目前国内上市的家庭智能路由产品大都采用了该公司的方案。该公司已被联发科收购。

此前，本书的作者们其实心里一直在有个疑问：为什么这类处理器20多年来从来没有一个国产厂商能设计？众所周知，MIPS体系结构的芯片其实国内也有厂商在出，并且有一些厂商还是大名鼎鼎的，不过大部分是雷声大雨点小。

不得不说MIPS与ARM在书本之外没有写明白的区别：ARM芯片由于制造厂商众多，因此其设计研发难度目前远低于MIPS体系结构芯片。而ARM芯片专利费高于MIPS体系结构芯片专利费、发热量高于MIPS体系结构芯片发热量、单指令集的CPU性能弱于MIPS体系结构芯片单指令集的CPU性能。但是由于名气响亮，如今国内很多公司都在设计ARM芯片，而这部分由于有补贴，可以获取利益。

很多读者可能会有疑问，既然设计芯片，为什么要给国外交专利费？我们先看看专业分工。

- 指令集：这是由公司、团队或组织提出的。
- IP：将方案付诸实施的经过测试的标准方法，并且得到业界的认可。
- 芯片设计：将IP根据产品需求设计成芯片。
- 流片：这是难度最高的工艺，是沙子变成晶圆的过程。
- 封测：封装及测试，将芯片封装成不同规格，并且测试是否是有效芯片。

以上专业分工都是由多家公司完成的，但是整个市场的风险都在芯片设计上，而芯片设计只是芯片生产的一个环节。如果你今天自己设计指令集或IP方案，那你也没必要去做芯片设计，原因是做这样一系列的事情，代价是非常大的，一般厂商难以承受。同时，最重要的是你的指令集和IP必须得到行业的广泛认可，不然就是闭门造车。

目前我们所熟知的RISC-V这一开源芯片解决方案便是如此，我们当然知道它是开源的，可是当你要做产品时，你是准备设计IP还是设计一颗芯片？当然，目前大部分RISC-V芯片其实也是购买IP方案来进行设计的。其实，我们也希望国内能出现IP方案厂商，但绝对不希望这样一个厂商同时也设计芯片。

1.4 本书背景介绍

1.4.1 自主、可控、开源

自主、可控是国家发展芯片的重点。本书在自主、可控上又加上了开源。很简单，有一位互联网知名人士曾说过什么是开源："开源就是这个时代集中力量办大事的体现。"是的，作为推广开源20多年的人，我们非常了解这句话的意义与分量。

自主：我们认为自主的芯片，研发团队一定要在国内。在国际化的今天，我们不可能闭门造车，不与国外交流，但是我们的芯片一定是要在国内研发的。

可控：Wi-Fi的安全性是老生常谈的问题，说实话本书的作者们并不十分在乎这件事，因为借助手机App里的后门，黑客不需要破解Wi-Fi就能盗取你的资料。但是战略安全是非常重要的，我们必须要保证在任何情况下，都有足够的芯片可以使用。这也是可控所带来的安全性问题。

开源：任何一个产品如果不开放，被大厂垄断，都将导致产业倒退。开源可以集中各类力量一起攻破难关，一起对全球产业起到引导作用。并且开源可以帮助很多创客、创业者实现创业梦想。

1.4.2 SF16A18芯片设计团队

上海矽昌通信技术有限公司的董事长和创始人李兴仁博士，2000年在复旦大学获得博士学位，曾领导开发中国首款北斗二代卫星导航芯片；2008年初以技术领军人、创始人的身份创办了上海盈方微电子有限公司，用3年时间研发出国内首款平板电脑CPU芯片；2014年12月二次创业，创办上海矽昌通信技术有限公司，专注于国产无线和有线网络通信集成电路芯片的研发和产业化，立志创办中国的"博通半导体"。

当前中国的互联网用户数全球第一，但宽带运营商的家庭宽带接入设备、Wi-Fi无线路由器及物联网智能电子产品，因无国产同类芯片，都只能采用海外的集成电路芯片产品。据统计，全球Wi-Fi无线路由器年需求量约5亿台，其中80%以上由中国制造，进口此类芯片，国家每年都要花费数十亿美元。

有鉴于此，上海矽昌通信技术有限公司首先从家庭宽带接入设备着手，布局智能家居的中枢——家庭智能网关，整体投入超8000万元人民币，历经5次试验投片，在2018年第一季度，成功实现了无线路由处理器芯片（代号SF16A18，简称A18）的量产。

A18芯片是国内首款自主开发的2.4GHz/5GHz双频无线路由处理器集成电路芯片，可广泛应用于中国电信光纤宽带入户设备、广电（如东方有线）同轴电缆宽带入户设备、4G转Wi-Fi宽带入户设备，以及家用Wi-Fi无线路由器等领域。同时，由于它集成了双核4处理器高性能CPU，也可以用于无线智能网关、无线智能路由器及智能音箱等产品；后续在公司资金充裕和市场发展迅捷的前提下，只需对接入方案、集成电路板重新设计，就可应用于物联网的各种终端。

矽昌公司凝聚和培养了一支80人的"攻坚克难，能打硬仗，能打胜仗"的优秀研发团队。经过3年多艰苦奋斗，A18芯片成功量产。在射频SoC设计、有线和无线网络通信算法及协议栈开发和优化、通信系统的稳定性和兼容性方面，矽昌公司突破了技术壁垒，完成了技术积累。目前，矽昌公司已经开始研发更高端的无线路由处理器芯片产品（代号SF19A28，简称A28）。A18芯片和后续的A28芯片如果能大规模商业应用，将替代海外公司的同类进口产品，为国家节省大量外汇，确保产业安全、网络安全、数据安全。

1.5 AIoT的技术应用

AIoT（AI+IoT，智能物联网）现在已经迅速发展，截至2021年，全球有100亿台以上的设备接入物联网。物联网的应用已经开始落地，其中智能家居已经深入千家万户，而在办公环境中，办公设备也日趋智能化，传感器带来的办公场景改变也越来越常见，正常采购的各种办公设备，如投影机、大屏、打印机、空气净化器、空调、照明设备、门禁等大都具有接入互联网的能力，每个设备都可以单独进行网络控制，所以办公室的物联网时代已经到来，多设备统一管理和权限分配都需要一套完整的智能办公软件来进行管控。

办公场景的变革，需要物联网满足高安全性、大并发、边缘计算、异构集成、企业软件集成以及大脑网关总控等要求，并且能够兼容当前的各种设备和弱电、强电要求，降低后续运维成本。在这些要求下，我们深入实际，研究提出了一个基于智能办公场景的物联网脑图构架来建立企业级物联网，创造办公场景的革命。

当前物联网在办公场景的应用大部分源于智能家居的应用，把很多家庭级的要求放入企业办公中势必存在很多问题。

1. 安全性缺失

企业的安全要求较高，相对于家用来说，物联网的数据传输需要进行私钥加密，网络也要节点隔离，并且不能将所有服务内容放入公共云端。需要强电改造的部分，要达到工业级电气标准要求，防止在工业用电的电涌下出现安全隐患。

2. 并发性不够

在家庭环境中，使用人数通常不会超过10人，使用设备不会超过100台。但是在办公场景中，一层楼人数为500人左右，物联网设备（包括照明设备、门禁、电动窗帘、其他电器等）超过1000台，整栋楼宇人数、设备数将会更多，所以并发性不够是当前办公场景的一个基本问题，其原因也是采用家庭物联网的部署结构无法承载日常的基础使用，也无法进行扩展。

3. 纯云端技术，可靠性不够

大部分智能家居架构是前端智能设备直接联网云端，然后通过云端进行物联网控制，比如手机远程和智能场景管理。但是在企业环境中，不能完全依赖纯云端，因为在日常工作环境中，照明设备、空调、门禁都需要即时响应，不能因为网络不好而无法调整、无法使用。

4. 无法兼容现有设备

在当前智能硬件过剩的情况下，不缺乏任何需要的前端设备，比如智能灯泡、智能门禁、智能投影机；但是在企业环境中，不可能将所有的照明设备都换成智能灯泡，将所有的门禁都换成智能门禁，这样成本巨大，并且维护成本也巨大。因此企业级物联网需要兼容当前的办公设备，而不是全面替换。

5. 无法打通企业级硬件

除去传统的电器，企业中还存在大量企业级硬件，这些硬件有的已经是智能化设备，有的是网络设备，通常遵循了一定的技术标准，比如摄像头支持RTSP协议，通过UDP协议传输；网络音箱支持DLNA协议；空调支持Modbus串口协议；打印机支持TCP/IP协议。

6. 无法打通企业软件

企业内部使用这些智能化应用的功能，通常需要和已经存在的OA或者会议预约等企业软件集成，使用其统一的组织人员结构和账号体系，这样可以在人员离职后统一注销或者管理物联网的权限，达到安全的使用标准。

7. 没有中心大脑，无法自定义场景

企业内部需要更多的场景变化，所以需要自身有一个中心网关来控制和协调各种设备的协作运行，在传统的智能家居架构中无此结构，更多的是靠云端一个中心来完成，除去可靠性不足以外，还无法做到更细粒度的场景定义，物联网数据也无法在本地存储和共享。

1.5.1 智能办公物联网脑图

基于上述问题，我们需要提出一个全新的企业级物联网架构来实现办公的智能化。这个全新的物联网架构名为"智能办公物联网脑图"，其实现思路是将企业办公智能化看作一个完整的智能机器人，有大脑、眼睛、嘴巴、鼻子、手、骨骼、神经网络和血管。虽然它不是人形机器人，但是结构一致，可以为企业办公提供真正的AI和IoT能力。

● **物联网网关（大脑）**

这是整个构架中最为重要的一个部分，它负责将所有的物联网设备、电气设备、其他智能设备联通，通过物联网的ZigBee、LoRa、NB-IoT、蓝牙协议，音频的VOIP、DLNA、Qplay等协议，视频的RTSP协议，电器类的modbus串口协议，以及强电控制类和传感器类的协议进行互联互通，实现对所有设备的协同控制，并且做了协议转化，可以通过软件API的方式轻松调用和控制各类设备，和云端AI引擎互通，调用AI能力实现可听、可懂、可执行的智能化大脑。另外，它还可以和企业级软件，如OA、HR、ERP对接，使用同一套组织架构和账号体系。同时，它也是一台可以发射物联网信号的网关，和AP一起就可以覆盖整个办公区。

● **摄像头（眼睛）**

网关可以接管企业中已经存在的安防摄像头，通过网关访问云端AI引擎的能力将普通安防摄像头无缝升级为智能人脸识别摄像头。有了这些"眼睛"，我们可以实现很多新的智能化场景，比如人脸识别门禁、人脸识别考勤、人脸识别会议室签到、人脸识别访客路径、人脸识别在岗人员统计等。

● **屏幕、音箱、话机（嘴巴）**

随着智能电视机的普及，企业内部开始安装很多智能电视机来播放相应的企业内容，比如欢迎语、企业中标通知、访客排队情况、企业概况、宣传视频、宣传图片等，这也是我们整个智能化办公的展示窗口。我们可以通过网关统一编辑企业节目列表，然后把它下发到不同的电视机、大屏幕上，不用通过遥控器进行单台管理，内容也可以升级为定时播放和插播模式。同时整个智能化的全景展示也可以发布到各个屏幕上，展示整个空间实时的人员在岗情况、会议室使用情况、设备照明使用情况、访客信息等。企业级音箱系统和电话系统可以通过网关对接云端AI引擎，将自动语音播报集成进来，对企业环境中的广播通知、背景音乐、点对点电话通知场景进行变革，充当智能化的"嘴巴"，实现更多的人性化场景应用，比如员工通过电话机拨号到虚拟行政助手，通过AI智能语音的应答和交付完成大部分与行政相关的服务和引导。

● **传感器（鼻子）**

该架构可以对接各种厂商的传感器，如温/湿度传感器、PM2.5传感器、有害气体传感器、人体热释电传感器、红外传感器、光照传感器、电量传感器等，这些传感器都是用来实现在不同场景下控制智能设备的数据采集器，通常按照以下方法来使用：传感器通过网关支持的协议接入网关，该网关通过Wi-Fi或者有线连接接入智能网络，通过网关的接口适配，在网关中可以直接操作和管理，并且将实时的数据展现在大屏或者手机终端中。

● **控制器（手）**

对于强电控制的照明设备、电子门禁、电动窗帘、电动幕布等设备，我们可以加装强电控制器完成控制。控制器充当了智能化场景中的"手",接收网关发出的指令,自动完成开关等操作。这类设备只需要加装在传统的灯、门、窗帘上,不改变传统设备的采购选型,就可以无缝地集成进整个智能化网络中。

- **办公家具（骨骼）**

目前一些办公家具已经带有智能化或者自动化的功能,这类办公家具包括升降办公桌、智能化座椅、智能工位、智能会议桌等。这些办公家具将越来越多地成为现代办公的标配,但是它们自有一套系统标准,需要通过网关集成起来,进行联动操作。

- **人工智能云端引擎（神经网络）**

AI能力已经非常普遍地使用在智能家居场景中,包括视觉识别、语音识别、语义翻译、语音翻译等。这些能力需要在云端获取,因此我们将用网关来对接这些能力,可以选择阿里、百度、腾讯等AI云服务商的引擎,通过适配后,跟本地的其他设备进行互动,实现人脸识别门禁、会议室同声传译、会议室语音记录、语音命令开关设备等操作。

- **无线IoT网络（血管）**

所有的无线网络和物联网络将同时集成在网关中,通过AP的扩展覆盖整个办公空间,这个网络将Wi-Fi、ZigBee、LoRa、蓝牙、NB-IoT融合在一起,统一覆盖,统一管理,实现整个智能化的"血管"功能,在不同的网络标准中传递统一的指令和数据。

1.5.2 应用实例解析：企业VIP接待

接待人员首先在网关程序中预约接待时间、来访人员数量和姓名,并且录入VIP人员照片,在访客到来前30分钟,系统自动在所有屏幕上播放欢迎语,同时将公司大门门禁改为常开模式,并发信息提醒接待人员就位。

（1）VIP到达公司所在大楼1楼门闸时,系统自动通过人脸识别开启门闸,并通知企业接待人员贵宾已经到达1楼,同时协同电梯,将楼层自动设置为目的地楼层。

（2）访客到达公司大门时,系统通过人脸识别确认身份后自动语音播报欢迎语。

（3）在访客参观过程中,不同屏幕可以通过插播的方式播放需要讲解的内容。

（4）待访客参观完成,系统在大门口自动通过人脸识别进行合影留念。

（5）访客离开后,系统自动收回门禁权限,屏幕恢复播放之前的内容,公司大门门禁进入日常状态。

在上述流程中,智能网关协调了屏幕控制、门闸控制、门禁控制、人脸识别、语音播报等功能,形成了一个新的智能化场景,提升了接待效率,很好地展现了企业形象。

1.5.3 应用实例解析：智能下班控制

在一个面积有1000平方米以上、规模在500人左右的公司,每天下班关闭各种电气设备已经变成一件非常麻烦的事情,通常需要行政人员巡场1个小时才能完成。并且在加班情况下,很多设备无法关闭。

应用智能下班控制功能后,管理员首先在网关中设置下班时间,在下班半小时后,系统能够通过物联网传感器发现企业每个区域还有多少人,在没有人的区域和公共区域自动关闭照明设备、空调、空气净化器等电气设备,在有人的区域延迟关闭;在到达设置的延迟关闭时间(晚上8点)后强制关闭电气设备,并向还在区域中的员工发送单独的通知,让他们自行在手机上开启自己需要的设备,如果该员工已经在OA中申请过加班,那么在加班期间不会强制关闭设备;系统在每晚凌晨2点也会再次检查是否有没关闭的设备,如果有则强制关闭。

在这个场景中,智能网关协调了照明设备、空调、人体红外热释电传感器、空气净化器等设备并和OA系统打通以获取加班信息。

1.5.4 应用实例解析:智能会议室

中大型企业对会议室占用非常难以控制,我们可以将会议室预约和门禁、会议室内的电气设备互动起来,实现使用效率的提升。

首先,使用者需要通过预约来使用会议室,在到预约时间的时候,才可以携带自己的手机进行扫码或者通过人脸识别开门。开门后,人体红外热释电传感器通知网关,人已经到达,并且自动签到和打开照明设备、幕布、电动窗帘、电视机、投影机等设备。在人员提前结束会议离开的时候,系统检查到一定时间内没有人员,将自动释放会议室资源给其他用户使用;如果到预定时间后系统检查到仍有人使用,将会提醒开会人员及时续订,如果无法续订,将在下一场会议开始之前将照明设备、幕布、电视机、投影机自动关闭,实现不恶意占用会议室。

在这个场景中,智能网关协调了照明设备、空调、门禁、幕布、电动窗帘、电视机、投影机、人体红外热释电传感器等并通过OA会议预定系统获取预定信息。

1.6 表达约束

本书涉及多个环境下的交叉操作,为了避免读者在使用上出现混淆,我们对路径进行一定约束。这些约束可能与实际操作看起来不一样,如表1-7、表1-8所示。

表1-7 本书相关环境

类型	作用
Windows	用于学习和测试的计算机安装的操作系统
Ubuntu 16.04	使用VirtualBox运行于Windows系统下,用作交叉编译环境
OpenWrt	使用在开发板上的操作系统,本书的重点内容

表1-8 路径约束

系统类型	真实提示符	本书提示符	作用
开发板	DF1A:$	DF1A:$	表示在开发板上操作
Windows	C:	C:	没有变化
虚拟机	xxx@xxx-VirtualBox	HOST:$	表示在虚拟机中进行操作

第二篇　SF16A18芯片的OpenWrt系统

2 环境与工具准备

2.1 SF16A18芯片的规格

SF16A18是一颗高性能SoC芯片,适用于边缘计算及网络通信产品。它在一个芯片上集成了所有你需要的接口。它采用了4核MIPS32 interAptiv处理器,内部集成了256KB的L2缓冲区,可以进行网络数据路由和转发,支持SPI、eMMC、SD多种启动方式。芯片内部采用TSMC 28nm数模混合设计工艺,创新性地集成了2.4GHz和5GHz并发的射频模块以及射频前端处理模块,如发射放大器(PA)、低噪声接收放大器(LNA)和收发切换开关(SWITCH)。同时,芯片集成5口100Mbit/s以太网接口以及RGMII高速以太网接口(见表2-1、图2-1)。

表2-1 SF16A18 芯片的特性

处理器子系统	4核MIPS32 InterAptiv指令集	双物理核心,4路处理单元,12线程
		16KB L1 I-Cache和16KB L1 D-Cache(每个物理核心)、256KB L2 Cache
		主频1~1.2GHz
	16位 DDR2/DDR3内存	
	8线SD/eMMC接口	
	4线SD/SDIO接口	
	8通道DMA	
	4路UART串口	
	2路SPI串口	
	3路I^{2}C串口	
	4个通用定时器	
	1个看门狗	
	2个16位PWM	
	2个I^{2}S音频接口	
	2个PCM音频接口	
	1个USB 2.0 OTG接口	
	大量可编程GPIO接口	
Wi-Fi 子系统	2.4GHz 802.11b/g/n射频,20/40MHz频宽,最大速率150Mbit/s	
	5GHz 802.11a/n/ac射频,20/40/80MHz频宽,最大速率433Mbit/s	
	1×1 SISO 2.4GHz/5GHz双频射频收发器,内置PA、LNA	
	两个嵌入式独立CPU专门实现Wi-Fi部分运算(纯硬件Wi-Fi基站模式)	
封装	TFBGA 360Pin封装,0.65mm Pitch	
	操作温度:工业级,-40~125℃	

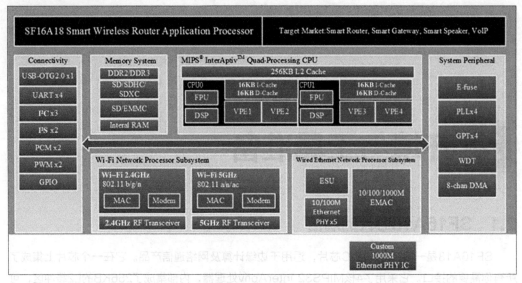

图2-1 SF16A18芯片框图

SF16A18芯片的关键创新如下。

硬件安全加速引擎： 硬件支持的加密算法种类齐全，广泛符合上层加解密软件要求，硬件加密从流程的最小粒度上保证操作的安全性。

性能强劲： CPU有4路处理单元、12线程，主频为1GHz，performance模式下可以达到1.2GHz，Coremark跑分为3.8 iteration/MHz，性能大幅领先竞争对手同类产品。芯片支持DDR2/DDR3外部大容量内存扩展。在对用户网络数据检查过程中，芯片不但要持续不断地对网络数据进行充分备份，还需要进行深度的数据解析和挖掘，这些软件处理对CPU的处理能力以及内存的容量都有很高要求，SF16A18的强大性能能够对此进行有力支持。

安全启动技术： 芯片通过固有的程序校验U-Boot，保证U-Boot的安全；通过U-Boot校验Linux内核，保证文件系统安全。芯片级的安全技术从根本上保证运行软件的安全性，保证在芯片上运行的系统无法被破解和更改。

SF16A18是国内首款自主知识产权且符合IEEE 802.11a/n/ac、802.11b/g/n标准的芯片，也是业界集成度最高的单芯片2.4GHz/5GHz双模Wi-Fi无线路由器SoC芯片，单一芯片同时集成了2.4GHz、5GHz双模RF收发器及PA、LNA，外围元器件依赖量减少50%。

SF16A18芯片与市面上主流ARM处理器和传统路由芯片的性能对比如图2-2、图2-3所示（请注意：在对比时，SF16A18芯片的温度为40~50℃，而ARM芯片的温度为70℃左右）。对比数据来自第三方，仅供参考。

第二篇 SF16A18芯片的OpenWrt系统

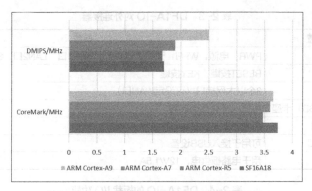

图2-2　SF16A18与主流ARM处理器的性能对比

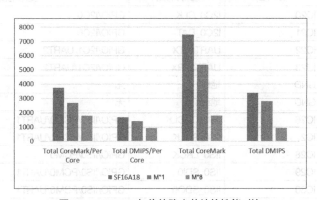

图2-3　SF16A18与传统路由芯片的性能对比

2.2　DF1A开发板介绍

DF1A是由Freeiris团队与SF16A18芯片厂合作，专门针对本书推出的全开源开发板。该板将硬件设计资料全部在线开源，同时可以完成本书全部内容的学习和测试。DF1A详细资料可以访问Freeiris官网获取。

开发板由两部分组成。

■ DF1A：核心板，包含处理器等核心部件（见表2-2）。

■ DF1A-IO：接口板，包含各类接口与板载部件（见表2-3、表2-4）。

表2-2　DF1A核心板

配置	参数
主处理器	SF16A18
辅处理器	FPU浮点运算单元
运行内存	DDR3 1024Mbit（128MB）
存储器	SPI 128Mbit（16MB）
调试接口	板载TTL串口
Wi-Fi	2.4GHz+5GHz双硬件Wi-Fi
天线	IPEX标准天线接口

表 2-3 DF1A-IO 对外连接器

方向	配置	参数
前面板	指示灯	PWR：电源、Wi-Fi：无线、WAN口、LAN1口、LAN2口、SYS（可编程控制）
前面板	按键	RESET按键、KEY按键
后面板	以太网接口	3个以太网接口（可设定WAN口）
后面板	TF（micro SD）卡座	可插入TF存储卡扩容
后面板	RS-485接口	可用于工业互联网和物联网
后面板	USB 2.0接口	可用于接入USB设备
后面板	DC接口	用于电路板供电，12V/1.5A

表 2-4 DF1A-IO 的板载 I/O 功能

分组编号	GPIO	默认功能	复用功能
PG1	IO20	I2C0_CLK	GPIO/I2C0
	IO21	I2C0_DAT	GPIO/I2C0
	IO22	UART2_TX	GPIO/I2C1/UART2
	IO23	UART2_RX	GPIO/I2C1/UART2
	GND	地线	无
	GND	地线	无
PG2	IO26	IIS0_CDCLK	GPIO/IIS0/PCM0/UART1
	IO27	IIS0_SCLK	GPIO/IIS0/PCM0/UART1
	IO28	IIS0_LRCK	GPIO/IIS0/PCM0
	IO29	IIS0_SDI0	GPIO/IIS0/PCM0/UART1
	IO30	IIS0_SDO0	GPIO/IIS0/PCM0/UART1
	GND	地线	无
PG3	IO33	GPIO	GPIO
	IO34	GPIO	GPIO
	IO35	GPIO	GPIO
	IO37	PWM1	GPIO/PWM1
	GND	地线	无
	GND	地线	无
PG4	USB_VBUS	USB_VBUS	无
	GND	地线	无
	DM	USB_DM	无
	DM1	USB_DM1	DM与DM1短接表示使用后面板USB 2.0接口
	DP	USB_DP	无
	DP1	USB_DP1	DP与DP1短接表示使用后面板USB 2.0接口
PG5	VCC3V3	无	3.3V对外供电
	GND	无	地线
	VCC3V3	无	3.3V对外供电
	GND	无	地线
	5V_IO	无	5V对外供电
	GND	无	地线
	5V_IO	无	5V对外供电
	GND	无	地线
	12V	无	12V对外供电
	GND	无	地线

2.3 U-Boot网页刷机

这一节，我们来看刷机的功能，这一节的内容可以不用实践（建议学习过编译固件后再进行本节的操作），日后若将系统弄出问题，使用本方法可安全、可靠地恢复系统。

BootLoader在操作系统内核运行之前运行，可以初始化硬件设备、建立内存空间映射图，从而将系统的软硬件环境带到一个合适状态，以便为最终调用操作系统内核准备好正确的环境。嵌入式系统中通常没有像BIOS那样的固件程序，因此整个系统的加载、启动任务就完全由BootLoader来完成。

在我们的开发板上，操作系统启动前，BootLoader要对一些基本的电子设备进行检测，同时完成CPU所要求的初始化，再启动Flash上的Linux内核。

开发板所使用的引导程序名字叫U-Boot，U-Boot是德国DENX小组开发的用于嵌入式CPU的引导程序（也支持x86体系结构），支持10多种不同的操作系统。

很多书籍都详细讲解了U-Boot是如何引导操作系统的，其实U-Boot如何引导都是CPU厂家提供的SDK里要做的事。

所谓刷机就是将新的系统刷写到板载Flash上，使用U-Boot刷写是很安全的，即使刷写坏了也可以重新刷写，一直到好用为止。第一次刷机要从U-Boot操作开始，开发板上提供了两个U-Boot以及两种刷机方法，一种是U-Boot的TFTP刷机，另一种是U-Boot的Web刷机。由于各厂商的内存映射不同，TFTP刷机方法存在一定风险，本书详细介绍Web刷机。

固件下载地址：freeiris官网/resource/sf16a18-df1a/

固件文件名：openwrt-siflower-sf16a18-fullmask-squashfs-sysupgrade.bin

准备步骤：

（1）将开发板WAN口与计算机网口直连，将计算机其他网卡全部禁用，开发板暂时不上电。

（2）打开Windows系统的"控制面板"→"网络连接"，本实例中使用的是"以太网2"（各计算机所使用的网口命名不同）。

（3）双击"以太网2"，设定TCP/IP协议的固定IP地址为"192.168.4.10"，子网掩码为"255.255.255.0"，其他均不设置（见图2-4）。

（4）单击"确认"退出设定。

图2-4　设定计算机的IP地址

上电步骤：

（1）先按住开发板上的按键"KEY"，再给开发板上电。

（2）保持按键按住8秒后松开按键。

（3）此时应该看到网卡已经有网络连接（见图2-5）。

图2-5　网卡已有网络连接

开始刷机：

（1）准备好刷机固件文件，将其放入计算机D盘中。

（2）用浏览器访问IP地址"http://192.168.4.1"，进入刷机界面。

（3）单击"选择文件"，选择刚才的固件文件（见图2-6）。

图2-6　访问Web刷机界面

（4）确认选择的文件是否正确，然后单击"Update firmware"开始更新固件。

刷机确认：

（1）更新过程需要2分30秒到3分钟来完成，这一过程中如果断电，需要重新开始刷机步骤。

（2）更新完成后，用浏览器再次访问"http://192.168.4.1"，可以看到主界面，表示刷机完成（见图2-7）。

图2-7　更新完成后用浏览器访问http://192.168.4.1

（3）刷机是有一定出错风险的，一旦出错请重新执行刷机步骤。

2.4 TTL串口调试

当我们进行系统调试时，串口必不可少。开发板上电前，一定要把核心板插入底板，然后接上串口调试线，再上电。OpenWrt在编译时已经默认打开串口作为本地终端了，通过串口终端，使用者可以像操作自己的计算机一样来操作开发板，实现SSH或Telnet做不了的事情，比如查看内核的一些信息、控制引导系统刷机等。开发板固件的默认登录账号为"admin"，密码也为"admin"，请在提示login时输入。

操作开发板的注意事项如下。
- 注意身上的静电要释放掉，触摸接地的物品，甚至触摸墙壁都有一些效果。
- 在连接好需要的调试线路之前不能给开发板上电。
- 注意开发板不要出现短路情况，不要将其他导体放在开发板上面或下面。

2.4.1 USB转TTL调试工具

串口调试工具我们一般采用USB转TTL调试工具，这类转换芯片大致有几种型号：CH340、CP2102、FT232、PL2303。从价格上来说，FT232>CP2102>PL2303>CH340。大家选择的时候，要注意兼容性，Linux一般兼容性比较好；Window只要安装对应版本的驱动，一般来说也没有什么问题。我们也遇到过某款芯片的开发板只在FT232芯片下不出现乱码，如果使用其他转换芯片，串口无论怎么设置均显示乱码的情况。

调试前先确定串口物理连接成功：将串口TTL端连接到开发板串口对应接线端，USB端插入宿主机的USB接口，连接线序如表2-5所示（其他如3.3V或5V不要连接）。

表2-5　TTL线连接线序

USB 转 TTL 调试工具串口 TTL 端	开发板串口接线端
RX	TX
TX	RX
GND	GND

2.4.2 Linux上位机串口调试

如果你使用的是Windows系统，请移步后面。在Linux系统中，推荐使用minicom作为串口调试软件。

Ubuntu系统下的安装方法：
```
HOST:~$ sudo apt-get -y install minicom
```
CentOS系统下的安装方法：
```
HOST:~$ sudo yum -y install minicom
```
确认串口硬件已识别的方法：
```
HOST:~$ ls /dev/ttyUSB0
/dev/ttyUSB0
```
如果以上步骤都没问题，我们就可以使用minicom进行串口调试了。通过minicom help命令，查看帮助。

```
OST:~$ minicom -help
Usage: minicom [OPTION]... [configuration]
A terminal program for Linux and other unix-like systems.
  -b, --baudrate         : set baudrate (ignore the value from config)
  -D, --device           : set device name (ignore the value from config)
  -s, --setup            : enter setup mode
  -o, --noinit           : do not initialize modem & lockfiles at startup
  -m, --metakey          : use meta or alt key for commands
  -M, --metakey8         : use 8bit meta key for commands
  -l, --ansi             : literal; assume screen uses non IBM-PC character set
  -L, --iso              : don't assume screen uses ISO8859
  -w, --wrap             : Linewrap on
  -H, --displayhex       : display output in hex
  -z, --statline         : try to use terminal's status line
  -7, --7bit             : force 7bit mode
  -8, --8bit             : force 8bit mode
  -c, --color=on/off     : ANSI style color usage on or off
  -a, --attrib=on/off    : use reverse or highlight attributes on or off
  -t, --term=TERM        : override TERM environment variable
  -S, --script=SCRIPT    : run SCRIPT at startup
  -d, --dial=ENTRY       : dial ENTRY from the dialing directory
  -p, --ptty=TTYP        : connect to pseudo terminal
  -C, --capturefile=FILE : start capturing to FILE
  -F, --statlinefmt      : format of status line
  -R, --remotecharset    : character set of communication partner
  -v, --version          : output version information and exit
  -h, --help             : show help
  configuration          : configuration file to use
These options can also be specified in the MINICOM environment variable.
This variable is currently unset.
The configuration directory for the access file and the configurations
is compiled to /etc/minicom.
Report bugs to <minicom-devel@lists.alioth.debian.org>.
```

minicom的参数很多，如果是第一次设置，请使用下面的命令。

```
HOST:~$ sudo minicom -s /dev/ttyUSB0
```

我们将看到图2-8所示的主菜单，只需要设置"Serial port setup"即可。

图2-8　minicom的主菜单

用上、下方向键选择，当光标在"Serial port setup"上时，按下Enter键，进入图2-9所示的设置界面，我们通常只需要设置"Serial Device"（串口设备节点）和"Bps/Par/Bits"（通信速率/数据位/奇偶校验位/停止位）即可。当前光标在"Change which setting?"上，键入"A"，此时光标移到A项对应处，改为将其"/dev/ttyUSB0"；然后对通信速率、数据位、奇偶校验位和停止位进行配置，键入"E"，通信速率/数据位/奇偶校验位/停止位选为"115200 8N1"（115 200bit/s，8位数据位，无奇偶校验位，1位停止位）；硬件/软件流控制分别键入"F"和"G"，并且都选No；最后按下Enter键保存，返回主菜单。

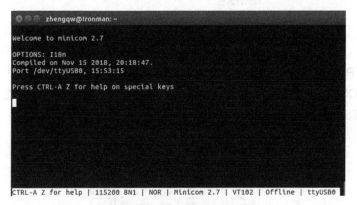

图2-9　Serial port setup配置界面

如果想让配置下次也生效，则配置完毕后，按下Enter键返回上级配置界面，并选择"Save setup as _dev_ttyUSB0"（倒数第4个选项）将其保存为默认配置。最后，选择"Exit from Minicom"命令退出，或选择"Exit"直接打开串口。

退出minicom后，下次可以通过下面的命令直接打开串口进行调试（见图2-10）。

```
HOST:~$ sudo minicom /dev/ttyUSB0
```

图2-10　minicom串口界面

在串口调试过程中，可以按下Ctrl+Z组合键显示帮助界面（见图2-11）。

图2-11　minicom帮助界面

所有的命令都是通过Ctrl+A <key>来组合的。退出minicom，需要先按下Ctrl+A组合键，再按下Q键，然后确认即可退出。

2.4.3　Windows上位机串口调试

Windows下串口调试工具非常多，大家可根据喜好和习惯进行选择，我们推荐使用PuTTY作为串口调试工具，它功能强大、简单易用。

（1）在计算机上安装USB转串口驱动程序。

（2）将USB转串口线接入计算机的USB接口，等待驱动程序安装完成，然后查看设备管理器，确认COM口的编号（见图2-12），因为每次USB接入计算机位置不同，COM口可能不一样。

图2-12　USB串口驱动程序安装成功，确认COM口的编号

（3）打开PuTTY软件（见图2-13），"Connection type"选择"Serial"。在"Serial line"中填写COM口编号，在"Speed"（速率）中填写115200，填写完毕后单击"Open"，如果没报错，直接出现黑色窗口，表示串口打开成功。

图2-13　PuTTY的配置界面

（4）将USB转串口线另外一端接入开发板底板的TTL接口，然后将底板上电，会看到如下信息。

```
Booting...
eth check connect ret=0 cost time=2000 ms
IROM DONE!
SiFlower SFAX8 Bootloader (Jul  9 2019 - 14:46:03)
ddr3 nt5cc128m16ip init start
MEM_PHY_CLK_DIV = 0x3
DDR training success
now ddr frequency is 400MHz!!!
ddr test
DR1BW a0000000 OK
DR1BR OK
DR2BW a0000000 OK
DR2BR OK
DR4BW a0000000 OK
DR4BR OK
DR8BW a0000000 OK
DR8BR OK
Boot from spi-flash
U-image: U-Boot 2016.07-rc2 for sfa18_p2, size is 310044
loaded - jumping to U-Boot 0xa0000000...
U-Boot 2016.07-rc2 (Jul 09 2019 - 14:46:10 +0800)
Board: MIPS sfa18 FULLMASK P20B
       Watchdog enabled
DRAM:  256 MiB
MMC:   emmc@7800000: 0sdio@7c00000: 1
SF: Detected EN25QH128A with page size 256 Bytes, erase size 4 KiB, total 16 MiB
In:    serial@8300000
Out:   serial@8300000
Err:   serial@8300000
Net:   Registering sfa18 net
Registering sfa18 eth
sf_eth0
Warning: sf_eth0 (eth0) using random MAC address - 4a:0a:ab:7c:96:2f
Hit any key to stop autoboot:  0
do_spi_flash----cmd = probe
SF: Detected EN25QH128A with page size 256 Bytes, erase size 4 KiB, total 16 MiB
do_spi_flash----cmd = read
device 0 offset 0xa0000, size 0x300000
SF: 3145728 bytes @ 0xa0000 Read: OK
## Booting kernel from Legacy Image at 81000000 ...
   Image Name:   MIPS OpenWrt Linux-3.18.29
   Image Type:   MIPS Linux Kernel Image (lzma compressed)
   Data Size:    1679584 Bytes = 1.6 MiB
   Load Address: 80100000
   Entry Point:  80105360
   Verifying Checksum ... OK
   Uncompressing Kernel Image ... OK
```

（5）信息会大量滚动，最后系统启动完毕，完成连接这一步骤。

USB转串口线看起来简单，但是你也可能会遇到一些问题。

● PuTTY打不开COM设备：驱动程序有问题，参数填写有问题，或TTL线有问题。
● 开发板上电以后屏幕什么都不显示：连接线有问题。
● 开发板烧了：赶紧联系硬件支持。
● 屏幕有显示，但显示的是乱码：可能是Speed（速率）没设置正确，也可能是线接触不良。重新连接所有设备（包括重插串口线），重新给开发板上电。
● 所有信息都能正常显示，但是按Enter键没反应并且无法输入信息：可能是线有问题，也可能是系统正在启动，还没到能输入的时候。

2.5 SSH远程登录

SSH（Secure Shell，安全外壳）是一种加密的网络传输协议。SSH通过在网络中创建安全的加密隧道来实现SSH客户端与服务器之间的安全连接。SSH是Linux系统中常用的远程命令管理协议，其作用类似我们通过串口登录，但是通过串口登录一定要有线连接，SSH可以通过网络直接访问，更方便我们进行远程调试。

默认情况下，OpenWrt系统的防火墙规则禁止通过WAN口访问SSH服务，只允许通过LAN口（有线或无线）访问，所以一定要通过LAN口进行连接，如果想通过WAN口访问SSH服务，则需要更改防火墙规则，而如何修改，请大家参考后面关于防火墙的章节。

2.5.1 SSH客户端常用软件

OpenWrt系统内运行了SSH服务器端程序，我们需要通过SSH客户端软件与SSH服务器端进行连接。SSH客户端非常多，比如Xshell、mobaxterm、SecureCRT等，但经常使用的就那几款，大家可以根据个人喜好和习惯进行选择。

如果你使用Linux，推荐使用OpenSSH Client作为SSH客户端。

Ubuntu用户可以通过apt命令进行安装：

```
HOST:~$ sudo apt-get -y install openssh-client
```

Centos用户可以通过yum命令进行安装：

```
HOST:~$ sudo yum -y install openssh-clients
```

如果你使用Windows，我们推荐使用PuTTY作为SSH客户端。PuTTY是一款集虚拟终端、系统控制台和网络文件传输于一体的自由及开放源代码（MIT License）的程序，由Simon Tatham为Windows平台开发，目前已经支持Linux。它支持多种协议，包括SSH、Telnet、Rlogin、Serial和Raw（原始的套接字连接），如图2-14所示。

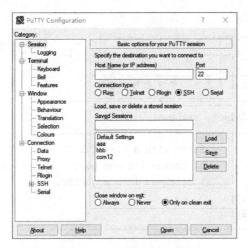

图2-14 PuTTY的操作界面

PuTTY默认只支持一个窗口会话，如果想要支持多个PuTTY窗口会话，那么推荐使用MTPuTTY工具（见图2-15）。它和PuTTY一样非常小巧，可以对PuTTY会话进行有效管理，支持无限多个PuTTY应用。一句话，你值得拥有。

图2-15 MTPuTTY的操作界面

2.5.2 SSH登录

将计算机通过有线网口与开发板LAN口连接，动态获取IP地址（192.168.4.X）。默认的开发板账号为admin，密码也为admin。

在Linux下直接使用ssh命令进行登录，命令格式如下。

```
ssh <远程用户名>@<目标IP> -p <端口,默认22>
HOST:~$ ssh admin@192.168.4.1
The authenticity of host '192.168.4.1 (192.168.4.1)' can't be established.
RSA key fingerprint is SHA256:h92x9Q9HhQ4lXx8tSFMB7cWo/4ENiP6ZSuhAA9Hs9cU.
Are you sure you want to continue connecting (yes/no)? yes
Warning: Permanently added '192.168.4.1' (RSA) to the list of known hosts.
admin@192.168.4.1's password:
BusyBox v1.23.2 (2019-06-24 10:12:04 CST) built-in shell (ash)
```

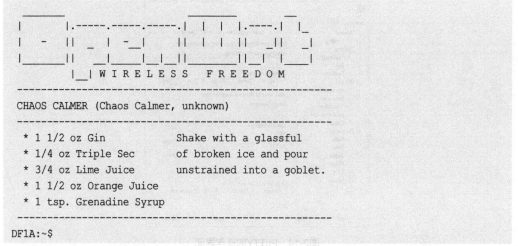

在Windows下，我们使用PuTTY工具，只需要打开一个PuTTY窗口（见图2-16），然后在"Host Name"处输入开发板的IP地址（192.168.4.1），在"Port"处输入端口号22，在"Connection type"处选择SSH，单击"Open"。

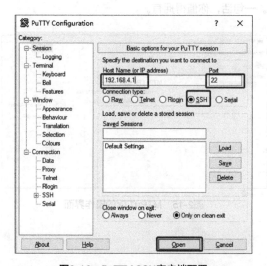

图2-16 PuTTY SSH客户端配置

第一次登录会提示警告信息（见图2-17），直接单击"是"确认就行，然后填写账户和密码。

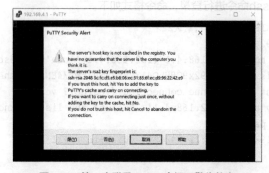

图2-17 第一次登录PuTTY会提示警告信息

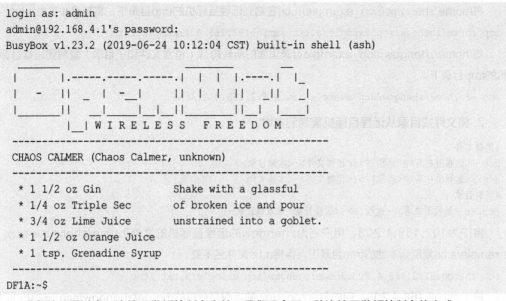

现在除了通过串口连接开发板控制台之外，我们又多了一种连接开发板控制台的方式。

2.6 SCP文件传输

我们已经学会了如何远程登录系统，但是在实际使用中，我们经常要实现文件的远程复制，虽然我们有多种方法可以选择，比如FTP、HTTP、Samba等，但是它们都不是特别便捷。我们可以直接用SSH附带的SCP协议实现文件的远程传输。接下来我们分别介绍如何在Linux下和Windows下将文件复制到OpenWrt目标系统中。

2.6.1 在Linux下使用scp命令

在Linux下，我们使用scp命令。scp是Secure Copy的缩写，是Linux系统下基于SSH登录进行安全的远程文件复制的命令。我们之前安装了OpenSSH客户端，就已经包含了scp命令了。

更详细的命令可以通过scp help查看。

```
HOST:~$ scp help
usage: scp [-12346BCpqrv] [-c cipher] [-F ssh_config] [-i identity_file]
           [-l limit] [-o ssh_option] [-P port] [-S program]
           [[user@]host1:]file1 ... [[user@]host2:]file2
```

下面介绍一下常用的命令格式。

1. 从本地将文件或目录复制到远程目标机

```
#复制文件
scp <本地文件> <远程用户名>@<远程IP>:<远程目录>
scp <本地文件> <远程用户名>@<远程IP>:<远程文件>
#复制目录
scp -r <本地目录> <远程用户名>@<远程IP>:<远程目录>
```

将/home/zhengqw/scp_example/a.txt复制到远程目标机的tmp目录下，保持a.txt文件名不变。

```
scp /home/zhengqw/scp_example/a.txt admin@192.168.4.1:/tmp
```

将/home/zhengqw/scp_example/a.txt复制到远程目标机的tmp目录下，将其重命名为b.txt。

```
scp /home/zhengqw/scp_example/a.txt admin@192.168.4.1:/tmp/b.txt
```

将/home/zhengqw/scp_example目录里面的所有内容（包含文件和子目录）复制到远程目标机的tmp目录下。

```
scp -r /home/zhengqw/scp_example admin@192.168.4.1:/tmp
```

2. 将文件或目录从远程目标机复制到本地

```
#复制文件
scp  <远程用户名>@<远程IP>:<远程文件> <本地目录>
scp  <远程用户名>@<远程IP>:<远程文件> <本地文件>
#复制目录
scp -r <远程用户名>@<远程IP>:<远程目录> <本地目录>
```

将IP为192.168.4.203、用户名为zhengqw的远程目标机的文件/home/zhengqw/scp_example/a.txt复制到本地的tmp目录下，保持a.txt文件名不变。

```
scp zhengqw@192.168.4.203:/home/zhengqw/scp_example/a.txt /tmp
```

将IP为192.168.4.203、用户名为zhengqw的远程目标机的/home/zhengqw/scp_example/目录里面的所有内容（包含文件和子目录）复制到本地的tmp目录下。

```
scp -r zhengqw@192.168.4.203:/home/zhengqw/scp_example/ /tmp
```

2.6.2 在Windows下使用SCP客户端

（1）先到WinSCP官方网站下载WinSCP，这是一个大名鼎鼎的SCP客户端软件。

（2）对下载包进行解压缩，然后双击进行"安装"，首先选择语言，单击"确定"（见图2-18）。

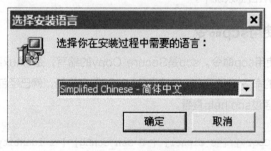

图2-18 选择语言

（3）单击"下一步"，选择同意许可协议，再单击"下一步"，安装类型选择"典型安装"，单击"下一步"（见图2-19）。

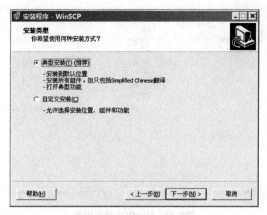

图2-19 选择"典型安装"

(4)用户界面选择"Commander界面",单击"下一步"(见图2-20)。

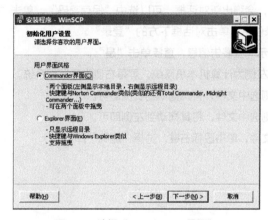

图2-20 选择"Commander界面"

(5)单击"安装",进入安装步骤,完成后单击"完成"。
(6)WinSCP首次启动后,会立即展现出登录创建窗口(见图2-21),填写以下登录信息。

- 主机名:填写开发板的IP地址,一般是192.168.4.1。
- 端口号:SSH的端口号。
- 用户名:填写admin。
- 密码:填写SSH登录用的密码。
- 文件协议:一定要选择SCP,否则登录无法完成。

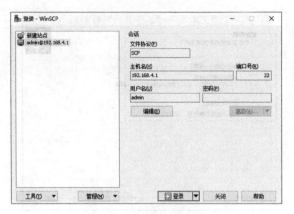

图2-21 WinSCP登录界面

（7）单击"保存"，会弹出新对话框，可以选中"保存密码"，单击"确认"，返回登录对话框。选中刚才创建的登录信息，单击对话框下方的"登录"。

（8）首次登录会提示一些警告信息，直接单击"是"。

（9）成功后，屏幕左侧为计算机本机系统，屏幕右侧为开发板系统。

- 上传文件：在左侧选中文件，将其拖动到右侧即可。
- 下载文件：在右侧选中文件，将其拖动到左侧即可。
- 删除文件：选中文件，单击鼠标右键，选择"删除"即可。

3 分区与软件包

3.1 SPI Flash分区原理

SPI是串行外设接口（Serial Peripheral Interface）的缩写。SPI是一种高速、全双工、同步的通信总线，它可以使CPU与外设通过串行方式进行通信，并且在芯片的引脚上只占用4根线即可完成传输：串行时钟线（SCLK）、主机输入/从机输出数据线（MISO）、主机输出/从机输入数据线（MOSI）和低电平有效的从机选择线（$\overline{SS}$）。SPI节约了芯片的引脚，同时为PCB布局节省空间提供方便。正是出于这种简单易用的特性，如今越来越多的芯片支持这种通信协议。

SPI Flash是具有串行通信接口的Flash。早期NorFlash的接口是并行的，即把数据线和地址线并排与IC的引脚连接。但是后来发现不同容量的Norflash数据线和地址线的数量不一样，不能在硬件上兼容，并且封装比较大，所以并行NorFlash接口后来逐渐被SPI NorFlash所取代。SPI NorFlash使用的引脚和封装更小，兼容性更好。所以现在提到SPI Flash就是指SPI NorFlash。

常见的SPI Flash芯片一般有8个引脚或16个引脚。本教程所用开发板DF1A采用是8引脚16MB SPI Flash，大家可以对着开发板找一下。

3.1.1 了解分区

我们知道磁盘可以进行分区，其实路由器Flash也是有分区的。当开发板刷机后，第一次启动时，通过串口可以看到以下信息。

```
[   12.037025] jffs2: notice: (478) jffs2_build_xattr_subsystem: complete building xattr subsystem, 0 of xdatum (0 unchecked, 0 orphan) and 0 of xref (0 dead, 0 orphan) found.
```

这句话的意思是：使用jffs2文件系统完成了格式化。Linux系统对闪存类存储器是采用MTD（Memory Technology Devices，内存技术设备）类设备驱动实现的，MTD是用于访问内存类设备（ROM、Flash）的Linux驱动子系统。它的主要目的是使闪存类设备更加容易被访问，为此它在硬件和上层提供了一个抽象的接口，使得在操作系统下，我们可以像操作硬盘一样操作这类设备。仔细观察Linux的启动信息，你会看到这么一段。

```
[    1.199720] 5 ofpart partitions found on MTD device spi0.0
[    1.205260] Creating 5 MTD partitions on "spi0.0":
[    1.210086] 0x000000000000-0x000000020000 : "spl-loader"
[    1.217105] 0x000000020000-0x000000080000 : "u-boot"
[    1.223582] 0x000000080000-0x000000090000 : "u-boot-env"
```

```
[    1.230375] 0x000000090000-0x0000000a0000 : "factory"
[    1.237000] 0x0000000a0000-0x000001000000 : "firmware"
[    1.296625] 2 uimage-fw partitions found on MTD device firmware
[    1.302596] 0x0000000a0000-0x00000023a120 : "kernel"
[    1.309260] 0x00000023a120-0x000001000000 : "rootfs"
[    1.315842] mtd: device 6 (rootfs) set to be root filesystem
[    1.321690] 1 squashfs-split partitions found on MTD device rootfs
[    1.327936] 0x000000940000-0x000001000000 : "rootfs_data"
```

第二句话的意思是,系统在SPI Flash(spi0.0)设备上创建了5个MTD分区。这5个分区以及它们的子分区的说明如表3-1所示。

表3-1 分区说明

分区ID		位置	容量	作用
spl-loader		0x000000000000~0x000000020000	128KB	第一个引导程序
u-boot		0x000000020000~0x000000080000	384KB	引导程序
u-boot-env		0x000000080000~0x000000090000	64KB	引导程序配置
factory		0x000000090000~0x0000000a0000	64KB	初始参数
firmware	整体	0x0000000a0000~0x000001000000	15.4MB	固件分区
	kernel子分区	0x0000000a0000~0x00000023a120	1.6MB	Linux内核
	rootfs子分区	0x00000023a120~0x000001000000	13.8MB	固件分区文件系统子集
	rootfs_data子分区	0x000000940000~0x000001000000	6.8MB	固件分区文件系统子集、可写分区子集

spl-loader,简称为SPL(部分进口芯片没有这一分区设计),主要用于DDR初始化。SPL从Flash的0地址开始,镜像约22KB,分区最小支持32KB。

u-boot是用于引导和启动内核程序的BootLoader。

u-boot-env是用于保存U-Boot使用的环境变量的分区,可以在U-Boot控制台中通过printenv命令查看其内容。如果U-Boot的配置固定,不需修改,可以去掉该分区。

factory用于保存系统信息、无线校准值和MAC地址信息,这一分区的内容格式,不同芯片厂商定义不同,以下内容均以SF16A18芯片进行说明。前面2KB为系统信息,后面2KB为无线校准值。后面具体开发应用时,我们会详细介绍这个分区的内容。SF16A18芯片的factory分区前2KB信息如表3-2所示。

表3-2 SF16A18芯片的factory分区前2KB信息说明

Index（字节）	Counts（字节）	内容	示例	用途
0~5	6	MAC起始地址	0xa8, 0x5a, 0xf3, 0, 0, 0	必须有一个唯一MAC地址
6~21	16	SN号	全0xff	App使用（云服务器将为每台设备分配一个SN号）
22	1	SN号设置确认码	0xff	App使用（由云服务器配置）
23~26	4	区分pcba和U-Boot	0	必须为0,在SPL中进行判断,跳转U-Boot还是pcba代码

续表

Index （字节）	Counts （字节）	内容	示例	用途
27~28	2	硬件版本号确认码（'hv'表示下面的硬件版本号是有效的）	'hv'	硬件版本号有效时才填写
29~60	32	硬件版本号	'A18_XC-LY801-C4_V3_20190412'	App使用，仅在网页上显示，可以不配置
61~62	2	国家和地区码	'CN'	影响Wi-Fi信道选择，如果为0xff，则为CN
63~64	2	产品型号确认码（'mv'表示下面的产品型号是有效的）	'mv'	产品型号有效时才填写
65~96	32	产品型号	'XC1200'	App使用，仅在网页上显示，可以不配置
97~100	4	硬件特性（默认为0xffffffff，每一位表示一个特性）	0xfffffffe	必须确认硬件规格后再填写
101~102	2	公司名称确认码（'vd'表示下面的公司名称是有效的）	'vd'	公司名称有效时才填写
103~118	16	公司名称	'siflower'	App使用，可以不配置
119~120	2	产品秘钥确认码（'pk'表示下面的产品秘钥是有效的）	'pk'	产品秘钥有效时才填写
121~152	32	产品秘钥（改成32字节）	'c51ce410c124a10e0db5e4b97fc2af39'	App使用，可以不配置（使用时秘钥需要根据不同设备向PM申请）
153~154	2	登录信息确认码（'li'表示下面的登录信息是有效的）	'li'	登录信息有效时才填写
155~158	4	登录信息	0xffffffff	控制Telnet Server、SSH Server、UART等（目前仅实现Telnet Server）

firmware包括整个OpenWrt系统和用户数据，对应镜像为openwrt-*.bin。其中kernel为内核镜像，rootfs为文件系统，包含了：

- 只读文件分区（SquashFS）放在首位置，大小随着固件编译内容不同有一定变化；
- 其余部分为可写分区（JFFS2文件系统），命名为rootfs-data。

DF1A开发板的Flash分区如图3-1所示。

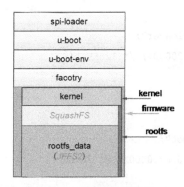

图3-1　SPI Flash的分区

（1）分区存在子分区，比如kernel和rootfs是firmware的子分区，而rootfs_data是rootfs的子分区。

（2）整个分区从0x000000000000开始，到0x000001000000结束，默认显示都是十六进制数值，可以通过计算器转换成十进制进行查看。

（3）kernel分区的容量计算公式如下，为什么这么写请看前面那个表的分区位置来琢磨。

0x0000000a0000-0x00000023a120=0x19a120,换算为十进制/1024/1024=1.6(MB)

（4）rootfs_data分区的容量计算公式如下。

0x000000940000-0x000001000000=0x6c0000,换算为十进制/1024/1024=6.8(MB)

分区的容量不是一成不变的，随着固件的变化，其firmware部分的容量也会改变，不过没关系，只要运用上面的计算方法，就可以计算分区容量是怎么变化的。

3.1.2 分区是怎么划分的

由于嵌入式系统的Flash容量都很小，而且不需要进行频繁调整，所以分区都是固定的。在路由器的Flash中，分区都是事先固化好的，基本上不用修改。

如果需要重新划分分区，需要在划分分区前进行功能规划，哪个分区做什么事，事先要设定好，每个分区的容量也要提前规划好，避免出现容量不够用的情况。分区的划分涉及的内容比较多，我们这里先进行简单介绍，等到后面的实战章节涉及相关内容，再详细展开。

分区信息定义在内核的DTS文件中，对于MIPS体系结构，DTS文件在内核源码的arch/mips/boot目录下。

DF1A开发板使用的是型号为w25q128（16MB SPI-Flash）的存储芯片兼容驱动程序，DTS分区信息如下。

```
w25q128@0 {
        compatible = "w25q128";
        reg = <0>;          /* chip select */
        spi-max-frequency = <33000000>;
        bank-width = <2>;
        device-width = <2>;
        #address-cells = <1>;
        #size-cells = <1>;
        partition@0 {
                label = "spl-loader";
                reg = <0x0 0x20000>; /* 128KB */
                read-only;
        };
        partition@20000 {
                label = "u-boot";
                reg = <0x20000 0x60000>; /* 384KB */
        };
        partition@80000 {
                label = "u-boot-env";
                reg = <0x80000 0x10000>; /* 64KB */
        };
```

```
        factory:partition@90000 {
            label = "factory";
            reg = <0x90000 0x10000>; /* 64KB */
        };
        partition@a0000 {
            label = "firmware";
            reg = <0xa0000 0xf60000>; /* 640KB~16MB */
        };
};
```

SPI Flash驱动程序会根据DTS中描述的分区信息进行Flash分区的划分。

查看MTD分区

```
DF1A:$ cat /proc/mtd
dev:    size     erasesize  name
mtd0: 00020000 00010000 "spl-loader"
mtd1: 00060000 00010000 "u-boot"
mtd2: 00010000 00010000 "u-boot-env"
mtd3: 00010000 00010000 "factory"
mtd4: 00f60000 00010000 "firmware"
mtd5: 0019a120 00010000 "kernel"
mtd6: 00dc5ee0 00010000 "rootfs"
mtd7: 006c0000 00010000 "rootfs_data"
```

查看MTD系统分区

```
DF1A:~$ cat /proc/partitions
major minor  #blocks  name
   31     0      128 mtdblock0
   31     1      384 mtdblock1
   31     2       64 mtdblock2
   31     3       64 mtdblock3
   31     4    15744 mtdblock4
   31     5     1640 mtdblock5
   31     6    14103 mtdblock6
   31     7     6912 mtdblock7
```

读取和备份分区

了解分区原理后，相信大家对固件升级或烧写原理很清楚了，烧写U-Boot或firmware（固件），实际上就是擦除和将新程序写入对应的分区。备份U-Boot或firmware，就是备份对应分区的内容。

下面列出几个常用备份、恢复命令。

1. 备份整个SPI Flash，将固件保存为dragonfly_1a_backup.bin

```
DF1A:~$ cat /dev/mtd0 /dev/mtd1 /dev/mtd2 /dev/mtd3 \
/dev/mtd4 > /tmp/dragonfly_1a_backup.bin
```

思考一下，为什么只要保存mtd0~mtd4的内容，就把整个SPI Flash给备份了？

2. 备份整个uboot分区，将固件保存为uboot_backup.bin

```
DF1A:~$ dd if=/dev/mtd1 of=/tmp/uboot_backup.bin
```

3. 备份整个factory分区，将固件保存为factory_backup.bin

```
DF1A:~$ dd if=/dev/mtd3 of=/tmp/factory_backup.bin
```

4. 从factory_backup.bin文件，恢复factory分区

```
DF1A:~$ dd if=/tmp/factory_backup.bin of=/dev/mtd3
```

factory分区没有文件系统，里面保存的是一些系统和应用的配置，比如当前这台机器的MAC地址。我们可以通过一个十六进制读取命令hexdump直接对其进行读取。

如前面所讲，factory分区对应的设备文件为/dev/mtd3或/dev/mtdblock3。

```
DF1A:~$ hexdump -C /dev/mtd3
00000000  10 16 88 14 3c fb ff ff  ff ff ff ff ff ff ff ff  |....<...........|
00000010  ff ff ff ff ff ff ff 00  00 00 00 68 76 41 31 38  |...........hvA18|
00000020  5f 45 56 42 5f 56 35 5f  32 30 31 39 30 34 32 34  |_EVB_V5_20190424|
00000030  00 00 00 00 00 00 00 00  00 00 00 00 00 43 4e 6d  |.............CNm|
00000040  76 45 56 42 5f 56 35 00  00 00 00 00 00 00 00 00  |vEVB_V5.........|
00000050  00 00 00 00 00 00 00 00  00 00 00 00 00 00 00 00  |................|
00000060  00 fe ff ff ff 76 64 73  69 66 6c 6f 77 65 72 00  |.....vdsiflower.|
00000070  00 00 00 00 00 00 00 70  6b 30 00 00 00 00 00 00  |.......pk0......|
00000080  00 00 00 00 00 00 00 00  00 00 00 00 00 00 00 00  |................|
00000090  00 00 00 00 00 00 00 00  6c 69 00 00 00 00 00 ff  |........li......|
000000a0  ff ff ff ff ff ff ff ff  ff ff ff ff ff ff ff ff  |................|
*
00000800  56 32 0e 0e 07 07 07 07  07 07 07 07 07 07 07 07  |V2..............|
00000810  07 07 07 07 07 07 07 07  07 07 07 07 07 07 07 07  |................|
*
00000970  06 06 06 06 06 06 06 06  06 06 06 06 06 06 06 06  |................|
*
00000990  06 07 07 07 07 07 07 06  06 06 06 07 07 07 07 07  |................|
000009a0  06 06 06 06 06 06 06 06  06 06 06 06 06 06 06 06  |................|
*
000009c0  06 06 06 06 06 06 07 07  07 07 07 06 06 06 06 06  |................|
000009d0  07 07 07 07 07 06 06 06  06 06 06 06 06 06 06 06  |................|
000009e0  06 06 06 06 06 06 06 06  06 06 06 06 06 06 06 06  |................|
000009f0  06 06 06 06 06 06 06 06  06 06 06 07 07 07 07 07  |................|
00000a00  06 06 06 06 06 06 07 07  07 07 06 06 06 06 06 06  |................|
00000a10  06 06 06 06 06 06 06 06  06 06 06 06 06 06 06 06  |................|
...
```

一般来说，芯片厂不会公开这部分参数，但是我们可以看看芯片厂的资料里哪个像MAC地址，这就是芯片出厂时的MAC地址。这部分内容不太容易理解，可以查看芯片厂提供的资料进行对比理解，例如对于DF1A开发板，参考表2-7可知factory分区的前6字节"10 16 88 14 3c fb"为开发板的MAC地址，通过ifconfig命令可以验证。

3.2 文件系统与透明挂载

通过前面的学习，我们知道firmware分区包含kernel和rootfs分区，rootfs又包含rootfs_data分区，其中固件原始的rootfs的文件系统是只读文件系统SquashFS，而rootfs_data的文件系统是可写文件系统JFFS2。

OpenWrt采用了基于Overlay技术的OverlayFS文件系统，将rootfs（SquashFS）和rootfs_data合并成一个逻辑分区，挂载为可见逻辑分区。而实际的rootfs（SquashFS）挂载在/rom下，rootfs_data挂载在/overlay下，这种技术叫作透明挂载。OpenWrt之所以有这些操作，主要是因为嵌入式系统的限制以及相关功能的需求导致使用单一的文件系统难以实现。OpenWrt根据所要实现的功能，选择不同的文件系统（SquashFS、JFFS2和OverlayFs）进行组合实现。

SquashFS文件系统

SquashFS是基于GPL协议开发的只读压缩文件系统，早期版本使用GZIP算法进行压缩，改进版本采用LZMA算法进行压缩。LZMA算法具有高压缩比、解压缩时仅需少量内存、解压和压缩速度快以及支持多线程等特点，非常适合用于嵌入式系统。SquashFS能够为文件系统内的文件、inode及目录结构进行压缩，并支持最大1024MB的区块，以提供更大的压缩比。OpenWrt中使用了采用LZMA算法的SquashFS文件系统，它具有以下特点：

- 空间占用更少（据统计，较之JFFS2文件系统可以节省20%~30%的空间）；
- 支持FailSafe功能，即系统出现问题可以恢复出厂配置；
- 浪费空间，当该文件系统中的某一文件被修改时，该文件会被复制到另一个分区（如JFFS2）中；
- 该文件系统在OpenWrt中为只读的，主要是用于存储系统编译后的默认系统文件，恢复出厂设置时会将默认系统文件通过Overlay技术呈现给用户。

JFFS2文件系统

JFFS2是一种可写的日志结构文件系统，可以使用LZMA算法（内核需要配置CONFIG_JFFS2_LZMA=y）进行压缩。它具有以下特点：

- 可写，是日志结构类型文件系统，且具有损耗平衡；
- 可压缩，所以程序占用空间更少；
- 读是以页（页大小为512字节）为单位进行的，而擦写是以擦写块（NOR:64~128KB，NAND:8~32KB）为单位操作的，该系统维护了几个链表来管理擦写块，根据擦写块上的内容，擦写块可能会在不同的链表上；
- 该文件系统在OpenWrt中主要用于存储可以更改的配置文件以及安装的软件包等，作为OpenWrt的可写部分提供了更新、升级软件不需要整体刷机的功能。

OverlayFS文件系统

OverlayFS由Miklos Szeredi开发，主要目的是在共同的基础文件系统上建立虚拟化系统。其主要的特性是：叠合两种不同的文件系统，一个只读，另一个可写。这样的特性对用户是透明的，使系统能支持恢复出厂配置等功能。OverlayFS文件系统在2014年12月3.18版本中被合并到Linux kernel主线中。该文件系统在OpenWrt中主要提供一种黏合机制，提高了OpenWrt采用不同文件系统实现不同功能的灵活性。

Overlay技术不只用于OpenWrt，Docker技术也采用Overlay技术，对于修改的image内容，Commit后都通过Overlay呈现，而image本身内容没有变化。

Overlay表示叠加和覆盖的意思。OverlayFS文件系统则表示一个文件系统覆盖在另一个文件系统上面。一个OverlayFS文件系统通常包含两个层（一般属于不同文件系统）——一个是upper层，另一个是lower层，对外呈现两个层叠加后的内容（见图3-2）。如果两个层所在文件系统目录树中存在相同名称的对象，那么叠加结果取决于对象的类型。

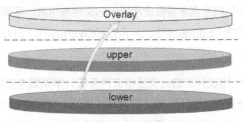

图3-2 Overlay示意图

- 文件：upper层中的文件呈现在Overlay中，lower层中的同名文件被隐藏。
- 目录：upper层中和lower层中的目录被合并到一个联合的目录，呈现在Overlay中。
- 一般来说，lower层文件系统为只读的，也可以是Overlay本身；而upper层文件系统通常为可写的。

3.2.1 Overlay实战

由于理解Overlay对于理解整个OpenWrt文件系统的原理至关重要，所以接下来我们重点讨论一下Overlay的相关内容。

Overlay的挂载命令如下（OpenWrt下不需要人工做这个操作）。

```
mount -t overlay overlay -o lowerdir=<lower>,upperdir=upper,workdir=work merged
```

其中"<lower>"表示lower层目录。Linux内核版本Linux 4.0及以上支持多lower层（最大支持500层）目录，Linux 3.18版本只支持一个lower层目录。不同的lower层目录使用":"分隔，如"lower1:lower2:lower3:……"，层次关系依次为lower1>lower2>……lowerX。upper和work目录分别表示upper层目录和文件系统挂载后用于存放临时和间接文件的工作基础目录（work base dir），最后的merged目录就是最终的挂载点目录。执行以上命令后，OverlayFS就成功挂载到merged目录下了。

- lowerdir=xxx：指定用户需要挂载的lower层目录。
- upperdir=xxx：指定用户需要挂载的upper层目录。
- workdir=xxx：指定文件系统的工作基础目录，挂载后内容会被清空，且在使用过程中其内容用户不可见。
- default_permissions：功能未使用。
- redirect_dir=on/off：开启或关闭redirect directory特性，开启后可支持merged目录和纯lower

层目录的rename/renameat系统调用。

- index=on/off：开启或关闭index特性。

如果你觉得Overlay的工作原理很不好理解，我们接下来通过实例来理解。

测试设备是否支持Overlay

（1）查看一下系统内核版本是不是大于等于3.18，并且是否支持OverlayFS文件系统。我们在开发板上进行测试，开发板内核版本为3.18.29，并且支持OverlayFS。

```
DF1A:~$ uname -r
3.18.29
DF1A:~$ cat /proc/filesystems|grep overlay
nodev  overlay
```

（2）创建overlay_example测试目录。

```
DF1A:~$ mkdir -p /overlay/overlay_example/
```

请思考一下，为什么要在/overlay目录下创建测试目录？

基本覆盖测试

（1）在overlay_example目录下创建OverlayFS文件系统的4个基础目录：lower、upper、work、merged。

（2）然后分别在两个lower目录和upper目录下创建同名文件foo、同名目录dir，并同时在dir目录下创建同名文件a，在lower/dir下创建文件b，在upper/dir下创建文件c。

（3）挂载OverlayFS文件系统后，在merged目录下能够看到foo，这就是OverlayFS的上下层合并。

（4）在merged/dir目录下看到来自lower层的文件b和来自upper层的文件a和c，位于最底层lower中的文件a被upper中的同名文件a覆盖（见图3-3），这就是OverlayFS的"上下层同名目录合并与同名文件覆盖"特性。

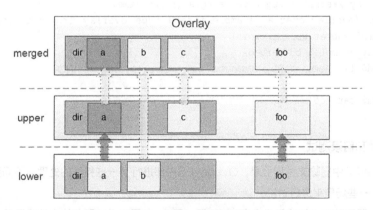

图3-3　Overlay合并示意图

以上操作的测试语法如下。

```
DF1A:~$ cd /overlay/overlay_example/
DF1A:overlay_example$ mkdir -p lower upper work merged
DF1A:overlay_example$ ls
```

```
lower    merged   upper    work
DF1A:overlay_example$ mkdir -p lower/dir upper/dir
DF1A:overlay_example$ touch lower/foo upper/foo
DF1A:overlay_example$ echo "i am lower foo"> lower/foo
DF1A:overlay_example$ echo "i am upper foo"> upper/foo
DF1A:overlay_example$ touch lower/dir/a upper/dir/a
DF1A:overlay_example$ touch lower/dir/b upper/dir/c
DF1A:overlay_example$ echo "i am lower dir a"> lower/dir/a
DF1A:overlay_example$ echo "i am upper dir a">upper/dir/a
DF1A:overlay_example$ mount -t overlay overlay -o lowerdir=lower,upperdir=upper,workdir=work merged
DF1A:overlay_example$ df
Filesystem           1K-blocks      Used Available Use% Mounted on
rootfs                    6656       548      6108   8% /
/dev/root                 7424      7424         0 100% /rom
tmpfs                    61348      1200     60148   2% /tmp
/dev/mtdblock7            6656       548      6108   8% /overlay
overlayfs:/overlay        6656       548      6108   8% /
tmpfs                      512         0       512   0% /dev
overlay                   6656       548      6108   8% /overlay/overlay_example/merged
DF1A:overlay_example$ mount
rootfs on / type rootfs (rw)
/dev/root on /rom type squashfs (ro,relatime)
proc on /proc type proc (rw,nosuid,nodev,noexec,noatime)
sysfs on /sys type sysfs (rw,nosuid,nodev,noexec,noatime)
tmpfs on /tmp type tmpfs (rw,nosuid,nodev,noatime)
/dev/mtdblock7 on /overlay type jffs2 (rw,noatime)
overlayfs:/overlay on / type overlay (rw,noatime,lowerdir=/,upperdir=/overlay/upper,workdir=/overlay/work)
tmpfs on /dev type tmpfs (rw,nosuid,relatime,size=512k,mode=755)
devpts on /dev/pts type devpts (rw,nosuid,noexec,relatime,mode=600)
debugfs on /sys/kernel/debug type debugfs (rw,noatime)
overlay on /overlay/overlay_example/merged type overlay (rw,relatime,lowerdir=lower,upperdir=upper,workdir=work)
DF1A:overlay_example$ cd merged/
DF1A:merged$ ls
dir  foo
DF1A:merged$ cat foo
i am upper foo
```

删除文件和目录测试

在OverlayFS中删除文件和目录，OverlayFS内部进行了一些特殊的处理。下面我们根据实际使用中遇到的一些场景来进行讨论。

- 场景：要删除的文件或目录来自upper层，且lower层中没有同名的文件或目录

upper层的文件系统是可写的，所有在OverlayFS中的操作，可以直接体现在upper层所对应的文件系统中，因此直接删除upper层或merged目录中的文件或目录，Overlay呈现的结果都是删除。

这里在upper目录下创建了文件foo和目录dir，然后在挂载OverlayFS后，从merged目录下删除它们，upper目录下的文件foo和目录dir也同时被直接删除（见图3-4）。

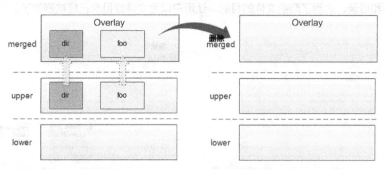

图3-4　仅upper层有文件或目录并执行删除操作

```
DF1A:merged$ cd /
DF1A:$ umount /overlay/overlay_example/merged
DF1A:$ rm -rf /overlay/overlay_example
DF1A:$ mkdir -p /overlay/overlay_example/
DF1A:$ cd /overlay/overlay_example/
DF1A:overlay_example$ mkdir lower upper work merged
DF1A:overlay_example$ ls
lower    merged  upper    work
DF1A:overlay_example$ touch upper/foo
DF1A:overlay_example$ mkdir upper/dir
DF1A:overlay_example$ mount -t overlay overlay -o lowerdir=lower,upperdir=upper,
workdir=work merged
DF1A:overlay_example$ ls -l upper/
drwxr-xr-x    2 admin     root             0 Jul 24 10:42 dir
-rw-r--r--    1 admin     root             0 Jul 24 10:42 foo
DF1A:overlay_example$ ls -l merged/
drwxr-xr-x    2 admin     root             0 Jul 24 10:42 dir
-rw-r--r--    1 admin     root             0 Jul 24 10:42 foo
DF1A:overlay_example$ rm -rf merged/*
DF1A:overlay_example$ ls -l merged/
DF1A:overlay_example$ ls -l upper/
```

● 场景：要删除的文件或目录来自lower层，upper层不存在覆盖文件

由于lower层中的内容对于OverlayFS来说是只读的，所以这里的删除并不是真正删除了文件，而是Overlay进行特殊的处理，让用户误以为文件被删除了。OverlayFS针对这种场景通过创建特殊whiteout文件设计了一套"魔术障眼法"。

whiteout概念存在于联合文件系统（UnionFS）中，代表某一类占位符形态的特殊文件，当用户目录与系统目录的共通部分联合到一个目录（例如bin目录）时，用户可删除归属于自己的某些系统文件副本，但归属于系统级的原件仍保留于同一个联合目录中，此时系统将产生一份whiteout文件，表示该文件在当前用户目录中已被删除，但在系统目录中仍然保留。

whiteout文件在用户删除文件或目录时在upper层创建，用于屏蔽底层的同名文件或目录，同时该whiteout文件在merged层是不可见的，所以用户就看不到被删除的文件或目录了（见图

3-5）。whiteout文件并非普通文件，而是主、次设备号都为0的字符设备，当用户在merged层通过ls命令检查父目录的目录项时，OverlayFS会自动过滤掉和whiteout文件自身以及和它同名的lower层文件和目录，达到了隐藏文件的目的，让用户以为文件或目录已经被删除了。

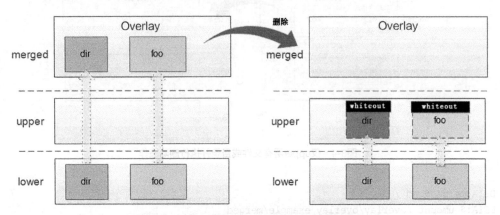

图3-5 仅lower层有文件或目录并执行删除操作

```
DF1A:overlay_example$ cd ..
DF1A:overlay$ umount /overlay/overlay_example/merged
DF1A:overlay$ rm -rf /overlay/overlay_example/
DF1A:overlay$ mkdir -p /overlay/overlay_example/
DF1A:overlay$ cd /overlay/overlay_example/
DF1A:overlay_example$ mkdir lower upper work merged
DF1A:overlay_example$ touch lower/foo
DF1A:overlay_example$ mkdir -p lower/dir
DF1A:overlay_example$ mount -t overlay overlay -o lowerdir=lower,upperdir=upper,
workdir=work merged
DF1A:overlay_example$ ls merged/
dir  foo
DF1A:overlay_example$ ls upper/
DF1A:overlay_example$ ls lower/
dir  foo
DF1A:overlay_example$ rm -rf merged/*
DF1A:overlay_example$ ls lower/
dir  foo
DF1A:overlay_example$ ls -l upper
c---------   1 admin    root         0,    0 Jul 23 17:08 dir
c---------   1 admin    root         0,    0 Jul 23 17:08 foo
DF1A:overlay_example$ ls -l lower/
drwxr-xr-x   2 admin    root              0 Jul 23 17:07 dir
-rw-r--r--   1 admin    root              0 Jul 23 17:07 foo
DF1A:overlay_example$ ls merged/
```

● 场景：要删除的文件是upper层覆盖lower层的文件，要删除的目录是上下层合并的目录

这是前两个场景的合并，OverlayFS既需要删除upper层对应文件系统中的文件或目录，也需要在对应位置创建同名whiteout文件，让upper层的文件和目录被删除后不至于让lower层的文件和目录被暴露出来。

```
DF1A:overlay_example$ cd /
DF1A:$ umount /overlay/overlay_example/merged
DF1A:$ rm -rf /overlay/overlay_example/
DF1A:$ mkdir -p /overlay/overlay_example/
DF1A:$ cd /overlay/overlay_example/
DF1A:overlay_example$ mkdir lower upper work merged
DF1A:overlay_example$ touch lower/foo upper/foo
DF1A:overlay_example$ mkdir -p lower/dir upper/dir
DF1A:overlay_example$ mount -t overlay overlay -o lowerdir=lower,upperdir=upper,
workdir=work merged
DF1A:overlay_example$ ls merged/
dir  foo
DF1A:overlay_example$ ls upper/
dir  foo
DF1A:overlay_example$ ls lower/
dir  foo
DF1A:overlay_example$ rm -rf merged/*
DF1A:overlay_example$ ls merged/
DF1A:overlay_example$ ls -l upper/
c---------  1 admin    root          0,   0 Jul 23 17:19 dir
c---------  1 admin    root          0,   0 Jul 23 17:19 foo
DF1A:overlay_example$ ls lower/
dir  foo
```

这里在upper目录和lower目录中都创建了文件foo和目录dir，在挂载OverlayFS后从merged目录删除它们，然后检查lower层中的文件和目录依然不变，同时upper层中的原有文件和目录已经被替换成两个同名的whitout文件了。操作流程及结果如图3-6所示。

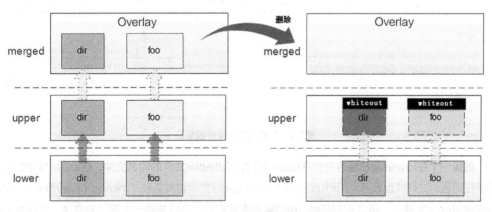

图3-6　upper和lower复合并执行删除操作

- 场景：创建文件和目录测试

创建文件和目录同删除类似，OverlayFS也需要针对不同的场景进行不同的处理。

创建一个全新的文件或目录：如果在lower层中和upper层中都不存在对应的文件或目录，那直接在upper层中对应的目录下创建文件或目录即可。

```
DF1A:overlay_example$ cd /
DF1A:$ umount /overlay/overlay_example/merged
```

```
DF1A:$ rm -rf /overlay/overlay_example/
DF1A:$ mkdir -p /overlay/overlay_example/
DF1A:$ cd /overlay/overlay_example/
DF1A:overlay_example$ mkdir lower upper work merged
DF1A:overlay_example$ mount -t overlay overlay -o lowerdir=lower,upperdir=upper,
workdir=work merged
DF1A:overlay_example$ ls
lower    merged   upper    work
DF1A:overlay_example$ ls merged/
DF1A:overlay_example$ ls lower/
DF1A:overlay_example$ ls upper/
DF1A:merged$ touch foo
DF1A:merged$ mkdir dir
DF1A:merged$ ls
dir  foo
DF1A:merged$ ls ../lower/
DF1A:merged$ ls -l ../upper/
drwxr-xr-x    2 admin    root            0 Jul 23 17:50 dir
-rw-r--r--    1 admin    root            0 Jul 23 17:50 foo
```

挂载后在merged目录中创建文件file和目录dir，它们直接被创建到了upper层对应的文件系统中，而lower层不受任何影响。操作流程及结果如图3-7所示。

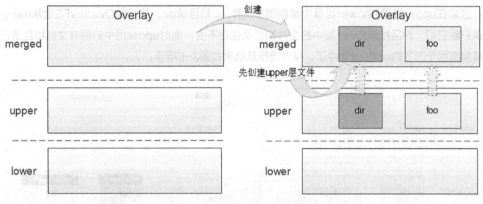

图3-7　创建全新的文件或目录

创建一个在lower层已经存在且在upper层有whiteout文件的同名文件： 在lower层中之前已经存在同名的文件或目录了，同时upper层也有whiteout文件将其隐藏（mknod upper/foo c 0 0，创建whiteout文件），所以用户在merged层看不到它们，可以新建一个同名的文件。在这种场景下，OverlayFS先删除upper层中的whiteout文件，然后创建新建的文件，这样在merged层中看到的文件就是来自upper层的新文件。

```
DF1A:overlay_example$ cd /
DF1A:$ umount /overlay/overlay_example/merged
DF1A:$ rm -rf /overlay/overlay_example/
DF1A:$ mkdir -p /overlay/overlay_example/
DF1A:$ cd /overlay/overlay_example/
```

```
DF1A:overlay_example$ mkdir lower upper work merged
DF1A:overlay_example$ touch lower/foo
DF1A:overlay_example$ mknod upper/foo c 0 0
DF1A:overlay_example$ mount -t overlay overlay -o lowerdir=lower,upperdir=upper,
workdir=work merged
DF1A:overlay_example$ ls merged/
DF1A:overlay_example$ ls -l upper/
crw-r--r--    1 admin      root             0,    0 Jul 23 18:13 foo
DF1A:overlay_example$ ls lower/
foo
DF1A:overlay_example$ touch merged/foo
DF1A:overlay_example$ ls -l merged/
-rw-r--r--    1 admin      root                   0 Jul 23 18:14 foo
DF1A:overlay_example$ ls -l upper/
-rw-r--r--    1 admin      root                   0 Jul 23 18:14 foo
```

这里先在lower目录中创建文件foo，然后在upper目录中创建同名的whiteout文件用于隐藏lower层中的文件，挂载OverlayFS文件系统后通过merged目录新建文件foo。lower目录中的原有foo文件不变，upper目录中的foo（whiteout文件）已经被替换成了新创建的文件foo，merged目录中的foo文件就是upper目录中的foo文件（见图3-8）。

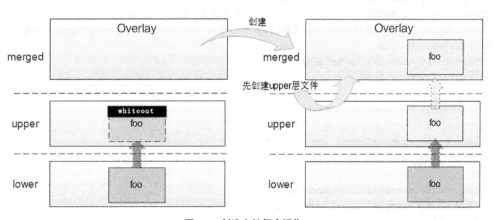

图3-8　创建文件复合操作

创建一个在lower层已经存在且在upper层有whiteout文件的同名目录：在lower层创建一个目录dir，在里面创建一个文件foo，在upper层创建一个同名的whiteout dir文件，用于隐藏lower层的dir目录，然后用户在merged层中又重新创建一个同名目录dir（见图3-9）。OverlayFS针对目录隐藏引入了一种opaque属性，它通过在upper层对应的目录上设置"trusted.overlay.opaque"扩展属性值为"y"来实现（upper层所在的文件系统需要支持xattr扩展属性），OverlayFS在读取上下层同名目录时，如果upper层的目录被设置了opaque属性，它将忽略这个目录下层的所有内容，以保证其是一个空目录。

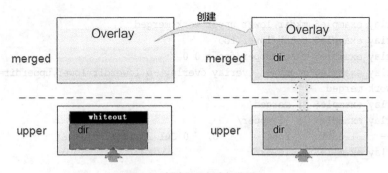

图3-9 创建目录复合操作图

```
DF1A:overlay_example$ cd /
DF1A:$ umount /overlay/overlay_example/merged
DF1A:$ rm -rf /overlay/overlay_example/
DF1A:$ mkdir -p /overlay/overlay_example/
DF1A:$ cd /overlay/overlay_example/
DF1A:overlay_example$ mkdir lower upper work merged
DF1A:overlay_example$ mkdir lower/dir
DF1A:overlay_example$ touch lower/dir/foo
DF1A:overlay_example$ mknod upper/dir c 0 0
DF1A:overlay_example$ ls upper/dir
upper/dir
DF1A:overlay_example$ ls -l upper/dir
crw-r--r--    1 admin    root       0,    0 Jul 24 12:07 upper/dir
DF1A:overlay_example$ mount -t overlay overlay -o lowerdir=lower,upperdir=upper,
workdir=work merged
DF1A:overlay_example$ ls merged/
DF1A:overlay_example$ ls upper/
dir
DF1A:overlay_example$ ls lower/
dir
DF1A:overlay_example$ mkdir merged/dir
DF1A:overlay_example$ ls merged/
dir
DF1A:overlay_example$ ls merged/dir/
DF1A:overlay_example$ ls -l upper/dir/
DF1A:overlay_example$ ls -l upper/
drwxr-xr-x    2 admin    root              0 Jul 24 12:08 dir
```

3.2.2 OpenWrt是怎么做的

我们先了解一下和Overlay相关的系统基本信息。

- OpenWrt系统是支持OverlayFS文件系统的（见图3-10）。其中没有nodev标志的输出，表示该文件系统不是虚文件系统，必须有实际的块设备。

```
nodev    sysfs
nodev    rootfs
nodev    ramfs
nodev    bdev
nodev    proc
nodev    tmpfs
nodev    debugfs
nodev    sockfs
nodev    pipefs
nodev    configfs
nodev    devpts
         squashfs
         vxfs
nodev    jffs2
nodev    overlay
nodev    mtd_inodefs
         ext3
         ext2
         ext4
         ntfs
         vfat
         fuseblk
nodev    fuse
nodev    fusectl
```

图3-10　系统支持的文件系统

● 通过mount命令，查看当前系统使用的文件系统（见图3-11）。Linux中的根目录以外的文件要想被访问，需要将其"关联"到根目录下的某个目录来实现，这种关联操作就是"挂载"，而挂载的目录就是"挂载点"，解除此关联关系的过程称为"卸载"。

```
rootfs on / type rootfs (rw)
/dev/root on /rom type squashfs (ro,relatime)
proc on /proc type proc (rw,nosuid,nodev,noexec,noatime)
sysfs on /sys type sysfs (rw,nosuid,nodev,noexec,noatime)
tmpfs on /tmp type tmpfs (rw,nosuid,nodev,noatime)
/dev/mtdblock7 on /overlay type jffs2 (rw,noatime)
overlayfs:/overlay on / type overlay (rw,noatime,lowerdir=/,upperdir=/ov
erlay/upper,workdir=/overlay/work)
tmpfs on /dev type tmpfs (rw,nosuid,relatime,size=512k,mode=755)
devpts on /dev/pts type devpts (rw,nosuid,noexec,relatime,mode=600)
debugfs on /sys/kernel/debug type debugfs (rw,noatime)
```

图3-11　查看当前分区挂载情况

● 查看系统启动参数（见图3-12）。内核在挂载根文件系统时，会根据rootfstype参数指定根文件系统类型为SquashFS和JFFS2。

```
DF1A:$ cat /proc/cmdline
memsize=128M console=ttyS0,115200n8 rootfstype=squashfs,jffs2 rdinit=/sb
in/init
```

图3-12　内核启动参数

启动过程

OpenWrt采用了基于Overlay技术的OverlayFS文件系统，将原始只读rootfs（SqushFS）分区和可写rootfs_data（JFFS2）分区合并成一个逻辑分区，挂载在根目录/。根据前面章节的内容，我们知道rootfs相当于Overlay的lower层，rootfs_data相当于Overlay的upper层，合并后挂载到根目录/。OpenWrt系统通过以下步骤实现这个过程。

（1）引导程序启动了内核后，由内核加载roofs_rom只读分区部分来完成系统的初步启动。

（2）rootfs_rom只读分区采用的是Linux内核支持的SquashFS文件系统，加载完毕后将其挂载到/rom目录，同时也挂载为根目录/。

（3）系统将使用JFFS2文件系统格式化的rootfs_data可写文件分区，并且将这部分挂载到/overlay目录。

（4）系统再将/overlay透明挂载为根目录/。

（5）最后将一部分内存挂载为/tmp目录。

（6）查看挂载情况：

```
DF1A:$ df
Filesystem           1K-blocks      Used Available Use% Mounted on
rootfs                    6976       964      6012  14% /
/dev/root                 7168      7168         0 100% /rom
tmpfs                   126308      1192    125116   1% /tmp
/dev/mtdblock7            6976       964      6012  14% /overlay
overlayfs:/overlay        6976       964      6012  14% /
tmpfs                      512         0       512   0% /dev
```

（7）透明挂载根目录/：系统先将rootfs_rom挂载为根目录/，这样就具备了一个完整的系统，然后将rootfs_data以透明方式挂载在根目录/上。OpenWrt透明挂载重叠之后的效果如下。

- 我们所看到的根文件系统是由rootfs_rom和rootfs_data两个分区合并在一起呈现的。
- 当我们修改一个任何位置的文件时，所做的修改在rootfs_data里会记录。
- 当我们删除一个文件时，所做的修改在rootfs_data里会记录。
- 当我们增加一个文件时，所做的修改在rootfs_data里会记录。
- 当我们读取文件时，首先检测rootfs_data里的内容，再检测rootfs_rom里的内容，一直到最后给你一个结果。

透明挂载的纷争

当对文件进行操作时，所占的空间是倍增的。比如我们修改了一个名字为abc的文件，那么在/rom里还有修改之前的那个abc，同时在/overlay里有修改之后的abc。

系统不论在任何时候，都能通过简单地删除掉/overlay里所有的文件达到复原的效果，这就是恢复出厂原理。一个简单的例子如下所示。

```
DF1A:$ cd /
DF1A:$ ls abc
ls: abc: No such file or directory
DF1A:$ ls /overlay/upper/abc
```

```
ls: /overlay/upper/abc: No such file or directory
DF1A:$ echo "hello world">abc
DF1A:$ ls abc
abc
DF1A:$ md5sum abc
6f5902ac237024bdd0c176cb93063dc4   abc
DF1A:$ cat abc
hello world
DF1A:$ ls /overlay/upper/abc
/overlay/upper/abc
DF1A:$ md5sum /overlay/upper/abc
6f5902ac237024bdd0c176cb93063dc4   /overlay/upper/abc
DF1A:$ cat /overlay/upper/abc
hello world
DF1A:$ rm abc
DF1A:$ ls abc
ls: abc: No such file or directory
DF1A:$ ls /overlay/upper/abc
ls: /overlay/upper/abc: No such file or directory
```

3.3 OPKG软件包管理

在OpenWrt下使用的软件包管理技术叫作OPKG（Open Package Management的缩写）。它是一款基于ipkg的轻量级包管理器，用C语言编写，类似于apt/dpkg，常用于路由器、交换机等嵌入式设备中，用来管理软件包的安装升级与下载，目前已成为开源界嵌入式系统软件包管理的标准。OpenEmbedded、OpenWrt、Yocto等项目都使用OPKG作为包管理软件，目前OPKG项目由Yocto项目维护。

通过OPKG管理软件包，可以轻松解决软件包依赖的问题。如果我们要安装软件包A，但是软件包A依赖于软件包B和C，即如果要安装软件包A，必须先安装软件包B和C，那么B和C如果又依赖其他软件包怎么办？解决这个问题的最好办法就是构建软件包的依赖关系，解决软件包依赖。在CentOS或Ubuntu上，大家安装软件包时，通过yum或者apt命令从软件包仓库中更新软件包源，然后通过命令很方便地实现软件包的依赖安装。在OpenWrt系统中，你可以通过OPKG包管理器的opkg命令完成软件包的依赖安装。OPKG试图在软件包仓库内解决依赖关系。如果失败了，它将会报告一个错误并停止安装该软件包。如果丢失第三方包的依赖关系、源码包依然可用的话，为了忽略依赖关系的错误，可以使用-force-depends选项。

OPKG管理的软件包扩展名为.ipk。关于软件包如何制作、OPKG又是如何解决软件包依赖的问题的，我们将在后续章节中给大家详细介绍。

3.3.1 OPKG语法格式

语法： opkg [选项…] 子命令 [参数…]
opkg必须有一个子命令（见表3-3、表3-4），选项（见表3-5、表3-6）和参数为可选内容。

表 3-3 与软件包管理相关的子命令

子命令	含义
update	从服务器上更新可用软件包列表：获取软件包安装列表，将其保存到内存分区的/tmp/opkg-lists/目录下，重启系统后列表会丢失，所以最好在每次安装软件包前先执行opkg update取得最新的软件包
upgrade <pkgs>	升级软件包：通常不推荐普通用户去升级软件包，因为一个典型的OpenWrt系统存储在只读的SquashFS文件系统中。同时，当升级工作做完后，使用了远超SquashFS文件系统或JFFS2文件系统默认安装基本软件包的容量。因此，推荐刷更新的OpenWrt固件来代替升级软件包
install <pkgs>	安装一个或多个软件包
configure <pkgs>	配置一个或多个未安装的包
remove <pkgs|regexp>	卸载一个或多个软件包，支持正则表达式
flag <flag> <pkgs>	标记一个或多个软件包。标记的软件包必须是已经安装过的。每次调用仅允许使用一个标记。可用标记有：hold、noprune、user、ok、installed、unpacked

表 3-4 与信息相关的子命令

子命令	含义
list	列出可用软件包。格式为：软件包包名-版本-描述信息
list-installed	列出已安装的软件包
list-upgradable	列出可升级的软件包
list-changed-conffiles	列出用户修改的配置文件
files <pkg>	列出软件包所属的文件及路径
search <file|regexp>	列出包含file文件的软件包
info [pkg|regexp]	显示pkg的软件包描述信息
status [pkg|regexp]	显示软件包pkg的状态
download <pkg>	将软件包pkg下载到当前目录
compare-versions <v1> <op> <v2>	使用关系<=、<、>、>=、=、<<、>>来比较<v1>和<v2>两个版本
print-architecture	列出可安装软件包的结构
depends [-A] [pkgname|pat]+	列出已安装软件包pkgname的所有依赖
whatdepends [-A] [pkgname|pat]+	列出所有依赖于pkgname的软件包，仅适用于已安装的软件包
whatdependsrec [-A] [pkgname|pat]+	递归列出所有依赖于pkgname的软件包，仅适用于已安装的软件包，显示信息比whatdepends命令多
whatprovides [-A] [pkgname|pat]+	列出某文件属于哪个pkgname的软件包
whatconflicts [-A] [pkgname|pat]+	列出pkgname软件包与其他软件包可能冲突的文件
whatreplaces [-A] [pkgname|pat]+	列出pkgname软件包中可能被其他软件包替代的文件

所有包含regexp的命令，表示正则表达式。

表 3-5 opkg 命令选项

短选项	长选项	含义
-A		查询全部软件包（不仅是已安装的，包含未安装的软件包列表）
-V[<level>]	--verbosity[=<level>]	设置verbosity级别为level。可用的级别有以下5种。 0：仅错误。 1：普通消息（默认）。 2：通知消息。 3：调试。 4：调试等级2
-f <conf_file>	--conf <conf_file>	设置opkg配置文件为conf_file。缺省为/etc/opkg.conf
	--cache <directory>	设置包缓存目录
-d <dest_name>	--dest <dest_name>	设置软件包安装、删除、升级的根目录为dest_name。dest_name应为已在配置文件中定义的目的名称
-o <dir>	--offline-root <dir>	指定离线安装软件包的根目录
	--add-arch <arch>:<prio>	注册架构为指定的优先级
	--add-dest <name>:<path>	注册目标名及其路径

表 3-6 opkg 强制选项

选项	含义
--force-depends	在安装、删除软件包时无视失败的依赖
--force-maintainer	覆盖已经存在的配置文件
--force-reinstall	重新安装软件包
--force-overwrite	覆盖其他软件包的文件
--force-downgrade	允许opkg降级软件包
--force-space	禁用可用空间检查
--force-remove	即使执行prerm脚本失败也强制删除软件包
--force-postinstall	离线模式下仍运行安装后脚本
--force-checksum	忽略校验和失配
--noaction	无操作-仅测试
--download-only	无操作-仅下载
--nodeps	不跟踪依赖
--force-removal-of-dependent-packages	删除软件包的同时，删除其所有依赖软件包
--autoremove	删除自动安装以满足依赖的软件包
--tmp-dir或-t	指定临时目录

3.3.2 OPKG实战

opkg的命令参数很多，大家不用一一记住，只要记住常用的命令即可，使用opkg help命令可随时查看命令帮助。接下来我们对常用命令和选项提供一些命令示例。

1. 查看系统已经安装的软件包

```
DF1A:$ opkg list-installed
01_network - 1-1
aclscript - 1-1
arp-scan - 1.9-40-g69b2f70-1
```

```
ate_cmd - 1-1
badblocks - 1.42.12-1
base-config - 1-1
base-files - 157.2-unknown
block-mount - 2016-01-10-96415afecef35766332067f4205ef3b2c7561d21
busybox - 1.23.2-1
check_net - 1-1
check_wds - 1-1
curl - 7.40.0-3.2
ddns-scripts - 2.7.6-13
devlistclean - 1-1
dnsmasq - 2.73-1
dropbear - 2015.67-1
e2fsprogs - 1.42.12-1
ethtool - 3.18-1
firewall - 2016-11-29-1
fixtime - 1-1
fstools - 2016-01-10-96415afecef35766332067f4205ef3b2c7561d21
qwifi - 1-1
hostapd-common - 2015-03-25-1
init-devlist - 1-1
ip6tables - 1.4.21-1
iperf - 2.0.10-1
ipset - 6.24-1
iptables - 1.4.21-1
iw - 4.3-1
iwinfo - 2015-06-01-ade8b1b299cbd5748db1acf80dd3e9f567938371
jshn - 2015-11-08-10429bccd0dc5d204635e110a7a8fae7b80d16cb
jsonfilter - 2014-06-19-cdc760c58077f44fc40adbbe41e1556a67c1b9a9
kernel - 3.18.29-1-98fc3c23a1455509383896be0e1234f4
kmod-button-hotplug - 3.18.29-3
kmod-cfg80211 - 3.18.29+2016-01-10-1
kmod-crypto-arc4 - 3.18.29-1
kmod-crypto-core - 3.18.29-1
kmod-crypto-hash - 3.18.29-1
kmod-fs-ext4 - 3.18.29-1
...
```

2. 查看某个软件包包含哪些文件及其安装目录

```
DF1A:$ opkg files opkg
Package opkg (9c97d5ecd795709c8584e972bfdf3aee3a5b846d-9) is installed on root and
has the following files:
/etc/opkg/customfeeds.conf
/lib/upgrade/keep.d/opkg
/bin/opkg
/etc/uci-defaults/20_migrate-feeds
/usr/sbin/opkg-key
/etc/opkg.conf
/etc/opkg/distfeeds.conf
```

3. 查看内核包软件包kmod-cfg80211包含的文件及其安装目录

```
DF1A:$ opkg files kmod-cfg80211
Package kmod-cfg80211 (3.18.29+2016-01-10-1) is installed on root and has the following \
files:
/lib/netifd/wireless/mac80211.sh
/lib/wifi/mac80211.sh
/lib/modules/3.18.29/compat.ko
/lib/modules/3.18.29/cfg80211.ko
```

4. 更新软件包列表

```
DF1A:$ opkg update
Downloading http://www.xxxx.xxx/resource/sf16a18-df1a/packages/base/Packages.gz.
Updated list of available packages in /var/opkg-lists/sf16a18-df1a_base.
Downloading http://www.xxxx.xxx/resource/sf16a18-df1a/packages/base/Packages.sig.
Signature check passed.
Downloading http://www.xxxx.xxx/resource/sf16a18-df1a/packages/packages/Packages.gz.
Updated list of available packages in /var/opkg-lists/sf16a18-df1a_packages.
Downloading http://www.xxxx.xxx/resource/sf16a18-df1a/packages/packages/Packages.sig.
Signature check passed.
```

5. 安装软件包

```
DF1A:$ opkg install minicom
Installing minicom (2.7-2) to root...
Downloading http://www.xxxx.xxx/resource/sf16a18-df1a/packages/base/minicom_2.7-2_mips_siflower.ipk.
Configuring minicom.
```

6. 仅下载软件包

```
DF1A:$ opkg download minicom
Downloading http://www.xxxx.xxx/resource/sf16a18-df1a/packages/base/minicom_2.7-2_mips_siflower.ipk.
Downloaded minicom as ./minicom_2.7-2_mips_siflower.ipk.
```

7. 卸载软件包

```
DF1A:$ opkg remove minicom
Removing package minicom from root...
```

8. 管道命令

```
DF1A:$ opkg list |grep netifd
netifd - 2015-12-16-245527193e90906451be35c2b8e972b8712ea6ab - OpenWrt Network Interface Configuration Daemon
```

9. regexp正则命令

```
DF1A:$ opkg info kmod-sched*
Package: kmod-sched-core
......
DF1A:$ opkg list-installed | awk '{print $1}'
01_network
aclscript
arp-scan
ate_cmd
badblocks
base-config
base-files
block-mount
...
```

3.3.3 OPKG源配置

OPKG也有类似于Ubuntu下apt管理软件的配置文件apt.conf的配置文件。

```
DF1A:$ ls /etc/opkg/
customfeeds.conf  distfeeds.conf   keys
```

customfeeds.conf和distfeeds.conf都可以作为源的配置，只是名称有区别，当执行opkg update时，会去配置文件中读取对应的源。

distfeeds.conf的内容示例如下。

```
DF1A:$ cat /etc/opkg/distfeeds.conf
src/gz sf16a18-df1a_base http://freeiris网站地址/resource/sf16a18-df1a/packages/base
src/gz sf16a18-df1a_telephony http://freeiris网站地址/resource/sf16a18-df1a/packages/telephony
src/gz sf16a18-df1a_packages http://freeiris网站地址/resource/sf16a18-df1a/packages/packages
src/gz sf16a18-df1a_routing http://freeiris网站地址/resource/sf16a18-df1a/packages/routing
src/gz sf16a18-df1a_luci http://freeiris网站地址/resource/sf16a18-df1a/packages/luci
src/gz sf16a18-df1a_management http://freeiris网站地址/resource/sf16a18-df1a/packages/management
```

- src/gz表示下载的源列表对应的格式为.gz。
- sf16a18-df1a表示下载的源列表对应的签名文件前缀。
- http://xxxx表示源的地址列表。

配置好软件包源文件后，就可以使用opkg update更新源列表。opkg会执行以下操作。

（1）根据opkg源的配置文件，从源的配置地址下载软件包列表压缩文件Package.gz。

（2）更新opkg-lists，保存的默认路径为/var/opkg-lists。

（3）下载对应源的Package.sig签名文件，并核对本地签名信息（/etc/opkg/keys/）。如果签名验证没有通过，提示更新失败；如果签名验证通过，本地/var/opkg-lists目录会存放对应的软件包的列表（二进制）。

（4）安装软件包时，会根据软件包描述的依赖关系，从源上下载对应的依赖软件包，先安装依赖软件包，再安装目标软件包。

OPKG常见问题

● 使用OPKG安装软件包时，有时会提示内核版本不匹配，大概是这样：

```
Collected errors:
 * satisfy_dependencies_for: Cannot satisfy the following dependencies for kmod-usb-storage:
 * 	kernel (= 3.3.8-1-6acd2a17c333f503dc86081b03fe73c0) *
kernel (= 3.3.8-1-6acd2a17c333f503dc86081b03fe73c0) *
 * opkg_install_cmd: Cannot install package kmod-usb-storage.
```

出现这个错误提示是因为服务器上的固件内核版本与设备上所安装的版本不一致，唯一的解决方法就是重新刷机。

● opkg update出现签名错误：这是因为固件内的OPKG签名文件和服务器上的源签名文件不一致。建议更新服务器上版本的固件。

```
DF1A:$ opkg update
Downloading http://××××/base/Packages.gz.
Updated list of available packages in /var/opkg-lists/dragonfly_1a_base.
Downloading http://××××/base/Packages.sig.
Signature check failed.
Remove wrong Signature file.
```

● 通过OPKG安装软件后，如果开发板的系统恢复了出厂设置，那么安装的软件包会全部消失。请思考一下为什么。

4 UCI统一配置

4.1 UCI介绍

嵌入式系统一般是各种软件包的集合,由于软件包大部分是开源的,不同的软件有自己的配置文件、配置参数和配置规则,有的软件是通过NVRAM存放配置(通常是在Flash中划分的某个分区),有的是通过不同目录下的配置文件存放配置,不同软件的配置文件的存放路径也各不相同,没有统一的标准,而且也没法统一。对于单个软件的配置,我们还能应付,但是对于像OpenWrt这种开源软件包的集合系统来说,就显得杂乱无章,也使得开发人员和使用者会产生困惑。

系统的配置应该简单、直接,这是OpenWrt配置接口的设计原则,为此OpenWrt引入了一套统一配置接口,这个配置接口就是UCI。

UCI是Unified Configuration Interface的缩写,即统一配置接口,是OpenWrt系统用户配置的统一接口。它提供了简单、容易、标准化的命令行和API接口,可以使用各种编程语言(如Lua、Python、PHP、C等)来执行命令和传递配置参数,从而达到修改系统参数的目的。UCI中已经包含了网络配置、无线配置、系统信息配置等基本路由器所需的主要配置参数。同时OpenWrt下默认的Web管理界面框架LuCI,也是利用UCI提供的API实现对UCI配置文件的管理。可以说整个OpenWrt的主要配置都是围绕UCI进行的。

UCI作为统一配置接口,更像一个黏合剂,每个软件原有的配置文件保持不变。想修改这个软件的配置文件时,我们通过uci命令修改这个软件的UCI配置文件,UCI系统会把UCI配置的参数和软件原有的配置文件和参数对应起来,然后重新装载或重启软件,完成修改配置。

虽然目前OpenWrt下已有大量软件包支持UCI配置管理,但这不是强制的,不是所有的软件包都实现了UCI配置管理。对于开发者,在OpenWrt下新移植和创建软件包时,要尽量符合OpenWrt的约定和规范。我们要先创建UCI配置文件,再把软件包的配置文件的配置参数与UCI文件的配置参数对应起来,实现参数的传递。

请不要担心,我们将在接下来的章节中深入学习UCI的具体细节。后面的LuCI开发和实战章节,也会涉及UCI相关的内容。

4.2 UCI的配置文件

UCI的配置文件被分割成/etc/config下的多个独立的文件,各个文件按名字含义对应系统的不同的功能配置，这些文件是格式遵循UCI规定的文件，可以通过文本编译器VI或者uci命令接口修改。使用uci命令修改配置文件会产生缓存，所以每次修改后要尽快确认保存，避免出现冲突。/etc/config下UCI配置文件的作用如表4-1所示。

表 4-1 /etc/config 下 UCI 配置文件的作用

文件	作用
基本配置	
dhcp	面向LAN口提供的IP地址分配服务配置
dropbear	SSH服务配置
firewall	路由转发、端口转发、防火墙规则
network	自身网络接口配置
system	时间服务器时区配置
wireless	无线网络配置
服务配置	
uhttpd	轻量级HTTP服务。LuCI 默认的HTTP服务器配置
luci	LuCI Web界面配置，包含语言、主题等
ucitrack	LuCI Web界面相关配置，描述LuCI修改功能配置后，影响哪些其他配置及其他配置需要执行的动作
fstab	系统启动时将要激活的静态文件系统和交换分区的配置
rpcd	ubus RPC daemon配置
upnp	upnp配置参数，即插即用网络协议
SF16A18芯片特有	
ap_groups	2.4GHz和5GHz AP基本配置
qos_cfg	网络QoS配置
siversion	路由器本地接口版本
basic_setting	路由器基本配置使能，是否启用向导、OTA参数、中继模式等
siwifi	路由器的SN号等配置
style	本地网页配置文件
sicloud	镜像服务器连接配置
wifi_info	SF16A18 Wi-Fi信息
devlist	LAN口连接过的设备信息列表
sidefault	访问控制列表默认配置
siserver	服务器配置
tsconfig	用户使用信息配置保存
widevlist	Wi-Fi连接设备信息列表

4.3 UCI配置文件语法

UCI文件的语法比较简单易懂。语法格式如下：

```
config 'section-type' 'section'
     option 'key' 'value'
     list 'list_key'  'list_value'
```

（1）config节点：以关键字config开始的一行用来代表当前节点。

● section-type：节点类型。

- section：节点名称。

（2）option选项：表示节点中的一个元素。
- key：键。
- value：值。

（3）list列表选项：表示列表形式的一组参数。
- list_key：列表键。
- list_value：列表值。

4.3.1 config节点

语法格式：

```
config 'section-type' 'section'
```

- config：表示节点，以关键字config开头，表示一个节点的开始。一般顶行写。
- section-type：表示节点的类型，可自定义节点类型名称，不能省略。
- section（可选）：表示节点名称，可自定义节点名称，可以省略，省略后表示匿名节点。

实际使用示例（/etc/config/network）如下。

```
config interface 'wan'
        option ifname 'eth0.2'
        option macaddr '10:16:88:14:3d:04'
        option force_link '0'
        option proto 'dhcp'
config alias
        option interface 'lan'
        option proto 'static'
        option ipaddr '192.168.5.251'
        option netmask '255.255.255.0'
```

config节点使用原则：
- UCI允许只有节点类型的匿名节点存在；
- 节点名称建议使用单引号包含，以免引起歧义；
- 节点中可以包含多个option选项或list列表选项；
- 节点遇到文件结束或遇到下一个节点代表完成。

4.3.2 option选项

语法格式：

```
option 'key' 'value'
```

- option：即选项，以关键字option开头，表示节点中的一个元素。
- key：表示关键字，可自定义名称，注意避免重名，不能省略。
- value：表示值，不能省略。

实际使用示例片段如下。

```
option ifname 'eth0.2'
option macaddr '10:16:88:14:3d:04'
```

```
option force_link '0'
option proto 'dhcp'
```

option选项使用原则：
- 选项的值建议使用单引号包含；
- 避免相同的选项键存在于同一个节点，否则只有一个生效。

4.3.3 list列表选项

语法格式：

```
list 'list_key' 'list_value'
```

- list：即列表，以关键字list开头，表示节点中的一个列表。
- list_key：表示列表键，不能省略。
- list_value：表示列表值，不能省略。

实际使用示例片段(/etc/config/firewall)如下。

```
config zone
        option name 'wan'
        list network 'wan'
        list network 'wan6'
        option output 'ACCEPT'
        option forward 'REJECT'
        option masq '1'
        option mtu_fix '1'
        option input 'REJECT'
```

list列表使用原则：
- 选项的值建议使用单引号包含；
- 列表键的名字如果相同，则相同键的值将会被当作数组传递给相应软件。

4.3.4 UCI的语法容错

允许的语法：

```
option example value
option 'example' value
option example "value"
option "example"   'value'
option   'example' "value"
```

不允许的语法：

```
option 'example" "value'
option example 一组以空格分隔的值
```

尽量使用常规字符去处理UCI，特殊字符有可能会破坏数据结构的完整性。

4.4　UCI命令行接口

UCI提供了命令行接口用于直接操作UCI配置文件。

4.4.1 uci命令语法

UCI配置文件是通过uci命令来进行操作的。uci命令的语法格式如下。

```
uci [<options>] <command> [<arguments>]
命令(command):
        batch
        export      [<config>]
        import      [<config>]
        changes     [<config>]
        commit      [<config>]
        add         <config> <section-type>
        add_list    <config>.<section>.<option>=<string>
        del_list    <config>.<section>.<option>=<string>
        show        [<config>[.<section>[.<option>]]]
        get         <config>.<section>[.<option>]
        set         <config>.<section>[.<option>]=<value>
        delete      <config>[.<section>[[.<option>][=<id>]]]
        rename      <config>.<section>[.<option>]=<name>
        revert      <config>[.<section>[.<option>]]
        reorder     <config>.<section>=<position>
选项(options):
        -c <path>:设置配置文件的搜索路径(默认值为/etc/config)。
        -d <str>:在uci show中设置列表值的分隔符。
        -f <file>:使用<file>而不是stdin作为输入。
        -m:导入时,将数据合并到现有包中。
        -n:在导出时命名未命名的部分(默认)。
        -N:不要命名未命名的部分。
        -p <path>:为配置更改文件添加搜索路径。
        -P <path>:为配置更改文件添加搜索路径,并将其用作默认值。
        -q:quiet模式(不打印错误消息)。
        -s:强制使用严格模式(停止解析器错误,默认)。
        -S:禁用严格模式。
        -X:不要在'show'上使用扩展语法。
```

uci命令的参数说明如表4-2所示。

表4-2 uci 命令的参数说明

命令	目标	描述
batch	-	执行一个多行UCI脚本
export	[<config>]	以人可读的格式导出配置
import	[<config>]	以UCI语法导入配置文件
changes	[<config>]	列出给定配置文件修改过的内容,如果没有参数,则列出所有修改过的配置文件
commit	[<config>]	将给定配置文件的更改(如果未给定,则将所有配置文件)写入文件系统。所有"uci set"、"uci add"、"uci rename"和"uci delete"命令的执行结果都被写入缓存,当使用"uci commit"命令后,才保存到文件系统中。在使用文本编辑器编辑配置文件之后,不需要这样做,但对于直接使用UCI文件的脚本、GUI和其他程序来说需要使用"uci commit"

续表

命令	目标	描述
add	<config> <section-type>	将匿名节点添加到指定配置中
add_list	<config>.<section>.<option>=<string>	将指定的字符串添加到现有列表选项中
del_list	<config>.<section>.<option>=<string>	从现有列表选项中删除指定的字符串
show	[<config>[.<section>[.<option>]]]	以程序化（压缩符号）方式显示指定的选项、节点或配置
get	<config>.<section>[.<option>]	获取指定选项的值或指定节点的类型
set	<config>.<section>[.<option>]=<value>	设置指定选项的值，或添加类型设置为指定值的新节点
delete	<config>[.<section>[.<option>]]	删除指定的节点或选项
rename	<config>.<section>[.<option>]=<name>	将指定的选项或节点重命名为指定的名称
revert	<config>[.<section>[.<option>]]	还原指定的选项、节点或配置文件
reorder	<config>.<section>=<position>	重新排序uci show节点的显示位置

4.4.2 UCI的匿名节点

一个节点可以有名称，也可以没有名称，有名称的节点叫作命名节点，没有名称的节点叫作匿名节点。

```
命名节点:config interface 'lan'
匿名节点:config alias
```

匿名节点会由UCI自动产生一个标识，这个标识的格式为"cfg<id>"，其中id会自动产生，名称为"@XXX[Y]"。

/etc/config/upnpd下面有个匿名节点alias，我们通过"uci export"以人可读方式显示配置。

```
DF1A:$ uci export upnpd
package upnpd
config upnpd 'config'
        option enabled '0'
        option enable_natpmp '1'
        option enable_upnp '1'
        option secure_mode '1'
        option log_output '0'
        option download '1024'
        option upload '512'
        option internal_iface 'lan'
        option port '5000'
        option upnp_lease_file '/var/upnp.leases'
config perm_rule
        option action 'allow'
        option ext_ports '1024-65535'
        option int_addr '0.0.0.0/0'
        option int_ports '1024-65535'
        option comment 'Allow high ports'
config perm_rule
        option action 'deny'
        option ext_ports '0-65535'
        option int_addr '0.0.0.0/0'
        option int_ports '0-65535'
```

```
        option comment 'Default deny'
```

我们通过"uci show"命令以压缩符号方式查看内容,@perm_rule[0]是匿名节点:

```
DF1A:$ uci show upnpd
upnpd.config=upnpd
upnpd.config.enabled='0'
upnpd.config.enable_natpmp='1'
upnpd.config.enable_upnp='1'
upnpd.config.secure_mode='1'
upnpd.config.log_output='0'
upnpd.config.download='1024'
upnpd.config.upload='512'
upnpd.config.internal_iface='lan'
upnpd.config.port='5000'
upnpd.config.upnp_lease_file='/var/upnp.leases'
upnpd.@perm_rule[0]=perm_rule
upnpd.@perm_rule[0].action='allow'
upnpd.@perm_rule[0].ext_ports='1024-65535'
upnpd.@perm_rule[0].int_addr='0.0.0.0/0'
upnpd.@perm_rule[0].int_ports='1024-65535'
upnpd.@perm_rule[0].comment='Allow high ports'
upnpd.@perm_rule[1]=perm_rule
upnpd.@perm_rule[1].action='deny'
upnpd.@perm_rule[1].ext_ports='0-65535'
upnpd.@perm_rule[1].int_addr='0.0.0.0/0'
upnpd.@perm_rule[1].int_ports='0-65535'
upnpd.@perm_rule[1].comment='Default deny'
```

通过"uci show upnpd.@perm_rule[0]"命令查看自动产生的匿名节点的cfg<id>。

```
DF1A:$ uci show upnpd.@perm_rule[0]
upnpd.cfg03ed70=perm_rule
upnpd.cfg03ed70.action='allow'
upnpd.cfg03ed70.ext_ports='1024-65535'
upnpd.cfg03ed70.int_addr='0.0.0.0/0'
upnpd.cfg03ed70.int_ports='1024-65535'
upnpd.cfg03ed70.comment='Allow high ports'
```

4.4.3 语法测试

使用Vim创建配置/etc/config/mytest,内容如下。

```
config system
        option device '5g'
        option system 'linux'
        option timezone 'UTC'
config server 'yotta'
        list server '××××1.com'
        list server '××××2.com'
        list server '××××3.com'
        option enabled '1'
        option start '100'
        option limit '150'
```

用"uci show"查看mytest的配置：

```
DF1A:$ uci show mytest
mytest.@system[0]=system
mytest.@system[0].device='5g'
mytest.@system[0].system='linux'
mytest.@system[0].timezone='UTC'
mytest.yotta=server
mytest.yotta.server='××××1.com' '××××2.com' '××××3.com'
mytest.yotta.enabled='1'
mytest.yotta.start='100'
mytest.yotta.limit='150'
```

取得有名节点yotta的列表server的值：

```
DF1A:$ uci get mytest.yotta.server
××××1.com ××××2.com ××××3.com
```

取得匿名节点@system[0]的选项device的值：

```
DF1A:$ uci get mytest.@system[0].device
5g
```

增加一个类型为push、名称为setting的命名节点：

```
DF1A:$ uci set mytest.setting=push
```

给setting节点增加enable选项：

```
DF1A:$ uci set mytest.setting.enable=1
```

增加一个类型为system的匿名节点：

```
DF1A:$ uci add mytest system
cfg06e48a
```

给新增的匿名节点cfg06e48a增加enable选项：

```
DF1A:$ uci set mytest.cfg06e48a.enable=0
```

修改类型为system的第一个匿名节点下的device选项：

```
DF1A:$ uci set mytest.cfg06e48a.device=smartrouter
```

删除掉列表mytest.yotta.server的一个值：

```
DF1A:$ uci del_list mytest.yotta.server=××××2.com
```

查看都修改了什么：

```
DF1A:$ uci changes mytest
mytest.setting='push'
mytest.setting.enable='1'
mytest.cfg06e48a='system'
mytest.cfg06e48a.enable='0'
mytest.cfg02e48a.device='smartrouter'
mytest.yotta.server-='××××2.com'
```

必须提交，才能使修改生效：

```
DF1A:$ uci commit mytest
```

用"uci export"查看mytest文件：

```
DF1A:$ uci export mytest
```

```
package mytest
config system
        option device '5g'
        option system 'linux'
     option timezone 'UTC'
config server 'yotta'
        list server '××××1.com'
        list server '××××3.com'
        option enabled '1'
        option start '100'
        option limit '150'
config push 'setting'
        option enable '1'
config system
        option enable '0'
        option device 'smartrouter'
```

用"uci show"查看mytest文件：

```
DF1A:$ uci show mytest
mytest.@system[0]=system
mytest.@system[0].device='5g'
mytest.@system[0].system='linux'
mytest.@system[0].timezone='UTC'
mytest.yotta=server
mytest.yotta.server='××××1.com' '××××3.com'
mytest.yotta.enabled='1'
mytest.yotta.start='100'
mytest.yotta.limit='150'
mytest.setting=push
mytest.setting.enable='1'
mytest.@system[1]=system
mytest.@system[1].enable='0'
mytest.@system[1].device='smartrouter'
```

4.4.4 举例：开启Wi-Fi

wireless文件中有个"wlan0-guest"接口，默认状态是关闭。用uci命令查看这个节点：

```
DF1A:$ uci show wireless.@wifi-iface[1]
wireless.cfg053579=wifi-iface
wireless.cfg053579.device='radio0'
wireless.cfg053579.ifname='wlan0-guest'
wireless.cfg053579.network='guest'
wireless.cfg053579.mode='ap'
wireless.cfg053579.ssid='SiWiFi-8868-2.4G-guest'
wireless.cfg053579.encryption='psk2+ccmp'
wireless.cfg053579.key='12345678'
wireless.cfg053579.isolate='1'
wireless.cfg053579.hidden='0'
wireless.cfg053579.group='1'
```

```
wireless.cfg053579.netisolate='0'
wireless.cfg053579.disable_input='0'
wireless.cfg053579.wps_pushbutton='0'
wireless.cfg053579.wps_label='0'
wireless.cfg053579.rps_cpus='2'
wireless.cfg053579.disabled='1'
```

可以看到wireless.cfg053579.disabled的值为1，表示禁用Wi-Fi，下面通过uci命令将其值设为0，使能节点。

```
DF1A:$ uci set wireless.cfg053579.disabled=0
DF1A:$ uci changes wireless
wireless.cfg053579.disabled='0'
DF1A:$ uci commit wireless
DF1A:$ uci show wireless.cfg053579.disabled
wireless.cfg053579.disabled='0'
```

使用wifi命令启动，使Wi-Fi配置生效：

```
DF1A:$ wifi
[ 3676.960930] IPv6: ADDRCONF(NETDEV_UP): br-guest: link is not ready
[ 3677.718510] lmac[0] vif_mgmt_register, vif_type : 2
[ 3677.725396] IPv6: ADDRCONF(NETDEV_UP): wlan0: link is not ready
[ 3677.735680] device wlan0 entered promiscuous mode
[ 3677.740684] br-lan: port 2(wlan0) entered forwarding state
[ 3677.746335] br-lan: port 2(wlan0) entered forwarding state
[ 3677.845039] lmac[1] vif_mgmt_register, vif_type : 2
[ 3677.851848] IPv6: ADDRCONF(NETDEV_UP): wlan1: link is not ready
[ 3677.860768] device wlan1 entered promiscuous mode
[ 3677.865885] br-lan: port 3(wlan1) entered forwarding state
[ 3677.871443] br-lan: port 3(wlan1) entered forwarding state
[ 3677.953594] br-lan: port 2(wlan0) entered disabled state
[ 3677.959808] br-lan: port 3(wlan1) entered disabled state
[ 3678.124363] br-lan: port 3(wlan1) entered forwarding state
[ 3678.129985] br-lan: port 3(wlan1) entered forwarding state
[ 3678.136227] IPv6: ADDRCONF(NETDEV_CHANGE): wlan1: link becomes ready
[ 3680.123336] br-lan: port 3(wlan1) entered forwarding state
[ 3687.220449] br-lan: port 2(wlan0) entered disabled state
[ 3687.283669] device wlan0 left promiscuous mode
[ 3687.288361] br-lan: port 2(wlan0) entered disabled state
[ 3687.738007] lmac[0] vif_mgmt_register, vif_type : 2
[ 3687.744976] IPv6: ADDRCONF(NETDEV_UP): wlan0: link is not ready
[ 3687.754316] device wlan0 entered promiscuous mode
```

4.5 UCI 的Lua接口

UCI除了提供命令行接口外，还提供了C、Lua等语言的应用程序接口。Web配置页面（如LuCI）就是利用了UCI所提供的API实现对UCI配置文件的修改的。在OpenWrt中，有3个与UCI相关的包：uci、libuci、libuci-lua。其中libuci-lua提供了UCI的Lua支持。熟悉Lua UCI API对后面学习LuCI非常有帮助。下面我们介绍一下如何使用Lua语言调用UCI接口。

4.5.1 libuci-lua API接口

libuci-lua为我们提供了一个Lua的动态库/usr/lib/lua/uci.so，在使用Lua UCI时，先要包含这个库。然后可以使用uci对象或uci对象实例调用Lua UCI API进行操作。

```
require("uci")
```

Lua UCI提供的API如下。

- cursor

含义：使用uci对象初始化一个uci上下文实例。后面可以通过这个实例调用UCI API。
语法：uci.cursor()
示例：local x = uci.cursor()

- load

含义：加载config文件。
语法：uci.load(config)
示例：uci.load("network")

- unload

含义：卸载config文件。
语法：uci.unload(config)
示例：uci.unload("network")

- get

含义：取得指定选项的值，返回值为选项值string或nil。
语法：x:get("config", "section", "option")或uci.get("config", "section", "option")
示例：
local x = uci.cursor()
local value = x:get("network", "lan", "proto")
或local value = uci.get("network", "lan", "proto")

- get_all

含义：取得所有config内容，返回一个table，每项内容包括以下几项。
[".type"]：节点类型。
[".name"]：节点名称，匿名节点会自动产生名称(格式为cfg<id>)。
[".index"]：列表索引(从1开始)。
[".anonymous"]：是否是匿名节点。
语法：x:get_all("config")或uci.get("config")
示例：
local x = uci.cursor()
local network_tbl =x:get_all("network")
或local network_tbl=uci.get_all("network")

- add

含义：添加一个类型为type的匿名节点，x:add方法会返回节点名称cfg<id>。
语法：x:add("config", "type")或uci.add("config", "type")
示例：
local x = uci.cursor()
local name = x:add("network", "switch")
或local name=uci.add("network", "switch")

● set

含义：创建命名节点或设置选项值/列表值，返回值为true或false。
语法：
添加类型为type的命名节点name：
x:set("config", "name", "type") 或uci.set("config", "name", "type")
设置选项值：
x:add("config", "section","option", "value")
或uci.add("config", "section","option", "value")
或add("config", "section.option", "value")
设置列表值：
x:set("config", "section", "list", { "value1", "value2" ,...})
或uci.set("config", "section", "list", { "value1", "value2",... })
或add("config", "section.list", { "value1", "value2",... })
示例：
设置添加命名节点：
uci.set("network", "tan", "interface")
设置选项值：
local x = uci.cursor()
x:set("network", "lan", "proto", "dhcp")
或uci.set("network", "lan", "proto", "dhcp")
或uci.set("network", "lan.proto", "dhcp")
设置列表值：
local x = uci.cursor()
x:set("system", "ntp", "server", {
  "0.openwrt.pool.ntp.org",
  "1.openwrt.pool.ntp.org",
  "2.openwrt.pool.ntp.org",
  "3.openwrt.pool.ntp.org"
})

● rename

含义：重命名config文件中option的名称。
语法：x:rename("config", "section","option","name") 或uci.rename("config", "section","option","name")
示例：
local x = uci.cursor()
x:rename("network", "lan", "proto", "netproto")

● delete

含义：删除选项或节点。
语法：
删除选项：
x:delete("config", "section", "option")
或uci.delete("config", "section", "option")
删除节点：
x:delete("config", "section")
或uci.delete("config", "section")
或delete("config", "section.option")
示例：
local x = uci.cursor()

```
x:rename("network", "lan", "proto", "netproto")
```

- **commit**

含义：提交config的修改内容。一般在执行所有API操作后，需要执行commit提交修改，否则修改的内容不被保存。
语法：`x:commit("config")` 或 `uci.commit("config")`
示例：`uci.commit("network")`

- **reorder**

含义：改变节点优先级，默认从0开始。
语法：`x:reorder("config","section","position")` 或 `uci.reorder("config","section","position")`
示例：
```
local x = uci.cursor()
x:reorder("wireless", "wireless.@wifi-iface[4]", 0)
```

- **revert**

含义：在提交前，恢复到修改之前的状态。
语法：`x:revert("config")` 或 `uci.revert("config")`
示例：`uci.revert("wireless")`

- **foreach**

含义：遍历config文件中所有的类型为type的节点，并调用回调函数，返回一个table，包含所有的选项，每项包含以下内容。
s['.type']:节点类型
s['.name']:节点名称
语法：
`x:foreach("config", "type", function(s) ... end)` 或 `uci.foreach("config", "type", function(s) ... end)`
示例：
```
x:foreach("wireless", "wifi-iface", function(s)
    for key, value in pairs(s) do
        print(key .. ': ' .. tostring(value))
    end
end)
```

- **get_confdir**

含义：返回config文件的路径，默认返回/etc/config。
语法：`x:get_confdir` 或 `uci.get_confdir`
示例：`uci.get_confdir()`

- **set_confdir**

含义：设置config文件的路径。
语法：`x:set_confdir("config path")` 或 `uci.set_confdir("config path")`
示例：`uci.set_confdir("/etc/config")`

4.5.2 Lua UCI实验

下面我们使用上面的API来做一个综合实验。我们使用之前的mytest配置文件，内容如下。

```
DF1A:$ uci export mytest
package mytest
```

```
config system
        option device '5g'
        option system 'linux'
        option timezone 'UTC'
config server 'yotta'
        list server '×××1.com'
        list server '×××3.com'
        option enabled '1'
        option start '100'
        option limit '150'
config push 'setting'
        option enable '1'
config system
        option enable '0'
        option device 'smartrouter'
```

test.lua代码如下。

```lua
require("uci")
function print_get_all(tbl)
 for k1,v1 in pairs(tbl) do
  print("---------------\n")
   if type(v1) == "table" then
    for k2, v2 in pairs(v1) do
     if type(v2) ~= "table" then
      print("\t" .. k2 .. ': ' .. tostring(v2))
     else
      print("\t" .. k2)
      for k3,v3 in pairs(v2) do
       print("\t\t[" .. k3 .. ']: ' .. tostring(v3))
      end
     end
    end
   end
 end
end
local x = uci.cursor()
--print
print("==============get all mytest before change====================\n")
local all_tbl = x:get_all("mytest")
print_get_all(all_tbl)
--add config section
x:set("mytest","solo","server")
local cfg_sid=x:add("mytest","server")
print(cfg_sid)
--add unnamed section option
x:set("mytest",cfg_sid,"serverlist",{"1.××1.com","2.××1.com","3.××1.com"})
x:set("mytest",cfg_sid,"proto","dhcp")
x:set("mytest",cfg_sid,"enabled","0")
x:set("mytest",cfg_sid,"ifname","eth1")
--modify
```

```
x:set("mytest","yotta","enabled","0")
x:set("mytest","solo","serverlist",{"1.××2.com","2.××2.com","3.××2.com"})
--rename option
x:rename("mytest", "yotta", "start", "end")
--delete
x:delete("mytest","yotta","limit")
local tbl_chg = x:changes()
x:commit("mytest")
--foreach
print("==============foreach server type====================\n")
x:foreach("mytest", "server", function(s)
    print("\n")
    for key, value in pairs(s) do
        if type(value) == "table" then
            print(key .. ': ')
            for k1,v2 in pairs(value) do
                print("\t[" .. k1 .. ']: ' .. tostring(v2))
            end
        else
            print(key .. ': ' .. tostring(value))
        end
    end
end)
--print
print("==============get all mytest after change====================\n")
local all_tbl = x:get_all("mytest")
print_get_all(all_tbl)
```

下面我们对代码内容进行说明。

（1）require("uci")加载UCI的Lua库，这个库是使用Lua UCI API必须加载的，放到最前面。

（2）使用local x=uci.cursor()获得上Lua UCI下文对象，后面的API操作直接使用x操作即可。

（3）执行x:get_all("mytest")，取得所有节点对象，节点对象包含配置文件中所有的匿名和命名节点，其返回值为一个table（见图4-1），然后使用print_get_all对其进行遍历。print_get_all()函数循环遍历所有config节点，打印节点的选项和列表值，还包含隐藏的.name（节点名称）、.type（节点类型）、.index（索引）、.anonymous（是否是匿名节点）。print_get_all()函数中包含3个循环，第一个循环遍历get_all()函数取得的所有节点对象table（包含匿名节点和命名节点），第二个循环遍历每个节点对象table中的属性、选项、列表（本身是一个table），第三个循环返回列表table中的索引和值。

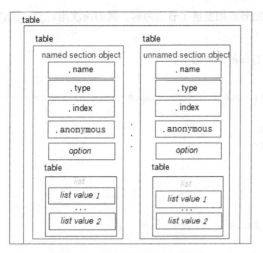

图4-1　get_all()返回的节点示意图

（4）x:set("mytest","solo","server")创建命名节点solo，类型为server。local cfg_sid=x:add("mytest","server")创建匿名节点，类型为server，匿名节点自动产生的cfg<id>赋值给cfg_sid，后面我们操作这个匿名节点，会用到这个cfg<id>。注意创建匿名节点虽然可以使用uci.add()方法，但是这个方法不会返回产生的cfg<id>，除非后面不使用这个cfg<id>，否则后面没办法操作这个匿名节点。

（5）给创建的匿名节点cfg<id>添加选项和列表内容，注意列表的格式。

```
x:set("mytest",cfg_sid,"serverlist",{"1.××1.com","2.××1.com","3.××1.com"})
x:set("mytest",cfg_sid,"proto","dhcp")
x:set("mytest",cfg_sid,"enabled","0")
x:set("mytest",cfg_sid,"ifname","eth1")
```

（6）修改mytest的命名节点yotta的enabled选项内容，修改刚才添加的solo节点的serverlist列表内容。

```
x:set("mytest","yotta","enabled","0")
x:set("mytest","solo","serverlist",{"1.××2.com","2.××2.com","3.××2.com"})
```

（7）用x:rename()函数修改节点选项的名称。修改yotta命名节点中的start选项名称为end。

```
x:rename("mytest", "yotta", "start", "end")
```

（8）用x:delete()函数删除节点选项。

```
x:delete("mytest","yotta","limit")
```

（9）x:changes()可以查看上面修改的内容，返回一个table，这里没有进行打印输出，大家可以自己打印一下，其实是遍历table。

（10）用x:commit("mytest")提交修改。这是进行所有修改之后必须执行的语句，否则修改的内容不会生效。它类似于前面我们学习的uci命令中的commit。

（11）提交修改后，我们就可以查看变更后的mytest内容。我们通过x:foreach()函数对"server"节点（包含匿名和命名节点）进行遍历，和之前的原则一样，遇到table就需要用循环进行遍历。

（12）因为前面增加节点和变更了节点内容，我们再次用x:get_all("mytest")对所有节点进行打印。

执行test.lua脚本，打印的结果如下。

```
DF1A:$ lua test.lua
==============get all mytest before change====================
---------------
        .name: yotta
        .type: server
        limit: 150
        enabled: 1
        start: 100
        .index: 1
        .anonymous: false
        server
                [1]: cn.××××1.com
                [2]: us.××××2.com
---------------
        .name: setting
        .type: push
        enable: 1
        .anonymous: false
        .index: 2
---------------



        .name: cfg06e48a
        .type: system
        enable: 0
        device: smartrouter
        .anonymous: true
        .index: 3
---------------
        .name: cfg02e48a
        .type: system
        timezone: UTC
        device: 5g
        .index: 0
        .anonymous: true
        system: linux
cfg09769c
==============foreach server type====================
.name: yotta
.type: server
end: 100
enabled: 0
.index: 1
.anonymous: false
```

```
server:
        [1]: cn.××××1.com
        [2]: us.××××2.com
.name: solo
.type: server
serverlist:
        [1]: 1.××2.com
        [2]: 2.××2.com
        [3]: 3.××2.com
.anonymous: false
.index: 4
.name: cfg09769c
.type: server
serverlist:
        [1]: 1.××1.com
        [2]: 2.××1.com
        [3]: 3.××1.com
ifname: eth1
enabled: 0
proto: dhcp
.anonymous: true
.index: 5
==============get all mytest after change=====================
---------------
        .name: yotta
        .type: server
        end: 100
        enabled: 0
        .index: 1
        .anonymous: false
        server
                [1]: ××××1.com
                [2]: ××××2.com
---------------
        .name: setting
        .type: push
        enable: 1
        .anonymous: false
        .index: 2
---------------
        .name: solo
        .type: server
        serverlist
                [1]: 1.××2.com
                [2]: 2.××2.com
                [3]: 3.××2.com
        .anonymous: false
        .index: 4
---------------
        .name: cfg02e48a
```

```
        .type: system
        timezone: UTC
        device: 5g
        .index: 0
        .anonymous: true
        system: linux
---------------
        .name: cfg09769c
        .type: server
        serverlist
                [1]: 1.××1.com
                [2]: 2.××1.com
                [3]: 3.××1.com
        ifname: eth1
        enabled: 0
        proto: dhcp
        .anonymous: true
        .index: 5
---------------
        .name: cfg06e48a
        .type: system
        enable: 0
        device: smartrouter
        .anonymous: true
        .index: 3
```

用"uci export"命令查看修改后的mytest：

```
DF1A:$ uci export mytest
package mytest
config system
        option device '5g'
        option system 'linux'
        option timezone 'UTC'
config server 'yotta'
        list server '××××1.com'
        list server '××××2.com'
        option end '100'
        option enabled '0'
config push 'setting'
        option enable '1'
config system
        option enable '0'
        option device 'smartrouter'
config server 'solo'
        list serverlist '1.××2.com'
        list serverlist '2.××2.com'
        list serverlist '3.××2.com'
config server
        list serverlist '1.××1.com'
        list serverlist '2.××1.com'
        list serverlist '3.××1.com'
```

```
option proto 'dhcp'
option enabled '0'
option ifname 'eth1'
```

大家可以对照代码和执行结果，看是否和自己的预期一样。相信大家学习完本章节，已经对UCI统一配置接口及Lua UCI API的使用方法有了初步认识。

5 网络配置

5.1 配置文件

路由器的基本功能是完成WAN口（连接互联网的接口）、LAN口（连接局域网的接口）、Wireless（无线接口）的基本配置，这些配置都是以UCI配置文件形式存在的，下面介绍几个配置文件。

- /etc/config/network：有线网络接口配置文件，可以对所有网口以及VLAN（虚拟局域网）进行配置。
- /etc/config/wireless：无线网络配置文件。该文件包含了无线网络协议、参数、速率等信息。

DF1A开发板底板上有3个网络接口：1个WAN口、2个LAN口。至于它们为什么可以成为WAN口和LAN口，那是因为芯片内部使用VLAN进行了划分。

UCI有线网络配置

```
DF1A:$ uci show network
network.loopback=interface
network.loopback.ifname='lo'
network.loopback.proto='static'
network.loopback.ipaddr='127.0.0.1'
network.loopback.netmask='255.0.0.0'
network.globals=globals
network.globals.ula_prefix='fd8d:78b8:b066::/48'
network.lan=interface
network.lan.ifname='eth0.1'
network.lan.force_link='1'
network.lan.macaddr='10:16:88:4f:88:66'
network.lan.type='bridge'
network.lan.proto='static'
network.lan.ipaddr='192.168.4.1'
network.lan.netmask='255.255.255.0'
network.lan.ip6assign='60'
network.lan.group='0'
network.lan.rps_cpus='2'
network.lan.xps_cpus='2'
network.wan=interface
network.wan.ifname='eth0.2'
network.wan.force_link='1'
```

```
network.wan.macaddr='10:16:88:4f:88:67'
network.wan.rps_cpus='1'
network.wan.xps_cpus='0'
network.wan.proto='dhcp'
network.wan6=interface
network.wan6.ifname='eth0.2'
network.wan6.proto='dhcpv6'
network.guest=interface
network.guest.ifname='guest'
network.guest.force_link='1'
network.guest.type='bridge'
network.guest.proto='static'
network.guest.ipaddr='10.0.0.1'
network.guest.netmask='255.255.255.0'
network.guest.group='0'
network.lease=interface
network.lease.ifname='lease'
network.lease.force_link='1'
network.lease.type='bridge'
network.lease.proto='static'
network.lease.ipaddr='10.2.0.1'
network.lease.netmask='255.255.255.0'
network.lease.group='0'
network.@switch[0]=switch
network.@switch[0].name='switch0'
network.@switch[0].reset='1'
network.@switch[0].enable_vlan='1'
network.@switch_vlan[0]=switch_vlan
network.@switch_vlan[0].device='switch0'
network.@switch_vlan[0].vlan='1'
network.@switch_vlan[0].ports='1 2 5t'
network.@switch_vlan[1]=switch_vlan
network.@switch_vlan[1].device='switch0'
network.@switch_vlan[1].vlan='2'
network.@switch_vlan[1].ports='0 5t'
```

● network.loopback：本地回环接口，如果在设备上访问loopback的IP地址，就是自己访问自己。

● network.lan：局域网接口。

● network.wan：外网接口。

● network.wan6：IPv6外网接口。

● network.guest：访客网络接口。

● network.lease：租赁网络接口。

● network.@alias[0]：保留调试网络接口。

UCI无线网络配置

```
DF1A:$ uci show wireless
wireless.radio0=wifi-device
```

```
wireless.radio0.type='mac80211'
wireless.radio0.channel='auto'
wireless.radio0.band='2.4G'
wireless.radio0.max_all_num_sta='40'
wireless.radio0.netisolate='0'
wireless.radio0.country='CN'
wireless.radio0.ht_coex='0'
wireless.radio0.noscan='0'
wireless.radio0.radio='1'
wireless.radio0.txpower_lvl='2'
wireless.radio0.path='10000000.palmbus/11000000.wifi-lb'
wireless.radio0.htmode='HT20'
wireless.radio0.hwmode='11g'
wireless.@wifi-iface[0]=wifi-iface
wireless.@wifi-iface[0].device='radio0'
wireless.@wifi-iface[0].ifname='wlan0'
wireless.@wifi-iface[0].network='lan'
wireless.@wifi-iface[0].mode='ap'
wireless.@wifi-iface[0].ssid='SiWiFi-8868-2.4G'
wireless.@wifi-iface[0].encryption='psk2+ccmp'
wireless.@wifi-iface[0].key='12345678'
wireless.@wifi-iface[0].isolate='0'
wireless.@wifi-iface[0].hidden='0'
wireless.@wifi-iface[0].macfilter='disable'
wireless.@wifi-iface[0].macfile='/etc/wlan-file/wlan0.allow'
wireless.@wifi-iface[0].group='1'
wireless.@wifi-iface[0].netisolate='0'
wireless.@wifi-iface[0].disable_input='0'
wireless.@wifi-iface[0].wps_pushbutton='1'
wireless.@wifi-iface[0].wps_label='0'
wireless.@wifi-iface[0].rps_cpus='2'
wireless.@wifi-iface[1]=wifi-iface
wireless.@wifi-iface[1].device='radio0'
wireless.@wifi-iface[1].ifname='wlan0-guest'
wireless.@wifi-iface[1].network='guest'
wireless.@wifi-iface[1].mode='ap'
wireless.@wifi-iface[1].ssid='SiWiFi-8868-2.4G-guest'
wireless.@wifi-iface[1].encryption='psk2+ccmp'
wireless.@wifi-iface[1].key='12345678'
wireless.@wifi-iface[1].isolate='1'
wireless.@wifi-iface[1].hidden='0'
wireless.@wifi-iface[1].group='1'
wireless.@wifi-iface[1].netisolate='0'
wireless.@wifi-iface[1].disable_input='0'
wireless.@wifi-iface[1].wps_pushbutton='0'
wireless.@wifi-iface[1].wps_label='0'
wireless.@wifi-iface[1].rps_cpus='2'
wireless.@wifi-iface[1].disabled='0'
wireless.@wifi-iface[2]=wifi-iface
wireless.@wifi-iface[2].device='radio0'
```

```
wireless.@wifi-iface[2].ifname='wlan0-lease'
wireless.@wifi-iface[2].network='lease'
wireless.@wifi-iface[2].mode='ap'
wireless.@wifi-iface[2].ssid='SiWiFi-租赁--2.4G8868'
wireless.@wifi-iface[2].encryption='none'
wireless.@wifi-iface[2].isolate='1'
wireless.@wifi-iface[2].hidden='0'
wireless.@wifi-iface[2].group='1'
wireless.@wifi-iface[2].netisolate='0'
wireless.@wifi-iface[2].maxassoc='40'
wireless.@wifi-iface[2].disable_input='0'
wireless.@wifi-iface[2].rps_cpus='2'
wireless.@wifi-iface[2].disabled='1'
wireless.radio1=wifi-device
wireless.radio1.type='mac80211'
wireless.radio1.channel='161'
wireless.radio1.band='5G'
wireless.radio1.max_all_num_sta='40'
wireless.radio1.netisolate='0'
wireless.radio1.country='CN'
wireless.radio1.ht_coex='0'
wireless.radio1.noscan='1'
wireless.radio1.radio='1'
wireless.radio1.txpower_lvl='2'
wireless.radio1.path='10000000.palmbus/11400000.wifi-hb'
wireless.radio1.htmode='VHT80'
wireless.radio1.hwmode='11a'
wireless.@wifi-iface[3]=wifi-iface
wireless.@wifi-iface[3].device='radio1'
wireless.@wifi-iface[3].ifname='wlan1'
wireless.@wifi-iface[3].network='lan'
wireless.@wifi-iface[3].mode='ap'
wireless.@wifi-iface[3].ssid='SiWiFi-886c'
wireless.@wifi-iface[3].encryption='psk2+ccmp'
wireless.@wifi-iface[3].key='12345678'
wireless.@wifi-iface[3].isolate='0'
wireless.@wifi-iface[3].hidden='0'
wireless.@wifi-iface[3].macfilter='disable'
wireless.@wifi-iface[3].macfile='/etc/wlan-file/wlan1.allow'
wireless.@wifi-iface[3].group='1'
wireless.@wifi-iface[3].netisolate='0'
wireless.@wifi-iface[3].disable_input='0'
wireless.@wifi-iface[3].wps_pushbutton='1'
wireless.@wifi-iface[3].wps_label='0'
wireless.@wifi-iface[3].rps_cpus='3'
wireless.@wifi-iface[4]=wifi-iface
wireless.@wifi-iface[4].device='radio1'
wireless.@wifi-iface[4].ifname='wlan1-guest'
wireless.@wifi-iface[4].network='guest'
wireless.@wifi-iface[4].mode='ap'
wireless.@wifi-iface[4].ssid='SiWiFi-886c-guest'
```

```
wireless.@wifi-iface[4].encryption='psk2+ccmp'
wireless.@wifi-iface[4].key='12345678'
wireless.@wifi-iface[4].isolate='1'
wireless.@wifi-iface[4].hidden='0'
wireless.@wifi-iface[4].group='1'
wireless.@wifi-iface[4].netisolate='0'
wireless.@wifi-iface[4].disable_input='0'
wireless.@wifi-iface[4].wps_pushbutton='0'
wireless.@wifi-iface[4].wps_label='0'
wireless.@wifi-iface[4].rps_cpus='3'
wireless.@wifi-iface[4].disabled='1'
```

● wireless.radio0：2.4GHz物理无线电设备。描述2.4GHz无线设备上无线接口的通用属性，如频道或天线选择，最终配置参数会传递给无线驱动模块。通常无线电设备支持多个无线网络。

● wireless.radio1：5GHz物理无线电设备。描述5GHz无线设备上无线接口的通用属性，如频道或天线选择，最终配置参数会传递给无线驱动模块。通常无线电设备支持多个无线网络。

● wireless.@wifi-iface[0]：wlan0接口，2.4GHz无线网络接口。

● wireless.@wifi-iface[1]：wlan0-guest接口，2.4GHz无线访客网络接口。

● wireless.@wifi-iface[2]：wlan0-lease接口，2.4GHz无线租赁网络接口。

● wireless.@wifi-iface[3]：wlan1接口，5GHz无线网络接口。

● wireless.@wifi-iface[4]：wlan1-guest接口，5GHz无线访客网络接口。

查看当前网络情况

```
DF1A:$ ifconfig
br-lan    Link encap:Ethernet  HWaddr 10:16:88:4F:88:66
          inet addr:192.168.4.1  Bcast:192.168.4.255  Mask:255.255.255.0
          inet6 addr: fe80::1216:88ff:fe4f:8866/64 Scope:Link
          inet6 addr: fd8d:78b8:b066::1/60 Scope:Global
          UP BROADCAST RUNNING MULTICAST  MTU:1500  Metric:1
          RX packets:1715 errors:0 dropped:0 overruns:0 frame:0
          TX packets:2010 errors:0 dropped:0 overruns:0 carrier:0
          collisions:0 txqueuelen:0
          RX bytes:321696 (314.1 KiB)  TX bytes:716146 (699.3 KiB)

eth0      Link encap:Ethernet  HWaddr 10:16:88:4F:88:66
          inet6 addr: fe80::1216:88ff:fe4f:8866/64 Scope:Link
          UP BROADCAST RUNNING MULTICAST  MTU:1500  Metric:1
          RX packets:7152 errors:0 dropped:0 overruns:0 frame:0
          TX packets:2419 errors:0 dropped:0 overruns:0 carrier:0
          collisions:0 txqueuelen:1000
          RX bytes:2049200 (1.9 MiB)  TX bytes:799273 (780.5 KiB)
          Interrupt:24

eth0.1    Link encap:Ethernet  HWaddr 10:16:88:4F:88:66
          UP BROADCAST RUNNING MULTICAST  MTU:1500  Metric:1
          RX packets:1829 errors:0 dropped:7 overruns:0 frame:0
          TX packets:1914 errors:0 dropped:0 overruns:0 carrier:0
          collisions:0 txqueuelen:0
          RX bytes:346906 (338.7 KiB)  TX bytes:706846 (690.2 KiB)

eth0.2    Link encap:Ethernet  HWaddr 10:16:88:4F:88:67
```

```
              inet addr:172.16.10.203  Bcast:172.16.10.255  Mask:255.255.255.0
              inet6 addr: fe80::1216:88ff:fe4f:8867/64 Scope:Link
              UP BROADCAST RUNNING MULTICAST  MTU:1500  Metric:1
              RX packets:5299 errors:0 dropped:7 overruns:0 frame:0
              TX packets:493 errors:0 dropped:0 overruns:0 carrier:0
              collisions:0 txqueuelen:0
              RX bytes:1673046 (1.5 MiB)  TX bytes:42929 (41.9 KiB)
lo            Link encap:Local Loopback
              inet addr:127.0.0.1  Mask:255.0.0.0
              inet6 addr: ::1/128 Scope:Host
              UP LOOPBACK RUNNING  MTU:65536  Metric:1
              RX packets:93 errors:0 dropped:0 overruns:0 frame:0
              TX packets:93 errors:0 dropped:0 overruns:0 carrier:0
              collisions:0 txqueuelen:0
              RX bytes:7764 (7.5 KiB)  TX bytes:7764 (7.5 KiB)
wlan0         Link encap:Ethernet  HWaddr 10:16:88:4F:88:6B
              inet6 addr: fe80::1216:88ff:fe4f:886b/64 Scope:Link
              UP BROADCAST RUNNING MULTICAST  MTU:1500  Metric:1
              RX packets:0 errors:0 dropped:0 overruns:0 frame:0
              TX packets:11 errors:0 dropped:0 overruns:0 carrier:0
              collisions:0 txqueuelen:1000
              RX bytes:0 (0.0 B)  TX bytes:1200 (1.1 KiB)
wlan1         Link encap:Ethernet  HWaddr 10:16:88:4F:88:6F
              inet6 addr: fe80::1216:88ff:fe4f:886f/64 Scope:Link
              UP BROADCAST RUNNING MULTICAST  MTU:1500  Metric:1
              RX packets:0 errors:0 dropped:0 overruns:0 frame:0
              TX packets:12 errors:0 dropped:0 overruns:0 carrier:0
              collisions:0 txqueuelen:1000
              RX bytes:0 (0.0 B)  TX bytes:1334 (1.3 KiB)
```

- br-lan：虚拟设备LAN口桥接设备，包含通过LAN口和无线口连入系统的设备统一桥接。
- eth0：真实设备，CPU中的交换机芯片。
- eth0.1：VLAN划分的LAN有线网口。
- eth0.2：VLAN划分的独立WAN口。
- lo：虚拟设备，回环设备。
- wlan0：真实设备，2.4GHz无线设备。启动Wi-Fi后将产生此无线设备。
- wlan1：真实设备，5GHz无线设备。启动Wi-Fi后将产生此无线设备。
- pppoe-wan：虚拟设备，PPPoE拨号上网成功后产生。

查看br-lan桥状态

```
DF1A:$ brctl show
bridge name     bridge id               STP enabled     interfaces
br-lan          7fff.1016884f8866       no              eth0.1
                                                        wlan1
                                                        wlan0
```

重启网络服务

```
/etc/init.d/network restart
```

重启无线网络服务

```
DF1A:$ wifi
```

5.2 WAN口配置

WAN口是路由器与外部连接的外网端口。在OpenWrt中，WAN口的UCI配置文件主要在/etc/config/network中。

5.2.1 WAN口配置参数

/etc/config/network配置文件的WAN口配置内容如下。

```
config interface 'wan'
        option ifname 'eth0.2'
        option force_link '1'
        option macaddr '10:16:88:4f:88:67'
        option rps_cpus '1'
        option xps_cpus '0'
        option proto 'dhcp'
```

通过UCI查看network.wan节点的内容：

```
DF1A:$ uci show network.wan
network.wan=interface
network.wan.ifname='eth0.2'
network.wan.force_link='1'
network.wan.macaddr='10:16:88:4f:88:67'
network.wan.rps_cpus='1'
network.wan.xps_cpus='0'
network.wan.proto='dhcp'
```

network.wan选项参数说明如表5-1所示。

表5-1 network.wan 选项参数说明

选项	说明	可选值及说明	必填
ifname	设备名称	eth0.2	是
proto	协议类型	static：静态IP地址 dhcp：动态获取IP地址 dhcpv6：动态获取IPv6地址 pppoe：拨号上网 ppp：点对点协议 pptp：远程VPN服务器 3g：连接3G/4G无线移动网络	是
macaddr	WAN口MAC地址，修改该地址即可 实现MAC地址克隆功能	首次数据根据factory分区内的参数自动生成	是
force_link	是否保留ip及gw	数值，1或0，Netifd引入的参数。 为1时，就算link是down，interface的ip及gw依然存在	否
rps_cpus、xps_cpus	为了降低进程切换产生的损耗，指定特定CPU内核来响应这一部分驱动操作	数值，CPU核心0~3。该参数不建议修改。 这是由芯片厂商增加的特定参数	否

静态IP

给WAN口设置静态的IP地址。如果设置的静态IP与上级路由或网络不在一个网段中，可能导致不能上网。可选的配置参数说明如表5-2所示。

表 5-2 静态 IP 配置参数说明

选项	说明	可选值及说明	必填
proto	协议类型	static	是
ifname	接口名称	eth0.2	是
macaddr	MAC地址	值根据factory分区自动生成	是
mtu	修改最大数据包大小，默认不用设置	数值	否
ipaddr	WAN口的IP地址	字符串	是
netmask	WAN口的子网掩码	字符串	是
gateway	默认网关	字符串	否
broadcast	广播地址	字符串	否
dns	DNS服务器地址	字符串	否
metric	路由默认的跃点数	整数	否

配置文件举例如下。

```
config interface 'wan'
        option ifname 'eth0.2'
        option proto 'static'
        option ipaddr '192.168.0.2'
        option netmask '255.255.255.0'
        option gateway '192.168.0.1'
        option dns '192.168.0.1'
```

动态获取IP

设置WAN口的IP为动态获取，由上级路由或网络提供DHCP服务，通过DHCP协议获取IP。可选的配置参数说明如表5-3所示。

表 5-3 动态获取 IP 配置参数说明

选项	说明	可选值及说明	必填
proto	协议类型	dhcp	是
ifname	接口名称	eth0.2	是
macaddr	MAC地址	值根据factory分区自动生成	是
mtu	修改最大数据包大小，默认不用设置	数值	否
reqopts	在向DHCP服务器发出请求时增加附加的DHCP信息	字符串	否
sendopts	要发送到服务器的其他DHCP选项的空间分隔列表	字符串	否
dns	使用指定的DNS服务器地址替代获得的DNS	字符串	否
force_link	是否保留ip及gw	数值，1或0	否

配置文件举例如下。

```
cconfig interface 'wan'
        option ifname 'eth0.2'
        option macaddr '10:16:88:14:3d:04'
        option force_link '0'
```

```
option proto 'dhcp'
```

PPPoE拨号上网

以PPPoE方式拨号，如果拨号成功，会获取ISP（因特网服务提供方）分配的IP地址。在设置PPPoE的时候要确保你的设备WAN口连接的外网具备PPPoE的服务器。PPPoE拨号上网配置参数说明如表5-4所示。

表5-4 PPPoE拨号上网配置参数说明

选项	说明	可选值及说明	必填
proto	协议类型	pppoe	是
ifname	接口名称	eth0.2	是
macaddr	MAC地址	根据factory分区自动生成的值	是
mtu	修改最大数据包大小，默认不用设置	数值	否
username	拨号用的账号	字符串	是
password	拨号用的密码	字符串	是
ac	使用指定的访问集中器进行连接	字符串	否
service	连接的服务名称	字符串	否
connect	连接时执行的外部脚本	字符串	否
disconnect	断开连接时执行的外部脚本	字符串	否
demand	等待多久没有活动就断开PPPoE连接	数字，单位为秒	否
dns	DNS服务器地址	字符串	否
pppd_options	用于pppd进程执行的附加参数	字符串	否

配置文件举例如下。

```
config interface 'wan'
        option ifname 'eth0.2'
        option proto 'pppoe'
        option username '280000000000'
        option password '21700000'
```

5.2.2 使WAN口配置生效

通过Vi编辑器、uci命令，或者UCI提供的其他语言接口修改WAN口配置参数，然后可以使用下面两种方法（任选其一）使网络配置生效。

● 通过/etc/init.d/network脚本重启：

```
DF1A:$ /etc/init.d/network restart
```

● 用ubus命令重启network服务：

```
DF1A:$ ubus call network restart
```

5.3 LAN口配置

LAN口下的设备可以通过WAN口接入网络，也可以直接访问设备上的各项功能（默认系统防火墙对LAN口不做任何拦截）。LAN口的配置主要在/etc/config/network中。network.lan选项参数说明如下所示。

/etc/config/network配置文件的内容如下。

```
config interface 'lan'
        option ifname 'eth0.1'
        option force_link '1'
        option macaddr '10:16:88:4f:88:66'
        option type 'bridge'
        option proto 'static'
        option ipaddr '192.168.4.1'
        option netmask '255.255.255.0'
        option ip6assign '60'
        option group '0'
        option rps_cpus '2'
        option xps_cpus '2'
```

UCI查看方法：

```
DF1A:$ uci show network.lan
network.lan=interface
network.lan.ifname='eth0.1'
network.lan.force_link='1'
network.lan.macaddr='10:16:88:4f:88:66'
network.lan.type='bridge'
network.lan.proto='static'
network.lan.ipaddr='192.168.4.1'
network.lan.netmask='255.255.255.0'
network.lan.ip6assign='60'
network.lan.group='0'
network.lan.rps_cpus='2'
network.lan.xps_cpus='2'
```

network.lan选项参数说明如表5-5所示。

表5-5 network.lan 选项参数说明

选项	说明	可选值及说明	必填
ifname	设备名称	eth0.2	是
proto	协议类型	static：静态IP地址	是
macaddr	LAN口MAC地址	首次数据根据factory分区内的参数自动生成	否
type	网络类型必须是桥模式，否则不具备交换机功能	bridge	是
ipaddr	LAN口的IP地址	字符串	是
netmask	LAN口的子网掩码	字符串	是
ip6assign	IPv6前缀长度。将指定长度的前缀委托给此接口	数字	否
group	0就代表LAN	数字	是

配置文件举例如下。

```
config interface 'lan'
        option ifname 'eth0.2'
        option type 'bridge'
        option proto 'static'
        option ipaddr '192.168.1.1'
        option netmask '255.255.255.0'
```

        option group '0'

    修改LAN口的配置后要重启网络服务：

```
DF1A:$ /etc/init.d/network restart
```

5.4 配置无线网络

5.4.1 配置文件

/etc/config/wireless配置文件的内容如下。

```
config wifi-device 'radio0'
        option type 'mac80211'
        option channel 'auto'
        option band '2.4G'
        option max_all_num_sta '40'
        option netisolate '0'
        option country 'CN'
        option ht_coex '0'
        option noscan '0'
        option radio '1'
        option txpower_lvl '2'
        option path '10000000.palmbus/11000000.wifi-lb'
        option htmode 'HT20'
        option hwmode '11g'
config wifi-iface
        option device 'radio0'
        option ifname 'wlan0'
        option network 'lan'
        option mode 'ap'
        option ssid 'SiWiFi-8868-2.4G'
        option encryption 'psk2+ccmp'
        option key '12345678'
        option isolate '0'
        option hidden '0'
        option macfilter 'disable'
        option macfile '/etc/wlan-file/wlan0.allow'
        option group '1'
        option netisolate '0'
        option disable_input '0'
        option wps_pushbutton '1'
        option wps_label '0'
        option rps_cpus '2'
config wifi-iface
        option device 'radio0'
        option ifname 'wlan0-guest'
        option network 'guest'
        option mode 'ap'
        option ssid 'SiWiFi-8868-2.4G-guest'
        option encryption 'psk2+ccmp'
```

```
                option key '12345678'
                option isolate '1'
                option hidden '0'
                option group '1'
                option netisolate '0'
                option disable_input '0'
                option wps_pushbutton '0'
                option wps_label '0'
                option rps_cpus '2'
                option disabled '1'
config wifi-iface
                option device 'radio0'
                option ifname 'wlan0-lease'
                option network 'lease'
                option mode 'ap'
                option ssid 'SiWiFi-租赁--2.4G8868'
                option encryption 'none'
                option isolate '1'
                option hidden '0'
                option group '1'
                option netisolate '0'
                option maxassoc '40'
                option disable_input '0'
                option rps_cpus '2'
                option disabled '1'
config wifi-device 'radio1'
                option type 'mac80211'
                option channel '161'
                option band '5G'
                option max_all_num_sta '40'
                option netisolate '0'
                option country 'CN'
                option ht_coex '0'
                option noscan '1'
                option radio '1'
                option txpower_lvl '2'
                option path '10000000.palmbus/11400000.wifi-hb'
                option htmode 'VHT80'
                option hwmode '11a'
config wifi-iface
                option device 'radio1'
                option ifname 'wlan1'
                option network 'lan'
                option mode 'ap'
                option ssid 'SiWiFi-886c'
                option encryption 'psk2+ccmp'
                option key '12345678'
                option isolate '0'
                option hidden '0'
                option macfilter 'disable'
                option macfile '/etc/wlan-file/wlan1.allow'
```

```
        option group '1'
        option netisolate '0'
        option disable_input '0'
        option wps_pushbutton '1'
        option wps_label '0'
        option rps_cpus '3'
config wifi-iface
        option device 'radio1'
        option ifname 'wlan1-guest'
        option network 'guest'
        option mode 'ap'
        option ssid 'SiWiFi-886c-guest'
        option encryption 'psk2+ccmp'
        option key '12345678'
        option isolate '1'
        option hidden '0'
        option group '1'
        option netisolate '0'
        option disable_input '0'
        option wps_pushbutton '0'
        option wps_label '0'
        option rps_cpus '3'
        option disabled '1'
```

配置分为两层，wifi-device对应具体的无线驱动设备，在我们的硬件中支持2.4GHz和5GHz两种device；wifi-iface对应了单个Wi-Fi接口，与单个SSID相对应。一个wifi-device下面可以配置多个wifi-iface。

wifi-device配置选项说明如表5-6所示。

表5-6 wifi-device 配置选项说明

选项	说明	可选值及说明	必填
channel	信道	默认为auto，需要根据country来选择，中国2.4GHz频段可用信道为1~13，5GHz频段可用信道为36、40、44、149~165	是
band	频段	选择2.4GHz或者5GHz	是
max_all_num_sta	该device下所有station个数最大值	数值	否
netisolate	是否独立	数值，1或0，如果配置为1，则从该device下的设备无法访问同一bridge中其他bssid的设备	否
country	国家和地区编码，跟支持的信道有关	中国默认为CN	是
ht_coex	共存模式	20MHz/40MHz	否
noscan	不扫描周围信道	数值，1或0	否
radio	目前未使用	数值，1或0，默认配置为1	否
path	对应了驱动在/sys/bus下面的节点	字符串	是
htmode	带宽模式	2.4GHz支持20MHz/40MHz，5GHz支持20MHz/40MHz/80MHz	是
hwmode	Wi-Fi工作模式	2.4GHz支持802.11b/g/n，5GHz支持802.11n/a/ac，最终的模式是由hwode和htmode共同决定的。要注意hwmode的取值范围问题，2.4GHz只有802.11b/11g，5GHz只有802.11a，配置错误可能导致Wi-Fi启动失败	是

不同的hwmode决定了htmode可以配置的选项。表5-7、表5-8列举了htmode与hwmode在不同模式下的对应关系。

表5-7　2.4GHz htmode 与 hwmode 对应关系

模式	配置	说明
802.11b	option hwmode 11b	该模式下必须删除option htmode、option ht_coex
802.11g	option hwmode 11g	该模式下必须删除option htmode、option ht_coex
802.11n	option hwmode 11g option htmode HT20/HT40 option ht_coex 1	

表5-8　5GHz htmode 与 hwmode 对应关系

模式	配置	说明
802.11a	option hwmode 11a	该模式下必须删除option htmode、option ht_coex
802.11n	option hwmode 11a option htmode HT20/HT40 option ht_coex 1	
802.11ac	option hwmode 11a option htmode VHT20/VHT40/VHT80	该模式下必须删除option ht_coex

配置文件举例如下。

```
config wifi-device 'radio0'
    option type 'mac80211'
    option channel '11'
    option band '2.4G'
    option max_all_num_sta '20'
    option netisolate '0'
    option country 'CN'
    option ht_coex '0'
    option noscan '0'
    option radio '1'
    option txpower_lvl '2'
    option path '10000000.palmbus/11000000.wifi-lb'
    option htmode 'HT40'
    option hwmode '11g'
```

wifi-iface配置选项说明如表5-9所示。

表5-9　wifi-iface 配置选项说明

选项	说明	可选值及说明	必填
device	对应的wifi-device节点	默认2.4GHz为radio0，5GHz为radio1	是
ifname	网卡的iface的名称，没有限制		是
network	对应的bridge的名称，如果需要给Wi-Fi加入LAVN口则需要配置该值为lan		是
mode	iface的模式	支持ap、sta、monitor，ap对应热点，sta对应station，monitor对应监听模式	是
ssid	热点的名称	最大不能超过32位	是
encryption	加密方式	默认为psk2+ccmp，如果无密码则配置为none	是
key	Wi-Fi密钥	psk2下必须8位及以上	是

续表

选项	说明	可选值及说明	必填
isolate	无线station内部隔离	1或0	否
hidden	Wi-Fi热点隐藏节点	1或0	否
macfilter	Wi-Fi黑白名单使能开关	enable或disable，默认为disable	否
macfile	Wi-Fi黑白名单文件		否
group	bridge中的分组，各个不同的group之间在bridge中是不能互相访问的		是
netisolate	如果配置为1，则从该bssid下的设备无法访问同一bridge中其他bssid的设备		否

配置文件举例如下。

```
config wifi-iface
    option device 'radio0'
    option ifname 'wlan0'
    option network 'lan'
    option mode 'ap'
    option ssid 'MyTest-2.4G'
    option encryption 'psk2+ccmp'
    option key 'helloworld'
    option isolate '0'
    option hidden '0'
    option macfilter 'disable'
    option macfile '/etc/wlan-file/wlan0.allow'
    option group '1'
    option netisolate '0'
```

修改无线配置后，需要使用wifi命令使配置生效。

```
DF1A:$ wifi
```

5.4.2 无线网络查看命令

查看无线网络状态：

```
DF1A:$ iwinfo wlan0 info
wlan0     ESSID: "SiWiFi-8868-2.4G"
          Access Point: 10:16:88:4F:88:6B
          Mode: Master  Channel: 12 (2.467 GHz)
          Tx-Power: 20 dBm  Link Quality: unknown/70
          Signal: unknown  Noise: unknown
          Bit Rate: unknown
          Encryption: WPA2 PSK (CCMP)
          Type: nl80211   HW Mode(s): 802.11bgn
          Hardware: unknown [Generic MAC80211]
          TX power offset: unknown
          Frequency offset: unknown
          Supports VAPs: yes  PHY name: phy0
```

显示在左面的wlan0为无线连接的设备名字，这个名字是由匿名节点wireless.@wifi-iface[0]创建的。

通过wlan0搜索范围内其他无线设备：

```
DF1A:$ iwinfo wlan0 scan
Cell 01 - Address: 68:DB:54:DA:FD:4E
          ESSID: "TECSOON_K3C"
          Mode: Master   Channel: 1
          Signal: -53 dBm   Quality: 57/70
          Encryption: mixed WPA/WPA2 PSK (TKIP, CCMP)
Cell 02 - Address: 4C:AB:FC:5C:11:15
          ESSID: "CMCC-YeDg"
          Mode: Master   Channel: 3
          Signal: -82 dBm   Quality: 28/70
          Encryption: mixed WPA/WPA2 PSK (TKIP, CCMP)
...
```

通过wlan0查看radio0支持的国家和地区编码：

```
DF1A:$ iwinfo wlan0 countrylist
   00 00
   AD AD
   AE AE
   AF AF
   AG AG
   AI AI
   AL AL
   AM AM
   AN AN
...
```

5.4.3 实现无线中继

无线中继可以实现以开发板为客户端，远程连接另外一个无线路由器，主要是通过增加wifi-iface实现的。增加无线中继后，系统就有两个匿名的wifi-iface配置，其中一个用来解决其他设备接入路由器，另外一个用来解决无线中继。

（1）为网络增加WWAN类型接口设备：

```
DF1A:$ uci set network.wwan=interface
DF1A:$ uci set network.wwan.proto=dhcp
DF1A:$ uci commit
DF1A:$ uci show network.wwan
network.wwan=interface
network.wwan.proto='dhcp'
```

（2）将WWAN设置到防火墙WAN区域范围。由于WAN区域属于匿名节点，首先找到WAN口节点的ID（下面示例中cfg0cdc81为节点ID，这个值是随机的，每台机器操作时可能不同）。

```
DF1A:$ uci show -X firewall|grep "name='wan'"
firewall.cfg0cdc81.name='wan'
```

修改区域范围：

```
DF1A:$ uci set firewall.cfg0cdc81.network='wan wan6 wwan'
DF1A:$ uci commit
DF1A:$ /etc/init.d/firewall reload
```

（3）新建wifi-iface匿名节点实现连接。搜索要中继的另外一个无线网络（请注意搜索所使用的wlan0为2.4GHz的），如ESSID为LELE的无线网络。

```
DF1A:$ iwinfo wlan0 scan
Cell 01 - Address: 68:DB:54:DA:FD:4E
          ESSID: "TECSOON_K3C"
          Mode: Master  Channel: 1
          Signal: -52 dBm  Quality: 58/70
          Encryption: mixed WPA/WPA2 PSK (TKIP, CCMP)
Cell 02 - Address: 4C:AB:FC:5C:11:15
          ESSID: "CMCC-YeDg"
          Mode: Master  Channel: 3
          Signal: -67 dBm  Quality: 43/70
          Encryption: mixed WPA/WPA2 PSK (TKIP, CCMP)
Cell 03 - Address: 1C:40:E8:15:59:27
          ESSID: "LELE"
          Mode: Master  Channel: 5
          Signal: -33 dBm  Quality: 70/70
          Encryption: WPA2 PSK (CCMP)
...
```

（4）根据获得的值在配置文件/etc/config/wireless中增加一个类型为wifi-iface的匿名节点，其中ssid填写获得的ESSID，bssid填写获得的Address，而encryption要和对方对应，举例如下。

```
config wifi-iface
        option device      radio0
        option ifname      wlan0-wwan
        option network     wwan
        option mode        sta
        option ssid        LELE
        option encryption  psk2+ccmp
        option key         12345678
        option bssid       1C:40:E8:15:59:27
        option isolate     '0'
        option hidden      '0'
        option macfilter   disable
        option macfile     /etc/wlan-file/wlan0.allow
        option group       1
        option netisolate  0
        option disable_input 0
        option wps_pushbutton '0'
        option wps_label   '0'
        option disabled    '0'
```

（5）使用wifi命令使设置生效，之后检测，如果有wlan0、wlan0-wwan两个设备，则表示生效，其中wlan0为原来的无线网络，wlan0-wwan表示远程无线连接到其他路由器上的无线网络。wlan0-wwan已经被上级路由器LELE分配IP地址172.16.10.116；如果没有被分配IP地址，肯定是配置有问题，要重复上面前两步操作。

```
DF1A:$ /etc/init.d/network restart
DF1A:$ ifconfig wlan0
wlan0     Link encap:Ethernet  HWaddr 10:16:88:14:3C:FF
          inet6 addr: fe80::1216:88ff:fe14:3cff/64 Scope:Link
          UP BROADCAST RUNNING MULTICAST  MTU:1500  Metric:1
          RX packets:0 errors:0 dropped:0 overruns:0 frame:0
          TX packets:16 errors:0 dropped:0 overruns:0 carrier:0
          collisions:0 txqueuelen:1000
          RX bytes:0 (0.0 B)  TX bytes:2486 (2.4 KiB)
DF1A:$ ifconfig wlan0-wwan
wlan0-wwan Link encap:Ethernet  HWaddr 10:16:88:14:3C:FC
          inet addr:172.16.10.116  Bcast:172.16.10.255  Mask:255.255.255.0
          inet6 addr: fe80::1216:88ff:fe14:3cfc/64 Scope:Link
          UP BROADCAST RUNNING MULTICAST  MTU:1500  Metric:1
          RX packets:149 errors:0 dropped:0 overruns:0 frame:0
          TX packets:23 errors:0 dropped:0 overruns:0 carrier:0
          collisions:0 txqueuelen:1000
          RX bytes:60754 (59.3 KiB)  TX bytes:2964 (2.8 KiB)
```

（6）测试中继后的网络是否能够联网。

```
DF1A:$ route -n
Kernel IP routing table
Destination     Gateway         Genmask         Flags Metric Ref    Use Iface
default         172.16.10.1     0.0.0.0         UG    0      0        0 wlan0-wwan
172.16.10.0     *               255.255.255.0   U     0      0        0 wlan0-wwan
172.16.10.1     *               255.255.255.255 UH    0      0        0 wlan0-wwan
192.168.4.0     *               255.255.255.0   U     0      0        0 br-lan
192.168.5.0     *               255.255.255.0   U     0      0        0 br-lan
DF1A:$ ping 172.16.10.1
PING 172.16.10.1 (172.16.10.1): 56 data bytes
64 bytes from 172.16.10.1: seq=0 ttl=64 time=2.878 ms
^C
--- 172.16.10.1 ping statistics ---
1 packets transmitted, 1 packets received, 0% packet loss
round-trip min/avg/max = 2.878/2.878/2.878 ms
DF1A:$ ping www.xxxx.com
PING www.xxxx.com (xx.xx.66.14): 56 data bytes
64 bytes from xx.xx.66.14: seq=0 ttl=51 time=22.609 ms
64 bytes from xx.xx.66.14: seq=1 ttl=51 time=27.929 ms
^C
--- www.xxxx.com ping statistics ---
2 packets transmitted, 2 packets received, 0% packet loss
round-trip min/avg/max = 22.609/25.269/27.929 ms
```

注意事项

无线中继要修改的文件较多，请认真操作，如果遇到错误需要重复进行。

无线中继建立成功后具备的功能如下。

- 用ifconfig命令可以看到多了两个连接，一个是wlan0，另一个是wlan0-wwan，其中有一个带IP地址，那个带IP地址的就是无线中继。
- 开发板本身可以直接连接外网。
- 通过无线或LAN口连接到开发板上的手机或计算机，可通过开发板的无线中继连接外网。

无线连接失败：无线连接失败多半是配置信息写错引起的，请重新检查，重复操作"新建wifi-iface匿名节点实现连接"的步骤。

关闭无线访问模式：直接去掉另外那个wifi-iface，并且重启无线网络就可以了。

5.5 DHCP服务

默认情况下，OpenWrt使用dnsmasq和odhcpd对外提供DNS和DHCP服务，可以为LAN下的有线设备和无线设备提供自动分配IP地址服务，如表5-10所示。

表5-10 DHCP 相关包

软件包	服务	端口	配置
dnsmasq	DNS	53/TCP、53/UDP	/etc/config/dhcp
dnsmasq	DHCP	67/UDP	/etc/config/dhcp
odhcpd	DHCPv6	547/UDP	

/etc/config/dhcp配置文件的内容如下。

```
config dnsmasq
    option domainneeded '1'
    option boguspriv '1'
    option filterwin2k '0'
    option localise_queries '1'
    option rebind_protection '1'
    option rebind_localhost '1'
    option local '/lan/'
    option domain 'lan'
    option expandhosts '1'
    option nonegcache '0'
    option authoritative '1'
    option readethers '1'
    option leasefile '/tmp/dhcp.leases'
    option resolvfile '/tmp/resolv.conf.auto'
    option localservice '1'
    option dhcpscript '/lib/netifd/dhcplease'
config dhcp 'lan'
    option interface 'lan'
    option start '100'
    option limit '150'
    option leasetime '12h'
```

```
            option dhcpv6 'server'
            option ra 'server'
    config dhcp 'guest'
            option interface 'guest'
            option start '100'
            option limit '150'
            option leasetime '12h'
    config dhcp 'lease'
            option interface 'lease'
            option start '100'
            option limit '150'
            option leasetime '12h'
    config dhcp 'wan'
            option interface 'wan'
            option ignore '1'
    config odhcpd 'odhcpd'
            option maindhcp '0'
            option leasefile '/tmp/hosts/odhcpd'
            option leasetrigger '/usr/sbin/odhcpd-update'
```

通过UCI查看节点的内容：

```
DF1A:$ uci show dhcp
dhcp.@dnsmasq[0]=dnsmasq
dhcp.@dnsmasq[0].domainneeded='1'
dhcp.@dnsmasq[0].boguspriv='1'
dhcp.@dnsmasq[0].filterwin2k='0'
dhcp.@dnsmasq[0].localise_queries='1'
dhcp.@dnsmasq[0].rebind_protection='1'
dhcp.@dnsmasq[0].rebind_localhost='1'
dhcp.@dnsmasq[0].local='/lan/'
dhcp.@dnsmasq[0].domain='lan'
dhcp.@dnsmasq[0].expandhosts='1'
dhcp.@dnsmasq[0].nonegcache='0'
dhcp.@dnsmasq[0].authoritative='1'
dhcp.@dnsmasq[0].readethers='1'
dhcp.@dnsmasq[0].leasefile='/tmp/dhcp.leases'
dhcp.@dnsmasq[0].resolvfile='/tmp/resolv.conf.auto'
dhcp.@dnsmasq[0].localservice='1'
dhcp.@dnsmasq[0].dhcpscript='/lib/netifd/dhcplease'
dhcp.lan=dhcp
dhcp.lan.interface='lan'
dhcp.lan.start='100'
dhcp.lan.limit='150'
dhcp.lan.leasetime='12h'
dhcp.lan.dhcpv6='server'
dhcp.lan.ra='server'
dhcp.guest=dhcp
dhcp.guest.interface='guest'
dhcp.guest.start='100'
dhcp.guest.limit='150'
```

```
dhcp.guest.leasetime='12h'
dhcp.lease=dhcp
dhcp.lease.interface='lease'
dhcp.lease.start='100'
dhcp.lease.limit='150'
dhcp.lease.leasetime='12h'
dhcp.wan=dhcp
dhcp.wan.interface='wan'
dhcp.wan.ignore='1'
dhcp.odhcpd=odhcpd
dhcp.odhcpd.maindhcp='0'
dhcp.odhcpd.leasefile='/tmp/hosts/odhcpd'
dhcp.odhcpd.leasetrigger='/usr/sbin/odhcpd-update'
```

dhcp.@dnsmasq[0]选项参数说明如表5-11所示。dhcp.lan/dhcp.guest/dhcp.lease选项参数说明如表5-12所示。dhcp.wan选项参数说明如表5-13所示。dhcp.odhcpd选项参数说明如表5-14所示。

表5-11 dhcp.@dnsmasq[0] 选项参数说明

选项	说明	可选值及说明	必填
domainneeded	不向上级域名服务器转发无效格式域名	0：禁用 1：启用	是
boguspriv	不转发私有地址空间	0：禁用 1：启用	是
filterwin2k	不转发公共名称服务器无法应答的请求	0：禁用 1：启用	是
localise_queries	允许获得同时发生的请求的来源网络地址	0：禁用 1：启用	是
rebind_protection	拒绝来自上游服务器并且属于本地私有IP段的绑定请求	0：禁用 1：启用	是
rebind_localhost	允许本机网段的绑定请求	0：禁用 1：启用	是
local	从中查找此域的DNS条目/etc/hosts	字符串	是
domain	设定本地网域名	lan	是
expandhosts	将本地域部分添加到/etc/hosts中的名称中	0：禁用 1：启用	是
nonegcache	禁用缓存否定的"无此类域"响应	0：禁用 1：启用	是
authoritative	强制dnsmasq进入权威模式	0：禁用 1：启用	是
readethers	读取/etc/ethers文件中关于hosts的信息	0：禁用 1：启用	是
leasefile	存储客户端DHCP申请信息的文件名	字符串	是
resolvfile	上游DNS地址文件名	字符串	是

表 5-12　dhcp.lan/dhcp.guest/dhcp.lease 选项参数说明

选项	说明	可选值及说明	必填
interface	对应网络设备	lan、guest（访客网络）、lease（租赁网络）	是
start	分配IP地址的开始位置	数字，范围为0~255	是
limit	分配IP地址总量	数字，加上开始位置，不能超过255	是
leasetime	DHCP租期	数字时间，12h表示12小时	是
dhcp_option	DHCP选项	字符串列表，多个参数以逗号分隔，例如"mtu, 1470"或"26, 1470"	否
dhcp_option_force	DHCP强制选项	字符串列表，同dhcp_option	否
dynamicdhcp	动态分配客户端地址	0：禁用 1：启用	否
force	即使在同一网段上检测到另一台DHCP服务器，也强制在指定接口上提供DHCP服务	0：禁用 1：启用	
ignore	忽略这个网络设备	0：不忽略 1：忽略	否
dhcpv4	指定启用DHCPv4服务器还是禁用	字符串，server、disabled	否
dhcpv6	指定启用DHCPv6服务器、中继还是禁用	字符串，server、relay、disabled	否

表 5-13　dhcp.wan 选项参数说明

选项	说明	可选值及说明	必填
interface	对应网络设备	wan	是
ignore	忽略这个网络设备	0：不忽略 1：忽略	是

表 5-14　dhcp.odhcpd 选项参数说明

选项	说明	可选值及说明	必填
legacy	如果启动但未设置dhcpv4选项，则启用DHCPv4	0：不启用（默认） 1：启用	否
maindhcp	使用odhcpd作为主要的DHCPv4服务	0：不启用（默认） 1：启用	否
leasefile	DHCPv6租用/主机文件	字符串	是
leasetrigger	租赁触发器脚本	字符串	是
loglevel	syslog级别	整型，0~7（默认为6）	否

修改后要重启DHCP服务。

```
DF1A:$ /etc/init.d/dnsmasq restart
```

5.6　如何连接外网

路由器的基本功能就是连接外部网络，我们可以通过表5-15所示的方法连接外网。

表 5-15　连接外网的主要方式

联网方式	模型
DHCP获取上级路由器IP	计算机/手机→LAN有线/WiFi→开发板（WAN口）→外网路由器→外网
PPPoE拨号	计算机/手机→LAN有线/WiFi→开发板（WAN口）→外网
无线中继	计算机/手机→LAN有线/WiFi→开发板（无线中继）→外网路由器→外网
4G/5G上网	计算机/手机→LAN有线/WiFi→开发板（4G/5G联网芯片）→移动通信网络→外网

6 服务功能

6.1 防火墙

OpenWrt的防火墙软件包名为firewall。firewall基于iptables和libiptc（libip4tc、libip6tc）库实现防火墙功能（见图6-1）。netfilter是Linux内核中的包过滤框架，它允许数据包过滤、网络地址和端口转换（NAT）以及其他数据包操作。iptables是内核netfilter框架的用户层程序接口，使得插入、修改和删除信息包过滤表中的规则变得容易。iptables和firewall应用程序使用netfilter libiptc库与netfilter内核模块进行通信。

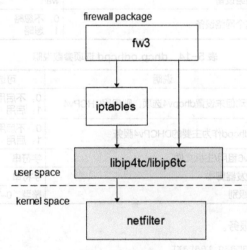

图6-1　firewall软件包与iptables、netfilter的关系

6.1.1 防火墙软件包

firewall软件包包含下面的文件。

```
DF1A:~$ opkg files firewall
Package firewall (2016-11-29-1) is installed on root and has the following files:
/etc/init.d/firewall
/sbin/fw3
/etc/config/firewall
/etc/firewall.user
/etc/hotplug.d/iface/20-firewall
```

● /etc/init.d/firewall：这是采用procd方式操作防火墙服务的启动、停止、重启、重新装载的脚本。

● /sbin/fw3：即firewall3，是OpenWrt下的netfilter/iptables规则构建器应用程序。它在用户空间中运行，将配置文件解析为一组iptables规则，并将每个规则发送到内核netfilter模块。通过fw3 print可以查看配置的iptables规则。

● /etc/config/firewall：防火墙UCI配置文件。

● /etc/firewall.user：用户自定义的防火墙脚本。

● /etc/hotplug.d/iface/20-firewall：防火墙的Hotplug脚本，描述内核通知应用层网络接口变化时，需要防火墙执行的操作。

6.1.2 防火墙配置

在OpenWrt中可以通过两种方式配置防火墙，一种是通过iptables工具进行配置，另一种是通过OpenWrt的UCI进行配置。使用UCI配置的优点是不用再学复杂的iptables命令。下面我们学习如何通过UCI方式来配置防火墙，具体就是通过配置 /etc/config/firewall这个文件来实现。

防火墙文件是OpenWrt中一个比较大的UCI文件，其内容大都是匿名的配置节点，这让配置变得稍显麻烦，建议使用UCI的-X参数查看或使用Vi编辑器查看防火墙文件。

```
DF1A:~$ uci show -X firewall
```

防火墙文件的内容分为6类节点。

● defaults（默认）类型匿名节点：只有一个配置节点，这是默认配置。

● zone（域）类型匿名节点：可以有多个zone，包含一个或多个interface，用作源或目的地的forwarding、rule和redirect。系统将LAN和WAN分为两个不同的zone，两个zone之间是隔离的。

● forwarding（转发）类型匿名节点：用于不同zone之间的转发。

● rule（规则）类型匿名节点：通常用来定义基本的接受、删除或拒绝规则，以允许或限制对特定端口或主机的访问。

● redirect（端口转发）类型匿名节点：用来配置实现具体端口转发功能。

● includes（引用其他脚本）类型匿名节点：在防火墙配置中指定一个或多个部分，可以包含自定义防火墙脚本。

一个最小的防火墙配置，通常包含一个defaults节点，至少包含两个zone（LAN和WAN）节点以及一个forwarding（允许数据从LAN发送到WAN）节点。

defaults（默认）节点配置

firewall.defaults是第一个配置节点，每个防火墙配置文件中至少包含一个defaults节点。

配置内容如下。

```
config defaults
    option syn_flood '1'
    option input 'ACCEPT'
    option output 'ACCEPT'
    option forward 'REJECT'
```

firewall.@defaults选项参数说明如表6-1所示。

表 6-1 firewall.@defaults 选项参数说明

选项	说明	可选值及说明	必填
input	设置filter表的INPUT链策略	ACCEPT: 允许 REJECT: 拒绝 (默认)	否
output	设置filter表OUTPUT链策略	ACCEPT: 允许 REJECT: 拒绝 (默认)	否
forward	设置filter表FORWARD链策略	ACCEPT: 允许 REJECT: 拒绝 (默认)	否
drop_invalid	丢弃无效数据包	REJECT: 拒绝 (默认)	否
syn_flood	是否启用防洪水攻击保护	0: 不启用 (默认) 1: 启用	否
synflood_rate	设置SYN数据包的速率限制,超过该值的数据被视为洪水数据	数字,默认值为25,单位为包/秒	否
synflood_burst	设置SYN数据包的突发限制,超过该值的数据被视为洪水数据	数字,默认值为25,单位为包/秒	否
disable_ipv6	启用IPv6防火墙	0: 不启用 (默认) 1: 启用	否
flow_offloading	设置是否启用软件流转移	0: 不启用 (默认) 1: 启用	否
flow_offloading_hw	设置是否启用硬件流转移	0: 不启用 (默认) 1: 启用	否
tcp_syncookies	可以抵御syn flood的技术	0: 不启用 (默认) 1: 启用	否
tcp_ecn	针对拥塞发生时包的控制	数字,默认值为50,单位为包/秒	否
tcp_westwood	拥塞和慢启动	0: 不启用 (默认) 1: 启用	否
tcp_window_scaling	TCP窗口缩放	0: 不启用 (默认) 1: 启用	否
accept_redirects	是否接受重定向的ICMP	0: 不启用 (默认) 1: 启用	否
accept_source_route	是否接收含有源路由信息的IP包	0: 不启用 (默认) 1: 启用	否
enabled	启用防火墙	0: 不启用 (默认) 1: 启用	否

zone(域)节点配置

在OpenWrt中,防火墙配置至少包含2个zone节点,一个用来描述WAN,另外一个用来描述LAN。一般情况下,除非是特别复杂的网络,否则不需要创建新的zone节点。

配置内容如下。

```
config zone
       option name 'lan'
       list network 'lan'
       option input 'ACCEPT'
       option output 'ACCEPT'
       option forward 'ACCEPT'
config zone
       option name 'guest'
       option input 'REJECT'
       option forward 'REJECT'
       option output 'ACCEPT'
       option network 'guest'
```

```
config zone
    option name 'lease'
    option input 'ACCEPT'
    option forward 'REJECT'
    option output 'ACCEPT'
    option network 'lease'
config zone
    option name 'wwan'
    list network 'wwan'
    option input 'REJECT'
    option output 'ACCEPT'
    option forward 'REJECT'
    option masq '1'
    option mtu_fix '1'
config zone
    option name 'wan'
    option output 'ACCEPT'
    option forward 'REJECT'
    option masq '1'
    option mtu_fix '1'
    option network 'wan wan6 wwan'
    option input 'ACCEPT'
```

firewall.@zone选项参数说明如表6-2所示。

表6-2　firewall.@zone 选项参数说明

选项	说明	可选值及说明	必填
name	名字，用来区分zone匿名节点	字符串，最大11个字符	是
network	接口列表，什么接口设备被绑定到这个zone上	字符串，一般是网络接口设备名称	否
masq	出站数据是否伪装，这通常在广域网上启用。如果是WAN口，此项必须为1	0：不启用（默认） 1：启用	否
masq_src	设定伪装的源子网源列表。允许多个子网	子网列表，默认为0.0.0.0/0，在子网前面加上！表示否定	否
masq_dest	设定伪装的目标子网列表。允许多个子网	子网列表，默认为0.0.0.0/0，在子网前面加上！表示否定	否
mtu_fix	为出站流量启用MSS。如果是WAN口请设为1	0：不启用（默认） 1：启用	否
input	默认情况下输入传输策略	ACCEPT：允许 REJECT：拒绝 DROP：抛弃（默认）	否
output	默认情况下输出传输策略	ACCEPT：允许 REJECT：拒绝 DROP：抛弃（默认）	否
forward	默认情况下转发传输策略	ACCEPT：允许 REJECT：拒绝 DROP：抛弃（默认）	否
family	iptables规则使用的协议	字符串，ipv4、ipv6或any，默认为any	否
conntrack	强制连接跟踪此区域	0：不启用 1：启用(当masq为1的时候)	否
log	在此区域创建拒绝或者丢弃的包的日志	0：不启用 1：启用	否
log_limit	限制每个日志的消息量	字符串	否
device	网络接口名称	网络接口列表	否

forwarding（转发）节点配置

我们通过转发节点配置可以实现两个不同zone之间的数据发送，默认需要定义一个LAN到WAN的转发。除非是特别复杂的网络，否则无须创建或修改转发节点配置。

配置内容如下。

```
config forwarding
        option src 'lan'
        option dest 'wan'
config forwarding
        option src 'guest'
        option dest 'wan'
config forwarding
        option src 'guest'
        option dest 'wwan'
config forwarding
        option src 'lease'
        option dest 'wan'
config forwarding
        option src 'lease'
        option dest 'wwan'
```

firewall.@forwarding选项参数说明如表6-3所示。

表6-3　firewall.@forwarding 选项参数说明

选项	说明	可选值及说明	必填
name	唯一转发名称	字符串	否
src	指定流量源zone的名称	字符串	是
dest	指定目的zone的名称	字符串	是
family	IP协议类型	字符串，ipv4、ipv6或any，默认为any	否
enabled	是否启动转发	0：不启用 1：启动（默认）	否

rule（规则）节点配置

rule是防火墙的许可规则机制，用于定义基本的接受、删除或拒绝规则，以允许或限制对特定端口或主机的访问。任何一个IP数据包都存在src（来源）、dest（目标）。

配置内容如下。

```
config rule
        option name 'DNSGuest'
        option src 'lease'
        option dest_port '53'
        option proto 'tcpudp'
        option target 'ACCEPT'
config rule
        option name 'DHCPGuest'
        option src 'lease'
```

```
        option src_port '67-68'
        option dest_port '67-68'
        option proto 'udp'
        option target 'ACCEPT'
config rule
        option name 'Allow-DHCP-Renew'
        option src 'wan'
        option proto 'udp'
        option dest_port '68'
        option target 'ACCEPT'
        option family 'ipv4'
config rule
        option name 'Allow-Ping'
        option src 'wan'
        option proto 'icmp'
        option icmp_type 'echo-request'
        option family 'ipv4'
        option target 'ACCEPT'
config rule
        option name 'Allow-IGMP'
        option src 'wan'
        option proto 'igmp'
        option family 'ipv4'
        option target 'ACCEPT'
config rule
        option name 'Allow-DHCPv6'
        option src 'wan'
        option proto 'udp'
        option src_ip 'fc00::/6'
        option dest_ip 'fc00::/6'
        option dest_port '546'
        option family 'ipv6'
        option target 'ACCEPT'
config rule
        option name 'Allow-MLD'
        option src 'wan'
        option proto 'icmp'
        option src_ip 'fe80::/10'
        list icmp_type '130/0'
        list icmp_type '131/0'
        list icmp_type '132/0'
        list icmp_type '143/0'
        option family 'ipv6'
        option target 'ACCEPT'
config rule
        option name 'Allow-ICMPv6-Input'
        option src 'wan'
        option proto 'icmp'
        list icmp_type 'echo-request'
```

```
        list icmp_type 'echo-reply'
        list icmp_type 'destination-unreachable'
        list icmp_type 'packet-too-big'
        list icmp_type 'time-exceeded'
        list icmp_type 'bad-header'
        list icmp_type 'unknown-header-type'
        list icmp_type 'router-solicitation'
        list icmp_type 'neighbour-solicitation'
        list icmp_type 'router-advertisement'
        list icmp_type 'neighbour-advertisement'
        option limit '1000/sec'
        option family 'ipv6'
        option target 'ACCEPT'
config rule
        option name 'Allow-ICMPv6-Forward'
        option src 'wan'
        option dest '*'
        option proto 'icmp'
        list icmp_type 'echo-request'
        list icmp_type 'echo-reply'
        list icmp_type 'destination-unreachable'
        list icmp_type 'packet-too-big'
        list icmp_type 'time-exceeded'
        list icmp_type 'bad-header'
        list icmp_type 'unknown-header-type'
        option limit '1000/sec'
        option family 'ipv6'
        option target 'ACCEPT'
config rule
        option src 'wan'
        option dest 'lan'
        option proto 'esp'
        option target 'ACCEPT'
config rule
        option src 'wan'
        option dest 'lan'
        option dest_port '500'
        option proto 'udp'
        option target 'ACCEPT'
```

firewall.@rule选项参数说明如表6-4所示。

表 6-4 firewall.@rule 选项参数说明

选项	说明	可选值及说明	必填
name	名称，用来区分rule匿名节点	字符串	是
target	规则动作	ACCEPT：允许 REJECT：拒绝 DROP：抛弃（默认） MARK：标记 NOTRACK	是
src	数据源的zone	字符串	否
src_ip	数据源的IP地址	字符串，IP地址	否
src_mac	数据源的MAC地址	字符串，MAC地址	否
src_port	数据源的端口，可以是一个端口或端口范围，但必须同时指定参数proto	数字，端口编号； 字符串，端口范围，如1000-2000	否
proto	数据源的协议类型	默认为tcpudp，all表示任意协议，具体协议支持tcp、udp、tcpudp、udplite、icmp、esp、ah、sctp	否
dest	目的地的zone	字符串	否
dest_ip	目的地的IP地址	字符串，IP地址	否
dest_port	目的地的端口，可以是一个端口或端口范围，但必须同时指定参数proto	数字，端口编号； 字符串，端口范围，如1000-2000	否
family	IP协议类型，默认为any	字符串，ipv4、ipv6、any	否
ipset	将流量与指定的ipset匹配	字符串，值前面加一个!来反转匹配	否
mark	将流量与指定的防火墙标记匹配	字符串	否
start_date	如果指定，则仅在指定日期（包括）之后匹配流量	日期（yyyy-mm-dd）	否
stop_date	如果指定，则仅在指定日期（包括）之前匹配流量	日期（yyyy-mm-dd）	否
start_time	如果指定，则仅在指定时间（包括）之后匹配流量	日期（yyyy-mm-dd）	否
stop_time	如果指定，则仅在指定时间（包括）之前匹配流量	日期（yyyy-mm-dd）	否
weekdays	如果指定，则仅在指定工作日间匹配流量	工作日列表	否
monthdays	如果指定，则仅在该月的指定日期匹配流量	日期列表	否
limit	最高平均匹配率	数字，可选/second、/minute、/hour、/day后缀，例如3/minute、3/min、3/m	否
limit_burst	要匹配的最大初始数据包数，允许短期平均值高于限制值	整数，默认为5	否
extra	要传递给iptables的额外参数	字符串	否
enabled	是否启用这个rule	0：不启用 1：启动（默认）	否

direction（端口转发）节点配置

端口转发是路由器中常见的功能，它允许访问者通过WAN口访问LAN口中的一个特定端口，并且将结果转发回给访问者。例如，将80端口开放到WAN口上。

firewall.@direction选项参数说明如表6-5所示。

表6-5 firewall.@direction 选项参数说明

选项	说明	可选值及说明	必填
name	名称，用来区分direction匿名节点	字符串	是
src	被转发来源zone	字符串，一般是wan	是
src_ip	被转发的IP地址	字符串，IP地址	否
src_mac	被转发的MAC地址	字符串，MAC地址	否
src_port	匹配端口或端口范围传入流量	数字，端口编号，如5000-5005	否
src_dport	对于DNAT，匹配目的端口的入站流量；对于SNAT，将源端口重写为指定的端口	数字，端口编号	否
proto	匹配指定的协议	all表示任意协议，具体协议支持tcp、udp、tcpudp、udplit、icmp、esp、ah、sctp	否
icmp_type	匹配icmp协议类型	默认为any	否
dest	指定流量目的zone	字符串	否
dest_ip	对于DNAT，将匹配的流量重定向到指定的主机；对于SNAT，匹配指定地址的流量	字符串，IP地址	否
dest_port	对于DNAT，将匹配的流量重定向到指定的端口；对于SNAT，匹配指定端口的流量	数字，端口编号	否
ipset	匹配指定ipset的流量	字符串	否
dest_mac	转发到哪个MAC地址	字符串，MAC地址	否
ipset	匹配指定ipset的流量	字符串	否
start_date	匹配指定日期后的流量	日期	否
stop_date	匹配指定日期前的流量	日期	否
start_time	匹配一天指定时间后的流量	时间	否
stop_time	匹配一天指定时间后的流量	时间	否
weekdays	匹配星期几的流量	字符串，星期列表，如mon、sun	否
monthdays	匹配几号的流量	字符串，日期列表，如1、3、4	否
utc_time	设置成UTC时间而不是本地时间	时间	否
target	产生rule时的NAT目标	字符串，默认为DNAT，可选DNAT、SNAT	否
family	IP协议类型	字符串，ipv4、ipv6、any，默认为any	否
reflection	激活NAT反射，适用于DNAT	0：不启用 1：启动（默认）	否
limit	最大平均匹配速率	指定为数字，具有可选的/second、/minute、/hour或/day后缀	否
limit_burst	最大初始包匹配数	数字	否
extra	给iptables的额外参数	字符串	否
enabled	是否启用	0：不启用 1：启动（默认）	否

include（引用其他脚本）节点配置

我们通过include在防火墙配置中指定一个或多个部分，可以包括自定义防火墙脚本。

配置内容如下。

```
config include 'miniupnpd'
    option type 'script'
    option path '/usr/share/miniupnpd/firewall.include'
    option family 'any'
    option reload '1'
```

firewall.@include选项参数说明如表6-6所示。

表6-6 firewall.@include 选项参数说明

选项	说明	可选值及说明	必填
enabled	是否使能该部分	0：不启用 1：启用（默认）	否
type	指定include的类型，可以是script传统Shell脚本包含的，也可以是iptables-restore格式的restore普通文件	字符串，默认为script	否
path	在引导或防火墙重新启动时执行的Shell脚本	文件名，默认为/etc/firewall.user	是
family	调用include的地址协议类型	字符串，默认为any，支持ipv4、ipv6、any	否
reload	是否应在重新加载时调用include	0：不启用（默认） 1：启用	否

6.1.3 防火墙命令

防火墙配置完毕后，可以通过命令行脚本使防火墙配置生效。

重置防火墙命令：

```
DF1A:~$ /etc/init.d/firewall reload
```

重启防火墙命令：

```
DF1A:~$ /etc/init.d/firewall restart
```

查看防火墙filter表完整策略：

```
DF1A:~$ iptables -L
```

也可以通过fw3命令操作：

```
fw3 [-q] {start|stop|flush|reload|restart}
```

重启防火墙命令：

```
DF1A:~$ fw3 restart
```

查看防火墙iptables配置：

```
DF1A:~$ fw3 print
```

6.1.4 防火墙案例

防火墙测试如图6-2所示。

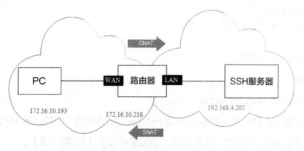

图6-2 防火墙测试示意图

允许某个IP通过WAN口访问路由器的22端口

（1）添加rule，设定允许172.16.10.193通过WAN口访问路由器22端口。

```
config rule
    option name 'allow-wan-ssh'
    option src 'wan'
    option 'dest_port' 22
    option target 'ACCEPT'
    option proto 'tcp'
    option src_ip '172.16.10.119'
```

（2）重启firewall：

```
DF1A:~$ /etc/init.d/firewall restart
```

（3）测试，在172.16.10.119上用SSH客户端通过WAN口访问路由器（见图6-3）。

图6-3　SSH登录

DNAT基于目标的网络地址转换

（1）添加redirect节点。外网访问路由器WAN口22端口，转发到LAN指定IP的5322端口：

```
config redirect
    option name 'dnat-tcpdup-22-to-5322'
    option src 'wan'
    option dest 'lan'
    option proto 'tcp udp'
    option src_port '22'
    option dest_ip '192.168.4.202'
    option dest_port '5322'
    option target 'DNAT'
```

（2）重启firewall：

```
DF1A:~$ /etc/init.d/firewall restart
```

（3）测试。在172.16.10.119上用SSH客户端通过WAN口（172.16.10.218）22端口访问路由器，路由器重定向到LAN下的IP为192.168.4.202的PC上（见图6-4）。

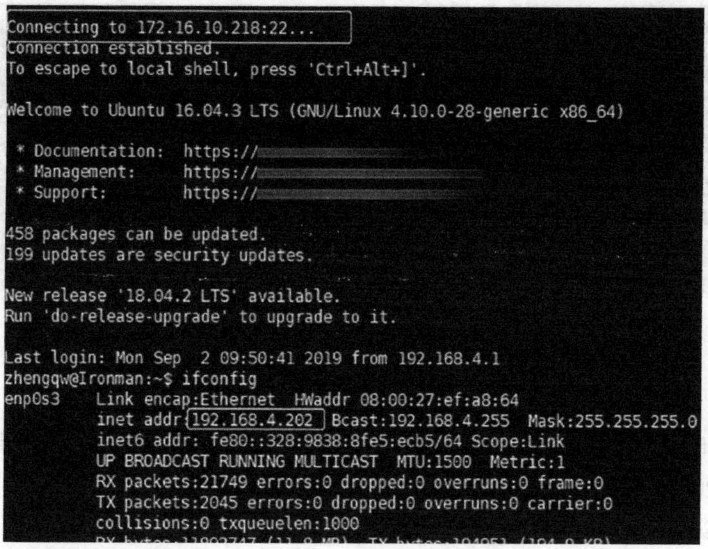

图6-4 DNAT结果

SNAT即源地址目标转换

（1）添加redirect节点。PC通过LAN口访问外网8080服务端口，修改PC的LAN口的IP为192.168.4.202：

```
config redirect
        option target          SNAT
        option name            "SNAT-tcp-8080"
        option src             lan
        option dest            wan
        option src_ip          192.168.4.202
        option src_dip         10.10.0.2
        option proto           tcp
        option dest_port       8080
        option enabled  1
```

（2）重启firewall：

```
DF1A:~$ /etc/init.d/firewall restart
```

（3）测试，在172.16.10.119上运行服务，监听8080端口：

```
HOST:~$ nc -l 8080
```

（4）在PC的LAN口（192.168.4.202）上连接服务器：

```
HOST:~$ nc -v 172.16.10.119 8080
```

（5）在172.16.10.119上查看连接情况。看到连接到172.16.10.119上的IP已经变为10.10.0.2（见图6-5）。但由于10.10.0.2是虚构的IP，只是建立了SYN连接，并没得到回复。如果修改src_dip为172.16.10.218（路由器WAN口IP），则能正确建立连接(ESTABLISHED)。大家可以自己测试。

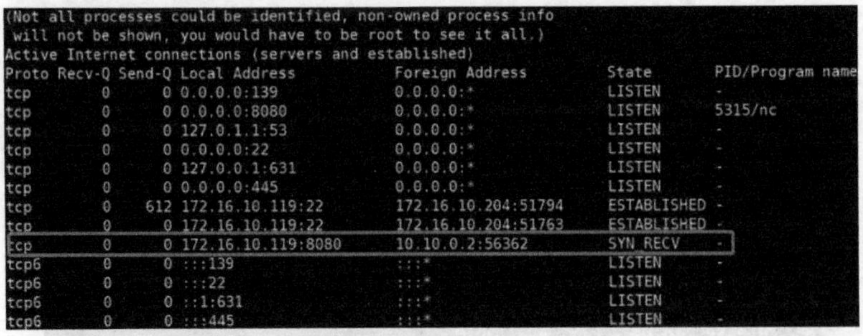

```
(Not all processes could be identified, non-owned process info
 will not be shown, you would have to be root to see it all.)
Active Internet connections (servers and established)
Proto Recv-Q Send-Q Local Address           Foreign Address         State       PID/Program name
tcp        0      0 0.0.0.0:139             0.0.0.0:*               LISTEN      -
tcp        0      0 0.0.0.0:8080            0.0.0.0:*               LISTEN      5315/nc
tcp        0      0 127.0.1.1:53            0.0.0.0:*               LISTEN      -
tcp        0      0 0.0.0.0:22              0.0.0.0:*               LISTEN      -
tcp        0      0 127.0.0.1:631           0.0.0.0:*               LISTEN      -
tcp        0      0 0.0.0.0:445             0.0.0.0:*               LISTEN      -
tcp        0    612 172.16.10.119:22        172.16.10.204:51794     ESTABLISHED -
tcp        0      0 172.16.10.119:22        172.16.10.204:51763     ESTABLISHED -
tcp        0      0 172.16.10.119:8080      10.10.0.2:56362         SYN_RECV    -
tcp6       0      0 :::139                  :::*                    LISTEN      -
tcp6       0      0 :::22                   :::*                    LISTEN      -
tcp6       0      0 ::1:631                 :::*                    LISTEN      -
tcp6       0      0 :::445                  :::*                    LISTEN      -
```

图6-5　SNAT结果

DMZ

（1）添加redirect节点，将所有来自WAN口对路由器的请求都转发到DMZ隔离区（192.168.4.202）上：

```
config redirect
        option src              wan
        option proto            all
        option dest_ip          192.168.4.202
```

（2）重启firewall：

```
DF1A:~$ /etc/init.d/firewall restart
```

（3）测试，在PC的LAN口（192.168.4.202）上运行服务，监听8080端口：

```
HOST:~$ nc -l 8080
```

（4）在172.16.10.119上连接服务器：

```
HOST:~$ nc -v 172.16.10.218 8080
Connection to 172.16.10.218 8080 port [tcp/http-alt] succeeded!
```

（5）在服务器上输入hello，客户端能够收到hello。

禁止LAN上主机访问WAN口80、443服务

（1）添加rule节点：

```
config rule
        option name 'Deny-lan-vist-wan-80'
        option src 'lan'
        option dest 'wan'
        option dest_port '80 443'
        option proto 'tcp'
        option target 'REJECT'
```

（2）重启firewall：

```
DF1A:~$ /etc/init.d/firewall restart
```

（3）测试，LAN口上任意设备都不能访问百度（HTTPS协议，443端口）。

6.2　UPnP与NATPMP

UPnP是一种通用即插即用（Universal Plug and Play，UPnP）网络协议，主要用于视频、音频领域的传输，对使用者来说，打开UPnP之后可以流畅地使用网络，提高NAT数据转换效率，加快P2P软件访问网络的速度（如观看在线视频和多点下载等），使网络更加稳定。但开启了UPnP会消耗一定CPU资源和内存。

NAT端口映射协议（NAT Port Mapping Protocol，NAT-PMP）是一个能自动创建网络地址转换（NAT）设置和端口映射配置而无须用户介入的网络协议。该协议能自动测定NAT网关的外部IPv4地址，并为应用程序提供与对等端通信的方法。NAT-PMP于2005年由苹果公司推出。NAT-PMP使用UDP协议，在5351端口运行。

6.2.1 安装配置UPnP

OpenWrt下的UPnP服务端软件名为miniupnpd，该软件不属于基本软件包，需要在线下载，请确保开发板可以连接上外网，然后按下面的步骤来下载、安装。

```
DF1A:$ opkg update
DF1A:$ opkg install miniupnpd
```

mniupnpd安装包包含以下文件。

- /usr/sbin/miniupnpd miniupnpd：主程序。
- /etc/config/upnpd miniupnpd：UCI配置文件。
- /etc/hotplug.d/iface/50-miniupnpd：接口启用时的Hotplug处理脚本。
- /user/share/miniupnpd/firewall.include：firewall包含的miniupnpd防火墙配置文件。
- /etc/uci-defaults/99-miniupnpd：会生成firewall配置文件的miniupnpd节点。
- /etc/init.d/miniupnpd：miniupnpd启动脚本。

UCI配置文件upnpd内容如下。

```
config upnpd 'config'
    option enabled '0'
    option enable_natpmp '1'
    option enable_upnp '1'
    option secure_mode '1'
    option log_output '0'
    option download '1024'
    option upload '512'
    option internal_iface 'lan'
    option port '5000'
    option upnp_lease_file '/var/upnp.leases'
config perm_rule
    option action 'allow'
    option ext_ports '1024-65535'
    option int_addr '0.0.0.0/0'
    option int_ports '1024-65535'
    option comment 'Allow high ports'
config perm_rule
    option action 'deny'
    option ext_ports '0-65535'
    option int_addr '0.0.0.0/0'
    option int_ports '0-65535'
    option comment 'Default deny'
```

miniupnpd的UCI配置文件，默认包含一个upnpd节点和若干个perm_rule（许可规则），

perm_rule规则按照它们在配置文件中出现的顺序依次应用。

upnpd节点

upnpd默认配置节点主要针对upnpd主程序进行配置。upnpd.config选项参数说明如表6-7所示。

表 6-7 upnpd.config 选项参数说明

选项	说明	可选值及说明	必填
enable_natpmp	开启NAT-PMP支持	0：禁用 1：启用（默认）	是
enable_upnp	开启UPnP支持	0：禁用 1：启用（默认）	是
secure_mode	安全模式，客户端只能给自己转发一个输入口	0：禁用 1：启用（默认）	是
log_output	日志输出级别，0表示不输出日志，如果设置了，日志将输出到syslog中	数字，范围为0~5，默认为0	是
download	允许来自WAN口的数据输入带宽	数字，单位为KB/s	是
upload	允许输出到WAN口的数据带宽	数字，单位为KB/s	是
external_iface	外网的zone	字符串，默认为wan	是
internal_iface	内网的zone	字符串，默认为wan	是
port	服务监听端口	数字	是
upnp_lease_file	UPnP客户端租用记录文件路径	字符串，文件路径	是

perm_rule节点

perm_rule主要进行端口授权许可配置，行为类似于防火墙的rule规则。perm_rule是匿名配置节点，允许多个匿名配置节点同时存在，当存在多个时，按照它们在配置文件中出现的顺序依次应用。upnpd.@perm_rule选项参数说明如表6-8所示。

表 6-8 upnpd.@perm_rule 选项参数说明

选项	说明	可选值及说明	必填
action	动作许可	allow：许可 deny：不许可（默认）	是
ext_ports	外部端口范围	字符串，开始端口-结束端口，例如0-65535	是
int_addr	IP地址，如果是0.0.0.0/0表示全部	字符串，默认为0.0.0.0/0，IP地址	是
int_ports	内部端口范围	字符串，开始端口-结束端口，例如0-65535	是
comment	备注信息，没有实际作用	字符串	是

6.2.2 UPnP命令

设置开机自动启动：

```
DF1A:$ /etc/init.d/miniupnpd enable
```

启动miniupnpd：

```
DF1A:$ /etc/init.d/miniupnpd start
```

端口1900用于UPnP发现，端口5351用于端口映射协议NAT-PMP（见图6-6）：

```
DF1A:$ netstat -tunpl|grep miniupnpd
```

```
tcp        0      0 :::5000                 :::*                    LISTEN      3180/miniupnpd
udp        0      0 192.168.4.1:43049       0.0.0.0:*                           3180/miniupnpd
udp        0      0 0.0.0.0:1900            0.0.0.0:*                           3180/miniupnpd
udp        0      0 192.168.4.1:5351        0.0.0.0:*                           3180/miniupnpd
udp        0      0 :::1900                 :::*                                3180/miniupnpd
udp        0      0 :::54239                :::*                                3180/miniupnpd
udp        0      0 :::5351                 :::*                                3180/miniupnpd
```

图6-6　miniunpd启动的端口

6.2.3 UPnP案例

使用UPnP能提高P2P软件（例如迅雷）的下载速度。在OpenWrt中启动miniupnpd，LAN客户端通过P2P软件进行下载操作，会在路由器中看到客户端信息被miniupnd记录到UPnP客户端租赁文件（/tmp/upnp.leases）中（见图6-7）。

```
TCP:21867:192.168.4.191:54321:0:MiniTP SDK
UDP:21867:192.168.4.191:12345:0:MiniTP SDK
```

图6-7　miniunpd客户端租赁文件

6.3　dropbear远程登录

SSH服务在OpenWrt下是通过一个名叫dropbear的软件包实现的。dropbear是在内存较小和处理器资源较少的嵌入式系统中替代OpenSSH的软件。

6.3.1　配置dropbear

dropbear软件包默认已经安装到开发板中。在开启SSH服务之前，我们可以通过串口访问开发板。

dropbear安装包包含以下文件。

● /usr/bin/dbclient：轻量级SSH2客户端。

● /usr/bin/scp：scp命令。

● /etc/dropbear/dropbear_rsa_host_key：RSA格式的SSH私钥。

● /etc/init.d/dropbear：dropbear启动脚本。

● /usr/bin/dropbearkey：创建私钥，生成RSA或DSS格式的SSH私钥，并将其保存到/etc/dropbear目录中。

● /etc/config/dropbear：UCI配置文件。

● /etc/dropbear/dropbear_dss_host_key authorized_keys：DSS格式的SSH私钥。

● /usr/sbin/dropbear：dropbear主命令。

● /usr/bin/ssh：ssh命令。

UCI配置文件dropbear的内容如下。

```
config dropbear
        option PasswordAuth 'on'
        option RootPasswordAuth 'on'
        option Port             '22'
#       option BannerFile       '/etc/banner'
```

用uci命令查看：

```
DF1A:$ uci show dropbear
dropbear.@dropbear[0]=dropbear
dropbear.@dropbear[0].PasswordAuth='on'
dropbear.@dropbear[0].RootPasswordAuth='on'
dropbear.@dropbear[0].Port='22'
```

dropbear.@dropbear匿名节点可配置选项参数如表6-9所示。

表6-9 dropbear.@dropbear 匿名节点可配置选项参数

选项	说明	可选值及说明	必填
enable	是否开启	0：禁用 1：启用（默认）	否
PasswordAuth	登录时提示输入密码	0：禁用 1：启用（默认）	否
RootPasswordAuth	允许root使用密码登录	0：禁用 1：启用（默认）	否
RootLogin	允许root登录	0：禁用 1：启用（默认）	否
BannerFile	登录后显示特定的欢迎信息	字符串，文件路径，默认为空	否
Port	SSH服务的端口号	数字，端口，默认为22端口	否
SSHKeepAlive	开启服务端心跳	数字，默认为300，单位为秒	否
IdleTimeout	空闲超时	数字，默认为0（表示不开启），单位为秒	否

6.3.2 dropbear命令

设置开机自动启动：

```
DF1A:$ /etc/init.d/dropbear enable
```

启动dropbear：

```
DF1A:$ /etc/init.d/dropbear start
```

查看端口22，确定SSH服务是否成功启动：

```
DF1A:$ netstat -tunpl|grep 22
tcp  0   0 0.0.0.0:22        0.0.0.0:*            LISTEN      15910/dropbear
tcp  0   0 :::22             :::*                 LISTEN      15910/dropbear
```

修改远程登录密码：确认要登录系统的用户已经设置了密码，如果没有设置密码，请先设置用户密码，否则SSH默认是拒绝登录的。系统默认的admin用户的密码为admin。我们也可以使用passwd命令重新给admin设置一个密码，密码要填写两次，屏幕不显示所填写内容。

```
DF1A:~$ passwd admin
Changing password for admin
New password:
Retype password:
Password for admin changed by admin
```

6.4 系统、时钟、日志

我们拿到开发板的第一件事，通常就是对开发板进行一些基本的设置，而这些设置大部分是通过修改UCI配置文件system来实现的。system配置文件包含对系统最基本操作的设置，例如主机名、时区以及将日志记录信息写入的方式和位置。

NTP（Network Time Protocol，网络时间协议）是用来使网络中的各个计算机时间同步的一种协议。我们采用NTP协议向网络上的时间授权服务器请求获得时间服务，把系统的时钟同步到UTC（世界协调时）时区，再根据本地时区的配置转换为本地时间。NTP精度在局域网内可达到0.1ms，在互联网上绝大多数情况下，其精度可以达到1~50ms。一些具有RTC硬件时钟的路由器可以通过NTP同步时钟，然后通过命令把获得的时钟信息写到RTC硬件时钟芯片里，这样即使网络断开也能提供准确时间。

6.4.1 配置system

UCI配置文件system的内容如下。

```
config system
        option zonename 'Asia/Shanghai'
        option timezone 'CST-8'
        option hostname 'SiWiFi3d04'
        option hostnameset '1'
config timeserver 'ntp'
        list server '0.××××.org'
        list server '1.××××.org'
        list server '2.××××.org'
        list server '3.××××.org'
        option enabled '1'
        option enable_server '0'
```

用uci命令查看：

```
DF1A:$ uci show system
system.@system[0]=system
system.@system[0].zonename='Asia/Shanghai'
system.@system[0].timezone='CST-8'
system.@system[0].hostname='SiWiFi3d04'
system.@system[0].hostnameset='1'
system.ntp=timeserver
system.ntp.server='0.××××.org' '1.××××.org' '2.××××.org' '3.××××.org'
system.ntp.enabled='1'
system.ntp.enable_server='0'
```

system.@system匿名节点选项参数说明如表6-10所示。system.ntp选项参数说明如表6-11所示。

表 6-10 system.@system 匿名节点选项参数说明

选项	说明	可选值及说明	必填
hostname	设备的主机名称	字符串，默认为OpenWrt	否
buffersize	内核信息输出的尺寸，默认由内核指定	数字	否
conloglevel	控制平台日志记录的级别，级别越高，记录的信息越多	数字，1~8，默认为7	否
cronloglevel	CRON服务写入系统syslog的日志级别	数字，默认为5 0：记录全部信息 8：记录命令被执行 9：只记录错误	否
klogconloglevel	内核向控制平台显示的信息级别	数字，默认为7	否
log_buffer_size	基于procd的系统日志的日志缓冲区的大小，可通过logread命令访问	数字	否
log_file	syslog日志文件路径	字符串，文件路径，默认为/var/log/messages	否
log_hostname	要发送到远程syslog的主机名	字符串	否
log_ip	syslog日志文件通过IP地址发送到另外一台机器上	字符串，IP地址，默认为空	否
log_port	另外那台记录syslog机器的接收端口	数字，端口号，默认为514	否
log_prefix	为通过网络发送的日志增加一个前缀	字符串，默认为空	否
log_proto	设置用于连接的协议	字符串，tcp或udp，默认为udp	否
log_size	日志文件的允许容量	数字，单位为KB，默认为16	否
timezone	时区，内容请参考时区表，默认为UTC	字符串，时区参数	否

表 6-11 system.ntp 选项参数说明

选项	说明	可选值及说明	必填
server	列表值，NTP服务器的地址	字符串，列表值可以重复存在，默认值为空	否
enable	是否开启NTP功能	0：禁用 1：启用	否
enable_server	启动模式，客户端表示向远端获取，服务端表示向内网计算机提供NTP服务	0：启动客户端和服务端 1：启动服务端	否

6.4.2 system命令

设置开机自动启动：

```
DF1A:$ /etc/init.d/system enable
```

启动system：

```
DF1A:$ /etc/init.d/system start
```

system案例

修改主机名称、时区、NTP服务器地址、日志：

```
DF1A:$ uci show -X system
system.cfg02e48a=system
system.cfg02e48a.zonename='Asia/Shanghai'
system.cfg02e48a.timezone='CST-8'
system.cfg02e48a.hostname='SiWiFi3d04'
system.cfg02e48a.hostnameset='1'
system.ntp=timeserver
```

```
system.ntp.server='0.××××.org' '1.××××.org' '2.××××.org' '3.××××.org'
system.ntp.enabled='1'
system.ntp.enable_server='0'
DF1A:$ uci set system.cfg02e48a.timezone=CST-8
DF1A:$ uci set system.cfg02e48a.zonename=Asia/Shanghai
DF1A:$ uci set system.cfg02e48a.log_file=/var/log/message
DF1A:$ uci set system.cfg02e48a.conloglevel=8
DF1A:$ uci set system.cfg02e48a.klogconloglevel=7
DF1A:$ uci delete system.ntp.server
DF1A:$ uci add_list system.ntp.server=ntp1.××.com
DF1A:$ uci add_list system.ntp.server=ntp2.××.com
DF1A:$ uci add_list system.ntp.server=ntp3.××.com
```

保存，确认修改：

```
DF1A:$ uci commit
```

重启使修改生效：

```
DF1A:$ reboot
```

查看时间：

```
DF1A:$ date
Tue Sep  3 15:10:15 CST 2019
```

查看log：

```
DF1A:$ cat /var/log/message
Tue Sep  3 15:13:41 2019 user.emerg syslog: Command failed: Not found
Tue Sep  3 15:13:41 2019 user.emerg syslog: Failed to parse message data
Tue Sep  3 15:13:41 2019 user.emerg syslog: up=
Tue Sep  3 15:13:41 2019 kern.warn kernel: [ 22.938763] cmd is 0, data is 0
Tue Sep  3 15:13:41 2019 kern.warn kernel: [ 22.941967] mac is 50:7b:9d:81:a9:67,key is 123
Tue Sep  3 15:13:41 2019 kern.warn kernel: [ 22.946689] have is 0
Tue Sep  3 15:13:41 2019 user.emerg syslog: /usr/bin/auto-check-ts-version.sh: line 41:
/usr/bin/pctl_upgrade: not found
Tue Sep  3 15:13:41 2019 kern.warn kernel: [ 23.046110] cmd is 0, data is 0
Tue Sep  3 15:13:41 2019 kern.warn kernel: [ 23.049323] mac is 6c:b7:49:d6:8f:24,key is 49
Tue Sep  3 15:13:41 2019 kern.warn kernel: [ 23.053980] have is 0
Tue Sep  3 15:13:44 2019 daemon.notice netifd: init DPS_NL ret=0
Tue Sep  3 15:13:44 2019 daemon.notice netifd: get dps group=4
Tue Sep  3 15:13:44 2019 kern.info kernel: [ 25.882340] sf_eth 10000000.ethernet eth0:
 INIT_INFO: read led-on-off-time form dts: 0xff 0x1ff
...
```

6.5 用命令刷固件

我们通过对OpenWrt系统固件进行升级，可以更换整个操作系统。在OpenWrt中有几种常见的固件升级方式：

- U-Boot+TFTP/Web；
- sysupgrade命令；
- mtd命令；
- 通过Web管理界面刷机；
- 用烧写器烧写。

每种方式都有自己的优点，有的方式是通过包装其他方式或命令来实现升级功能的，如用sysupgrade命令刷固件是通过mtd命令升级固件分区的；通过Web管理界面刷机，内部是通过sysupgrade命令来进行固件升级的；U-Boot+Web与U-Boot+TFTP都是通过U-Boot mtd命令进行升级的。

关于U-Boot刷机的功能，我们已经在2.3节学习过。本节我们主要讲解用sysupgrade命令和mtd命令进行刷机的方法。

固件准备已经在U-Boot刷机部分说明，请查阅。请将固件上传至/tmp/下（其他目录容量不足以保存）。

6.5.1 备份、恢复系统及配置

在刷机之前，首先要备份系统及其配置，包括系统分区、系统配置文件、工厂调校分区及系统配置分区等信息。

查看系统当前分区信息：

```
DF1A:$ cat /proc/mtd
dev:    size   erasesize  name
mtd0: 00020000 00010000 "spl-loader"
mtd1: 00060000 00010000 "u-boot"
mtd2: 00010000 00010000 "u-boot-env"
mtd3: 00010000 00010000 "factory"
mtd4: 00f60000 00010000 "firmware"
mtd5: 0019a2fe 00010000 "kernel"
mtd6: 00dc5d02 00010000 "rootfs"
mtd7: 006d0000 00010000 "rootfs_data"
```

备份整个固件：

```
cat /dev/mtd0 /dev/mtd1 /dev/mtd2 /dev/mtd3 /dev/mtd4 > /tmp/drogonfly-1a.bin
```

备份uboot分区：

```
dd if=/dev/mtd1 of=/tmp/uboot.bin
```

备份factory分区：

```
dd if=/dev/mtd3 of=/tmp/factory.bin
```

备份firmware分区/固件：

```
dd if=/dev/mtd4 of=/tmp/firmware.bin
```

备份路由器配置：

```
sysupgrade -b /tmp/back.tar.gz
```

6.5.2 mtd命令

mtd命令是由mtd软件包提供的通过命令行操作MTD设备的工具。mtd命令格式如下。

```
mtd [<options> ...] <command> [<arguments> ...] <device>[:<device>...]
```

mtd命令说明如表6-12所示，选项说明如表6-13所示。

表6-12 mtd 命令说明

命令	描述
unlock <dev>	解锁mtd分区
refresh <dev>	刷新mtd分区
erase <dev>	擦除设备上的所有数据
verify <imagefile>|-	验证imagefile（-用于标准输入）
write <imagefile>|-	将imagefile（-用于标准输入）写入设备
jffs2write <file>	将file附加到设备上的JFFS2分区

表6-13 mtd 选项说明

选项	描述
-q	安静模式
-n	在没有先擦除块的情况下写入
-r	命令执行成功后重启
-f	强制写入，没有trx检查
-e <device>	执行命令前擦除device
-d <name>	jffs2write的目录，默认为"tmp"
-j <file>	在写入image时将file集成到JFFS2数据中
-s <number>	将数据追加到JFFS2分区时跳过前number个字节，默认number为0
-p	从分区偏移量开始写入
-l <length>	要转储的数据的长度

mtd命令示例

下面我们通过一些示例，来学习常用的mtd命令的使用方法。

● 解锁mtd分区

执行mtd write命令时，通常会自动解锁对应的mtd分区，不用单独执行此命令。但是有些U-Boot需要先执行mtd unlock，才能执行mtd write。解锁mtd分区演示：

```
DF1A:$ mtd unlock u-boot
Unlocking u-boot ...
```

● 擦除分区

以firmware分区为例。测试前首先备份firmware分区，再擦除分区，擦除结束后重启系统：

```
DF1A:$ dd if=/dev/mtd4 of=/tmp/firmware.bin
31488+0 records in
31488+0 records out
DF1A:$ mtd -r write /tmp/firmware.bin firmware
Unlocking firmware ...
Writing from /tmp/firmware.bin to firmware ...   [w]
```

- 验证固件和分区

备份firmware分区，使用分区备份文件firmware.bin来验证备份的文件是否与分区内容一致：

```
DF1A:$ dd if=/dev/mtd4 of=/tmp/firmware.bin
31488+0 records in
31488+0 records out
DF1A:$ mtd verify /tmp/firmware.bin /dev/mtd4
Verifying /dev/mtd4 against /tmp/firmware.bin ...
37e51117e74eb9705821755b39e8325b - /dev/mtd4
37e51117e74eb9705821755b39e8325b - /tmp/firmware.bin
Success
DF1A:$ mtd verify /tmp/firmware.bin /dev/mtd5
Verifying /dev/mtd5 against /tmp/firmware.bin ...
Could not open mtd device: /dev/mtd5
Could not open mtd device: /dev/mtd5
```

- 恢复系统默认设置

使用mtd清除/overlay分区信息后重启即恢复默认设置：

```
DF1A:$ mtd -r erase rootfs_data
Unlocking rootfs_data ...
Erasing rootfs_data ...
Rebooting ...
```

- 直接删除/overlay分区并重启系统，与上面的命令具有相同效果：

```
rm -rf /overlay/* && reboot
```

6.5.3 sysupgrade命令

sysupgrade是OpenWrt最常用的升级固件脚本命令，推荐使用。在使用Web管理界面升级固件时，内部调用的就是sysupgrade命令。sysupgrade命令本身是一个Shell脚本，通过包含其他脚本和命令来实现升级固件（见图6-8）。

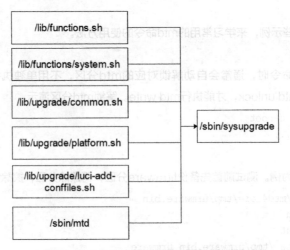

图6-8 sysupgrade命令包含其他脚本和命令

sysupgrade命令由OpenWrt基础软件包提供，sysupgrade命令的格式如下：
```
/sbin/sysupgrade [<upgrade-option>...] <image file or URL>
/sbin/sysupgrade [-q] [-i] <backup-command> <file>
```

sysupgrade参数说明如表6-14所示。

表6-14　sysupgrade 参数说明

参数	描述
升级支持的参数	
-d <delay>	在系统重启之前添加延迟
-f <config>	从.tar.gz文件恢复配置（文件或网址）
-i	交互模式
-c	尝试保留/etc/下所有修改的文件
-n	不保留配置，即更新固件后，Flash中不包含修改的配置文件
-T | --test	不执行升级，仅验证镜像和配置（.tar.gz文件）
-F | --force	强制升级，不论固件验证成功与否
-q	安静模式，不输出日志
-v	调试模式，输出日志
-h | --help	帮助
备份支持的参数	
-b | --create-backup <file>	备份/etc/和/etc/config下的文件。如果需要额外备份文件，需要在/etc/sysupgrade.conf文件中指定文件，备份格式为.tar.gz
-r | --restore-backup <file>	恢复sysupgrade -b创建的.tar.gz备份文件
-l | --list-backup	仅列出备份文件，即sysupgrade -b时要备份的文件

sysupgrade备份示例

● 查看sysupgrade默认备份的文件：

```
DF1A:$ sysupgrade -l
/etc/config/ap_groups
/etc/config/basic_setting
...
/etc/sysupgrade.conf
/etc/uhttpd.crt
/etc/uhttpd.key
/etc/url_list
```

sysupgrade默认备份的文件仅包含/etc/和/etc/config目录下的文件。/etc/sysupgrade.conf是sysupgrade配置文件，主要配置升级时需要备份的文件和目录。如果需要增加备份文件，则需要修改/etc/sysupgrade.conf配置文件。

我们添加一个目录（/lib/upgrade/）和一个文件（/sbin/sysupgrade）进行测试：

```
DF1A:$ vi /etc/sysupgrade.conf
DF1A:$ cat /etc/sysupgrade.conf
## This file contains files and directories that should
## be preserved during an upgrade.
# /etc/example.conf
```

```
# /etc/openvpn/
/sbin/sysupgrade
/lib/upgrade/
```

再次执行"sysupgrade -l"后，能看到我们添加的目录下的所有文件和添加的文件已经被包含进来了：

```
DF1A:$ sysupgrade -l
...
/lib/upgrade/common.sh
/lib/upgrade/keep.d/base-files
/lib/upgrade/keep.d/base-files-essential
/lib/upgrade/keep.d/opkg
/lib/upgrade/keep.d/ppp
/lib/upgrade/keep.d/sf-ts-config
/lib/upgrade/luci-add-conffiles.sh
/lib/upgrade/platform.sh
/sbin/sysupgrade
```

- 使用"sysupgrade -b"命令备份文件：

```
DF1A:$ cd /tmp
DF1A:tmp$ sysupgrade -b backup.tar.gz
```

默认备份格式为.tar.gz包，通过命令查看备份软件包的内容：

```
DF1A:tmp$ gzip -dc backup.tar.gz |tar -tvf -
-rw-rw-r-- 0/0       496 2019-07-17 09:30:51 etc/config/ap_groups
-rw-rw-r-- 0/0       390 2019-07-17 09:30:51 etc/config/basic_setting
...
-rw-rw-r-- 0/0      5942 2019-07-17 09:30:51 lib/upgrade/common.sh
-rw-r--r-- 0/0        63 2019-07-17 16:46:15 lib/upgrade/keep.d/base-files
-rw-rw-r-- 0/0       172 2019-07-17 09:30:51 lib/upgrade/keep.d/base-files-essential
-rw-r--r-- 0/0        16 2019-07-17 16:43:56 lib/upgrade/keep.d/opkg
-rw-r--r-- 0/0        68 2019-07-17 16:31:42 lib/upgrade/keep.d/ppp
-rw-r--r-- 0/0        15 2019-07-17 16:30:59 lib/upgrade/keep.d/sf-ts-config
-rw-rw-r-- 0/0       360 2019-07-17 09:30:52 lib/upgrade/luci-add-conffiles.sh
-rw-rw-r-- 0/0      1468 2019-07-17 09:30:51 lib/upgrade/platform.sh
-rwxrwxr-x 0/0      6760 2019-07-17 09:30:51 sbin/sysupgrade
-rw-r--r-- 0/0      1191 2019-07-17 16:32:14 etc/uhttpd.key
-rw-r--r-- 0/0       903 2019-07-17 16:32:14 etc/uhttpd.crt
```

- 使用"sysupgrade -r"命令恢复备份文件：

```
DF1A:tmp$ sysupgrade -r backup.tar.gz
```

大家可以先删除某个备份文件，然后恢复备份文件，验证备份文件是否被恢复。

sysupgrade升级固件示例

sysupgrade将自动重启系统，升级固件的整个过程会消耗一段时间。请将固件上传至/tmp目录下。

- 刷机并恢复系统配置：

```
#升级固件并在新固件中恢复备份配置backup.tar.gz
```

```
sysupgrade -f /tmp/backup.tar.gz -v /tmp/openwrt-siflower-sf16a18-fullmask-
squashfs-sysupgrade.bin
```

- 刷机不保留当前系统配置：

```
sysupgrade -n -v openwrt-siflower-sf16a18-fullmask-squashfs-sysupgrade.bin
```

- 刷机并保留当前配置：

```
sysupgrade -c -v /tmp/openwrt-siflower-sf16a18-fullmask-squashfs-sysupgrade.bin
```

升级固件后重新登录系统，查看/etc/config/system的内容，发现保留了配置：

```
DF1A:tmp$ uci export system
package system
config system
        option zonename 'Asia/Shanghai'
        option timezone 'CST-8'
        option hostname 'SiWiFi3d04'
        option hostnameset '1'
config timeserver 'ntp'
        option enabled '1'
        option enable_server '0'
        option server 'ntp1.×××.com ntp2.×××.com ntp3.×××.com'
```

6.6 域名劫持

域名劫持也称为DNS劫持，是指在劫持的网络范围内拦截域名解析请求，分析请求的域名，重定向指定域名到指定的IP地址。

域名劫持有很多应用，像很多商业路由器中，把路由器默认的LAN管理IP地址和公司域名对应，进行域名劫持，以便快速访问路由器Web管理界面。例如连接小米路由器，在浏览器中输入miwifi网址后，网页会自动转跳到小米路由器的Web管理页面。

在OpenWrt中，dnsmasq软件包提供DNS缓存和DHCP服务功能。作为域名服务器（DNS），dnsmasq可以通过缓存DNS请求来提高对访问过的网址的连接速度。作为DHCP服务器，dnsmasq可以为路由器LAN分配IP地址和提供路由。

域名劫持的实现

域名劫持有很多实现方法（例如修改/etc/hosts），本文通过修改dnsmasq配置文件和脚本来实现域名劫持。

（1）添加新节点：

```
#在/etc/config/dhcp配置文件中添加dhcp domain节点
config domain dnshijacking
        list name 'xxxx.com'
list name 'xxxx.cn'
option ip '192.168.4.1'
```

- name：域名列表，添加要劫持的域名。
- ip：劫持域名的IP地址，也可以不写。

（2）修改dnsmasq启动脚本：

```
#用Vi编辑/etc/init.d/dnsmasq脚本,修改dhcp_domain_ad()函数
dhcp_domain_add() {
      local cfg="$1"
      local ip name names record
      config_get names "$cfg" name "$2"
      [ -n "$names" ] || return 0
      config_get ip "$cfg" ip "$3"
      [ -n "$ip" ] || return 0
      for name in $names; do
              echo "$name"
              record="${record:+$record }$name"
      done
      [ "$cfg" == dnshijacking ] && {
              network_get_ipaddr lanaddr "lan"
              [ "$ip" != "$lanaddr" ] && {
                      uci_set dhcp $1 ip $lanaddr
                      uci_commit dhcp
                      ip=$lanaddr
              }
      }
      echo "$ip $record" >> $HOSTFILE
}
```

（3）重启dnsmasq：

```
DF1A:$ /etc/init.d/dnsmasq restart
```

（4）验证：修改dnsmasq启动脚本，实际上是在dnsmasq本地hosts文件（addn-hosts=/tmp/hosts）中增加域名和IP的对应关系。

```
#通过/tmp/hosts/dhcp文件查看域名和IP的对应关系
DF1A:$ cat /tmp/hosts/dhcp
# auto-generated config file from /etc/config/dhcp
192.168.4.1 xxxx.com xxxx.cn
192.168.4.1 SiWiFi3d04
```

（5）用浏览器通过http://xxxx.com或http://xxxx.cn域名来访问开发板Web管理界面（IP地址为192.168.4.1），如图6-9所示。

图6-9 开发板Web管理界面的域名劫持

6.7 服务与常用命令

6.7.1 服务命令

OpenWrt系统的/etc/init.d/目录下存放着许多系统服务的init脚本，这些脚本实现了不同服务的启动、停止、重启、重新加载配置等功能。关于init脚本的调用关系，我们在之前章节中介绍过，更详细的内容请参考前面章节。

init脚本都会引入一个/etc/rc.common文件，/etc/rc.common相当于init脚本的模板，定义了各种命令默认的处理函数。每一个init脚本都会包含下面的命令。

```
start：启动服务
stop：停止服务
restart：重启服务
reload：重新加载配置文件,如果失败,会重启服务
enable：设置开机自动启动
disable：禁止开机启动
```

启动服务：
```
/etc/init.d/XXX start
```

启动防火墙服务：
```
/etc/init.d/firewall start
```

关闭服务：
```
/etc/init.d/XXX stop
```

关闭防火墙服务：
```
/etc/init.d/firewall stop
```

重启服务：
```
/etc/init.d/XXX restart
```

重启防火墙服务：
```
DF1A:~$ /etc/init.d/firewall restart
```

重新加载服务配置文件：
```
/etc/init.d/XXX reload
```

重新加载防火墙配置文件：
```
/etc/init.d/firewall reload
```

设置服务开机启动：
```
/etc/init.d/XXX enable
```

设置防火墙开机启动：
```
/etc/init.d/firewall enable
```

禁止服务开机启动：
```
/etc/init.d/XXX disable
```

禁止防火墙服务开机启动：
```
/etc/init.d/firewall disable
```

6.7.2 系统常见命令

dmesg命令

dmesg的全称是display message，即显示信息。dmesg命令用来在系统中显示内核的相关信息。dmesg命令是从内核环形缓冲区中获取数据的。dmesg命令可以查看自系统启动后的内核消息、守护进程相关的信息等，对于我们快速识别内核和硬件问题、快速诊断和排除故障非常有帮助。dmesg命令非常简单，可以通过组合其他Shell命令来实现搜索和过滤功能。

dmesg命令的格式：

```
dmesg [-c] [-n LEVEL] [-s SIZE]
```

dmesg命令参数说明如表6-15所示。

表 6-15　dmesg 命令参数说明

参数	描述
-c	显示日志信息后，清空环形缓冲区
-n LEVEL	设定打印日志的级别，内核有8种级别（1~8），数值越大，显示的信息越多
-s SIZE	设置日志缓冲区大小，默认为8196字节

命令示例如下。

列出所有消息：

```
DF1A:$ dmesg
```

过滤消息：

```
DF1A:$ dmesg |grep -i usb
DF1A:$ dmesg |grep -i mtd
DF1A:$ dmesg |grep -i memory
DF1A:$ dmesg |grep -i sd
DF1A:$ dmesg |grep -i mtd
```

清空dmesg缓冲区日志：

```
DF1A:$ dmesg -c
```

设定日志级别：

```
DF1A:$ dmesg -n 8
```

logread命令

我们在OpenWrt中可通过logread命令查看运行时的消息日志。logread命令由logd软件包提供。logd是一个守护进程，通过与/dev/log通信转发内核消息，并提供ubus日志对象（提供读写方法）。logread命令是一个ubus读取消息的工具，与logd守护进程进行通信，以获取消息。

logread命令的格式：

```
logread [options]
选项：
    -s <path>
    -l <count>
    -e <pattern>
    -r <server> <port>
    -F <file>
```

```
    -S <bytes>
    -p <file>
    -h <hostname>
    -P <prefix>
    -f
    -u
    -0
```

命令示例如下。

列出所有日志：

```
DF1A:$ logread
```

列出最后10行日志：

```
DF1A:$ logread -l 10
```

过滤日志：

```
DF1A:$ logread -e netifd
```

不退出监控日志：

```
DF1A:$ logread -f
```

将指定log输出到文件中，修改UCI配置文件system：

```
config system
    option zonename 'Asia/Shanghai'
    option timezone 'CST-8'
    option hostname 'SiWiFi3d04'
    option hostnameset '1'
    option log_file '/var/log/mylog'
    option log_remote 0
```

重启服务：

```
DF1A:$ /etc/init.d/log restart
DF1A:$ /etc/init.d/system restart
```

将日志文件输出到文件中：

```
DF1A:$ cat /tmp/log/mylog
```

将指定log输出到远程端：

```
config system
    option zonename 'Asia/Shanghai'
    option timezone 'CST-8'
    option hostname 'SiWiFi3d04'
    option hostnameset '1'
    option log_ip '172.16.10.119'
    option log_port '8212'
    option log_proto 'tcp'
    option log_remote '1'
```

配置防火墙，添加rule节点：

```
config rule
    option target 'ACCEPT'
    option dest 'wan'
```

```
        option proto 'tcp udp'
        option dest_port '8212'
        option name 'ACCEPT-WAN-LOG'
```

在IP为172.16.10.119的服务器上监控日志输出：

```
nc -4 -l 8212
```

重启服务：

```
DF1A:$ /etc/init.d/log restart
DF1A:$ /etc/init.d/system restart
DF1A:$ /etc/init.d/network restart
```

输出服务器日志：

```
HOST:~$ nc -4 -l 8212
<30>Sep  5 11:21:35 logread[4247]: Logread connected to 172.16.10.119:8212
<29>Sep  5 11:24:22 netifd: Interface 'cfg094d8f' has link connectivity loss
<29>Sep  5 11:24:22 netifd: Interface 'lan' is now down
<6>Sep  5 11:24:22 kernel: [ 1363.929546] br-lan: port 3(wlan1) entered disabled state
<6>Sep  5 11:24:22 kernel: [ 1363.935160] br-lan: port 2(wlan0) entered disabled state
<6>Sep  5 11:24:22 kernel: [ 1363.954476] device eth0 left promiscuous mode
<6>Sep  5 11:24:22 kernel: [ 1363.960039] br-lan: port 1(eth0) entered disabled state
<13>Sep  5 11:24:22 relay: Reloading relay due to ifdown of cfg094d8f ()
<6>Sep  5 11:24:22 kernel: [ 1364.572203] IPv6: ADDRCONF(NETDEV_UP): eth0: link is not ready
<6>Sep  5 11:24:22 kernel: [ 1364.579787] device wlan0 left promiscuous mode
<6>Sep  5 11:24:22 kernel: [ 1364.584961] br-lan: port 2(wlan0) entered disabled state
<6>Sep  5 11:24:23 kernel: [ 1364.592922] device wlan1 left promiscuous mode
<6>Sep  5 11:24:23 kernel: [ 1364.598494] br-lan: port 3(wlan1) entered disabled state
<29>Sep  5 11:24:23 netifd: Interface 'lan' is disabled
<29>Sep  5 11:24:23 netifd: Bridge 'br-lan' link is down
<29>Sep  5 11:24:23 netifd: Interface 'lan' has link connectivity loss
<29>Sep  5 11:24:23 netifd: Interface 'loopback' is now down
<29>Sep  5 11:24:23 netifd: Interface 'loopback' is disabled
<29>Sep  5 11:24:23 netifd: Network device 'lo' link is down
<29>Sep  5 11:24:23 netifd: Interface 'loopback' has link connectivity loss
<29>Sep  5 11:24:23 netifd: Interface 'wan6' is now down
<28>Sep  5 11:24:23 dnsmasq[3588]: no servers found in /tmp/resolv.conf.auto, will retry
```

top命令

top是Linux下常用的性能分析工具，能够实时显示系统中各个进程的资源占用状况，方便我们实时跟踪进程，类似于Windows的任务管理器。默认的top命令是由BusyBox提供的。

top命令的格式：

```
top [-b] [-nCOUNT] [-dSECONDS] [-m]
```

在命令行中输入top，回车后的显示如图6-10所示。

```
Mem: 49408K used, 203212K free, 1248K shrd, 5148K buff, 17696K cached
CPU:   0.0% usr  0.1% sys  0.0% nic 99.8% idle  0.0% io  0.0% irq  0.0% sirq
Load average: 1.00 1.01 1.02 1/69 5887
  PID  PPID USER     STAT   VSZ  %VSZ CPU %CPU COMMAND
 5313     1 admin    S     2960   1.1   2  0.0 /usr/sbin/hostapd -P /var/run/wifi
 5887   727 admin    R     1576   0.6   0  0.0 top
 4247     1 admin    S     1260   0.5   0  0.0 /sbin/logread -f -r 172.16.10.119
 4871     1 admin    S     2960   1.1   3  0.0 /usr/sbin/hostapd -P /var/run/wifi
 4460     1 admin    S     1696   0.6   3  0.0 /sbin/netifd
 1847     1 admin    S     9532   3.7   1  0.0 /usr/sbin/ssst
 2056     1 admin    S     5192   2.0   0  0.0 /usr/sbin/new_device_listen
 5080     1 nobody   S     3324   1.3   0  0.0 /usr/sbin/dnsmasq -C /var/etc/dnsm
 1824     1 admin    S     3312   1.3   0  0.0 /usr/sbin/nmbd -F
 1823     1 admin    S     3208   1.2   2  0.0 /usr/sbin/smbd -F
 1716     1 admin    S     2292   0.9   3  0.0 /usr/sbin/uhttpd -f -h /www -r SiW
 1105     1 admin    S     2072   0.8   2  0.0 /sbin/rpcd
 1108     1 admin    S     1636   0.6   0  0.0 {first-check-ts-} /bin/sh /usr/bin
  727     1 admin    S     1580   0.6   1  0.0 /bin/ash --login
 2040     1 admin    S     1572   0.6   3  0.0 /usr/sbin/ntpd -n -S /usr/sbin/ntp
 1617     1 admin    S     1572   0.6   1  0.0 /usr/sbin/crond -f -c /etc/crontab
 4933  4460 admin    S     1568   0.6   3  0.0 udhcpc -p /var/run/udhcpc-eth1.pid
 5886  1108 admin    S     1560   0.6   1  0.0 sleep 780
    1     0 admin    S     1540   0.6   2  0.0 /sbin/procd
 1537     1 admin    S     1292   0.5   1  0.0 /usr/sbin/odhcpd
```

图6-10 用top命令显示进程信息

- 其中第一行显示内存的总量、空闲量、缓存容量。
- 第二行显示CPU的占用率、用户进程的CPU占用百分比、系统占用CPU的百分比、改变优先级进程的CPU占用百分比、CPU空闲率、I/O和硬软中断CPU占用百分比。
- 第三行显示系统1分钟前、5分钟前、15分钟前到现在的平均负载值。
- 从第五行起分别显示进程号、父进程号、进程创建者、进程状态（R：运行；S/D：睡眠；T：停止；Z：僵尸）、进程使用的虚拟内存总量、虚拟内存百分比、使用的CPU核心号、进程占用CPU百分比、进程名称。
- 在界面中可以使用N、M、P、T按键，来按照PID、MEM、CPU、Time进行排序。

free命令

free命令显示内存的使用情况，包括实体内存、虚拟的交换分区内存、共享内存区段，及系统核心使用的缓存等。

图6-11所示为执行free命令后显示的内存信息（单位为字节）。

```
              total        used        free      shared     buffers
Mem:         122700       49316       73384        1208       5336
-/+ buffers:              43980       78720
Swap:             0           0           0
```

图6-11 用free命令显示当前内存信息

- total：去掉为硬件和操作系统保留的内存后剩余的内存总量，与cat /proc/meminfo |grep MemTotal命令的执行结果相同。
- userd：当前已使用的内存总量。
- free：空闲的或可以使用的内存总量。

- shared：多进程共享的内存总量。
- buffers：缓存总量。
- Swap：交换分区虚拟内存总量。

其他实用命令

重启系统：

reboot

查看系统类型：

cat /proc/cpuinfo |grep 'system type' |awk -F ': ' '{print $2}'

查看CPU型号：

cat /proc/cpuinfo |grep 'cpu model'

查看设备型号：

cat /proc/cpuinfo |grep 'machine'

查看系统CPU核心数：

cat /proc/cpuinfo |grep processor|wc -l

7 存储器扩展

7.1 存储器的准备

经过了前几章的学习,你可能发现开发板的内存太小了、闪存也太小了,怎么办?本章将介绍如何扩展开发板存储器,包括存储器选型、初始化存储器和进行存储器扩容设置。

7.1.1 存储器选型

开发板具备两种存储器接口:一种是USB接口,另一种是TF(microSD)卡接口,这些接口可以满足我们对扩展存储器的需要,另外配合系统中的Swap交换分区技术可以实现内存扩展(交换分区类似于Windows下的虚拟内存)。

目前可选的存储器有以下几种。

USB移动硬盘或U盘: 需要注意的是供电稳定性,如果使用移动硬盘,推荐可以为移动硬盘另外提供一路外部供电。开发板USB接口为USB 2.0,最高速率为480Mbit/s,在这种速率下,既兼容USB 2.0接口的存储器,也兼容USB 3.0接口的存储器。

TF卡:芯片内置TF卡接口。目前TF卡的存储器单元技术有以下几种。

● **SLC:** Single Layer Cell,单级单元。SLC的特点是成本高、容量小、速度快,每个存储单元有1bit数据,每个单元可擦写次数达10万次。SLC一般使用在工业级产品上,其价格也贵很多。

● **MLC:** Multi-Level Cell,多级单元。MLC在与SLC相同的单元密度下,每个存储单元可存储2bit数据,目前已经较少使用。由于密度增加和算法原因,每个单元理论可擦写次数为3000~10 000次。

● **TLC/3D TLC:** Trinary-Level Cell,三级单元。TLC在与SLC相同的单元密度下,每个存储单元可存储3bit数据。TLC是目前低价位U盘、TF卡(尤其是高容量TF卡)常用的,其特点是密度大、成本低,缺点是速度慢、容易损坏,每个单元理论可擦写次数为500~1000次。

如果你使用开发板进行长期数据保存,可以采用外置带电源的移动硬盘;如果容量要求不大,也可以采用TF卡或U盘。

7.1.2 存储器识别

注意:在学习完本章节内容前,**不要插入存储器!**

系统可能存在未知设备自动挂载,这样会导致插入存储器后磁盘分区被自动挂载,影响我们操作。因此首先删除掉已有的挂载设定，后面的章节会专门介绍自动挂载。

```
DF1A:$ uci delete fstab.@mount[0]
DF1A:$ uci delete fstab.@mount[0]
DF1A:$ uci set fstab.@global[0].anon_mount='0'
DF1A:$ uci commit fstab
DF1A:$ /etc/init.d/fstab restart
```

安装相关软件包

安装包含字符集、工具的安装包（在开发板固件中已包含了驱动部分）：

```
DF1A:$ opkg update
DF1A:$ opkg install kmod-fs-ext4 kmod-fs-vfat kmod-nls-utf8 kmod-nls-cp437
DF1A:$ opkg install block-mount e2fsprogs
```

关于Hotplug自动挂载

● 在开启Hotplug自动挂载的状态下，插入任何存储器，存储器都会被自动挂载到/mnt/下的一个目录中（即使UCI配置fstab写了其他的挂载方法依旧这样）。

● 将外部存储挂载为Overlay分区时，依然会产生/mnt/xxx同时挂载的现象，这样会导致使用混乱。

● 如果不关闭Hotplug自动挂载，需要在插入设备后手工取消，每次都要操作比较麻烦。

关闭Hotplug自动挂载的方法：

```
#关闭方法
DF1A:~$ rm /etc/hotplug.d/block/10-mount
#恢复Hotplug自动挂载(如果以后有需要,可以这样操作)
DF1A:~$ cp /rom/etc/hotplug.d/block/10-mount /etc/hotplug.d/block/
#重新启动系统
DF1A:~$ reboot
```

不关闭Hotplug自动挂载，手动取消当前挂载：

```
#检测是否存在TF卡的分区
DF1A:~$ mount|grep mmcblk0
/dev/mmcblk0p1 on /mnt/mmcblk0p1 type ext4 (rw,relatime,data=ordered)
#取消挂载
DF1A:~$ umount /mnt/mmcblk0p1/
```

重启使修改生效：

```
DF1A:$ reboot
```

识别TF卡

将存储器插入TF卡槽内，将可看到如下信息，表示检测到TF卡。

```
[  168.784546] mmc_host mmc0: Bus speed (slot 0) = 150000000Hz (slot req 25000000Hz, actual 25000000HZ div = 3)
[  168.794661] mmc0: new high speed SDHC card at address aaaa
[  168.801568] mmcblk0: mmc0:aaaa SS08G 7.40 GiB
[  168.809643]  mmcblk0:
```

使用fdisk命令可以查看到TF卡已有分区和存储器的容量信息。

```
DF1A:~$ fdisk -l
```

```
Disk /dev/mmcblk0: 7948 MB, 7948206080 bytes
4 heads, 16 sectors/track, 242560 cylinders
Units = cylinders of 64 * 512 = 32768 bytes
        Device Boot       Start         End      Blocks   Id  System
```

● 在开发板中，TF卡为CPU原生支持，采用MMC驱动。所生成的设备标准为：mmcblk[编号]。

● 请记住设备文件的位置，这个位置将用于格式化和分区，例如上面通过fdisk查看到的/dev/mmcblk0。

识别U盘

将U盘插入USB 2.0接口，将会看到如下信息，表示检测到U盘。

```
[  483.987530] usb 1-1: new high-speed USB device number 2 using dwc2
[  484.206249] usb-storage 1-1:1.0: USB Mass Storage device detected
[  484.213448] scsi host0: usb-storage 1-1:1.0
[  485.219948] scsi 0:0:0:0: Direct-Access     Generic  Flash Disk       8.07 PQ: 0 ANSI: 4
[  485.231256] sd 0:0:0:0: [sda] 16498688 512-byte logical blocks: (8.44 GB/7.86 GiB)
[  485.239990] sd 0:0:0:0: [sda] Write Protect is off
[  485.245644] sd 0:0:0:0: [sda] Write cache: disabled, read cache: enabled, doesn't support DPO or FUA
[  485.259754]  sda: sda1
[  485.266455] sd 0:0:0:0: [sda] Attached SCSI removable disk
```

使用fdisk命令可以查看到U盘已有分区和存储器的容量信息。

```
DF1A:~$ fdisk -l
......................................................
Disk /dev/sda: 8447 MB, 8447328256 bytes
255 heads, 63 sectors/track, 1026 cylinders
Units = cylinders of 16065 * 512 = 8225280 bytes
     Device Boot     Start         End      Blocks   Id  System
/dev/sda1    *           1        1027     8248320    c  Win95 FAT32 (LBA)
```

● USB所生成的设备文件是以/dev/sd[X][N]的形式来编码的，其中X表示设备序列号，N表示分区编号。

● 上面演示的U盘设备为/dev/sda，其出厂已经有一个分区/dev/sda1。

可能出现的问题：如果U盘设备重启，顺序可能会产生变化，/dev/sda可能会变成/dev/sdb，我们将在后面的章节介绍使用UUID来解决该问题。

以下章节使用TF卡进行讲解，其操作原理与U盘相同。

7.1.3 存储器分区

我们使用的8GB TF卡计划按照如下方法分区。

● 分区一：容量7GB，用作存储器。
● 分区二：剩余空间全部分为Swap分区，用作虚拟内存扩展。

对mmcblk0的这个TF卡进行分区：

```
DF1A:~$ fdisk /dev/mmcblk0
```

```
The number of cylinders for this disk is set to 242560.
There is nothing wrong with that, but this is larger than 1024,
and could in certain setups cause problems with:
1) software that runs at boot time (e.g., old versions of LILO)
2) booting and partitioning software from other OSs
   (e.g., DOS FDISK, OS/2 FDISK)
Command (m for help):
```

输入d指令删除掉旧分区（如果有多个旧分区，删除程序会让你输入删除哪一个；如果只有一个分区，该分区会默认被删除掉）：

```
Command (m for help): d
No partition is defined yet!
```

用n指令创建分区：

```
Command (m for help): n
```

选择被创建分区的类型，p为主分区：

```
Partition type:
   p   primary (0 primary, 0 extended, 4 free)
   e   extended
Select (default p): p
```

再输入分区编号（主分区编号为1~4）：

```
Partition number (1-4, default 1): 1
```

设置被创建的分区尺寸，First cylinder不用填写，在Last cylinder填写+7G表示创建7GB大小的分区：

```
First cylinder (1-242560, default 1): Using default value 1
Last cylinder or +size or +sizeM or +sizeK (1-242560, default 242560): +7G
```

再用全部剩余容量创建第二个分区，语法与创建第一个分区相同（第二个分区不用输入Last cylinder编号，自动使用全部可用容量）：

```
Command (m for help): n
Command action
   e   extended
   p   primary partition (1-4)
p
Partition number (1-4): 2
First cylinder (213625-242560, default 213625): Using default value 213625
Last cylinder or +size or +sizeM or +sizeK (213625-242560, default 242560): Using default value 242560
```

用t指令将第二个分区设置为Swap交换分区类型，然后输入编号82：

```
Command (m for help): t
Partition number (1-4): 2
Hex code (type L to list codes): 82
Changed system type of partition 2 to 82 (Linux swap)
```

检查创建列表：

```
Command (m for help): p
Disk /dev/mmcblk0: 7948 MB, 7948206080 bytes
4 heads, 16 sectors/track, 242560 cylinders
```

```
Units = cylinders of 64 * 512 = 32768 bytes
        Device Boot      Start         End      Blocks   Id System
/dev/mmcblk0p1               1      213624     6835960   83 Linux
/dev/mmcblk0p2          213625      242560      925952   82 Linux swap
```

请记住以下两个文件的位置，格式化时要使用：
- 第一个分区，存放内容的设备文件为/dev/mmcblk0p1；
- 第二个分区，Swap交换分区，设备文件为/dev/mmcblk0p2。

用w指令将分区表写入存储器：

```
Command (m for help): w
The partition table has been altered.
Calling ioctl() to re-read partition table
[  464.050658] mmcblk0: p1 p2
```

7.1.4 存储器格式化

推荐采用EXT4作为文件系统格式，系统目前支持EXT2、EXT3、EXT4三种格式，EXT4格式适用于大容量存储器。

用EXT4格式格式化第一个分区：

```
DF1A:~$ mkfs.ext4 /dev/mmcblk0p1
mke2fs 1.42.12 (29-Aug-2014)
Creating filesystem with 1708990 4k blocks and 427392 inodes
Filesystem UUID: 9092ad2e-27fd-4b44-9916-32556ce705a8
Superblock backups stored on blocks:
        32768, 98304, 163840, 229376, 294912, 819200, 884736, 1605632
Allocating group tables: done
Writing inode tables: done
Creating journal (32768 blocks): done
Writing superblocks and filesystem accounting information: done
```

格式化交换分区：

```
DF1A:~$ mkswap /dev/mmcblk0p2
Setting up swapspace version 1, size = 948170752 bytes
```

格式化完成后，存储器准备就绪，在下一节中，我们将介绍存储器的使用方法。

7.2 存储器的使用

在这里我们以TF卡进行演示，介绍存储器的3种使用方式：
- 将EXT4格式分区挂载为Overlay透明分区，替代rootfs_data分区实现扩大可写分区容量；
- 将EXT4格式分区挂载为/mnt/extdisk目录，用来进行单独存储；
- 挂载Swap交换分区，增加内存容量。

7.2.1 挂载Overlay分区

开发板中SPI Flash的容量仅为16MB，其中rootfs_data分区部分作为Overlay分区使用，这样可写部分非常小，如果安装软件过多或用作存储显然是不够的。

查看SPI Flash的rootfs_data分区容量：

```
DF1A:$ cat /proc/mtd
dev:    size     erasesize  name
mtd0: 00020000 00010000 "spl-loader"
mtd1: 00060000 00010000 "u-boot"
mtd2: 00010000 00010000 "u-boot-env"
mtd3: 00010000 00010000 "factory"
mtd4: 00f60000 00010000 "firmware"
mtd5: 0019a321 00010000 "kernel"
mtd6: 00dc5cdf 00010000 "rootfs"
mtd7: 00630000 00010000 "rootfs_data"
```

将存储器挂载为Overlay分区，我们需要经过以下流程。

1. 将当前rootfs_data分区的内容复制到TF卡中。
2. 设定系统当前的UCI配置fstab，实现系统启动时自动挂载：

（1）这个修改还是在rootfs_data中完成的（因为当前系统还是将rootfs_data挂载为Overlay分区）；

（2）当系统启动时照样默认使用rootfs_data，不过在这一过程中一旦执行完fstab，将切换为TF卡的存储分区，完成后面的全部工作。

获取分区的UUID

我们通过分区的UUID可以避免设备重启后存储器设备文件变化，导致挂载不上，这是目前Linux类操作系统的通用做法，我们这里使用block命令来读取UUID：

```
DF1A:$ block info
/dev/mtdblock6: UUID="adf3e024-38314dc1-8d101492-393b784b" VERSION="1024.0" TYPE="squashfs"
/dev/mtdblock7: TYPE="jffs2"
/dev/mmcblk0p1: UUID="9092ad2e-27fd-4b44-9916-32556ce705a8" NAME="EXT_JOURNAL" VERSION="1.0" TYPE="ext4"
/dev/mmcblk0p2: VERSION="1" TYPE="swap"
```

如果你还记得上一节所介绍的分区格式的话，应该知晓/dev/mmcblk0p1整个分区作为存储器，而/dev/mmcblk0p2分区作为Swap交换内存分区。这里请记住UUID，后续章节要使用。

SPI中rootfs_data数据迁移

挂载存储器分区（请先确认是否已经被Hotplug挂载过，如果挂载过，请参考前面说明先取消掉）：

```
#创建临时挂载点
DF1A:$ mkdir /tmp/extdisk
#挂载分区
DF1A:$ mount -t ext4 /dev/mmcblk0p1 /tmp/extdisk
```

将整个/overlay中的内容复制到存储器：

```
DF1A:$ tar -C /overlay -cvf - . | tar -C /tmp/extdisk -xf -
```

关闭临时挂载，完成流程：

```
DF1A:$ umount /dev/mmcblk0p1
```

UCI配置文件fstab

UCI配置文件fstab是一个相比直接使用mount命令更为简便的进行开机自动挂载处理的软

件包。

UCI配置文件fstab的内容如下。

```
config global
        option anon_swap '0'
        option auto_swap '1'
        option auto_mount '1'
        option delay_root '5'
        option check_fs '0'
        option anon_mount '0'
config mount
option  target   '/mnt/sda1'
        option  uuid     '3cefc1eb-c9c9-4fc6-ba80-ff2714a6c2aa'
        option  enabled  '0'
```

fstab.@global匿名节点选项参数说明如表7-1所示。fstab.@mount匿名节点选项参数说明如表7-2所示。

表 7-1 fstab.@global 匿名节点选项参数说明

选项	说明	可选值及说明	必填
anon_swap	自动挂载任何一个外置磁盘上的交换分区	0：禁用（默认） 1：启用	否
anon_mount	当Hotplug程序执行到挂载的时候，自动将没有配置的存储器分区挂载到/mnt/下	0：禁用（默认） 1：启用	否
auto_swap	启用自动挂载Swap的设置	0：禁用 1：启用（默认）	否
auto_mount	启动自动挂载存储器分区	0：禁用 1：启用（默认）	否
delay_root	挂载之前等待多少秒，等待存储器连接上线	数字，默认为5，单位为秒	否
check_fs	在挂载时对ext类型的分区进行检测	0：禁用（默认） 1：启用	否

表 7-2 fstab.@mount 匿名节点选项参数说明

选项	说明	可选值及说明	必填
target	挂载到什么路径	字符串，路径，默认为空	是
device	被挂载的设备符号，如果有UUID可以不填此参数	字符串，设备符号，默认为空	否
uuid	被挂载的设备UUID	字符串，设备UUID编码，默认为空	否
fstype	被挂载的分区类型	Auto（自动）、ext4、ext3、fat32，默认为auto	否
options	挂载的命令参数	字符串，默认为空	否
enabled	启用自动挂载	0：禁用（默认） 1：启用	否
enabled_fsck	启用在挂载时对分区进行文件系统检测	0：禁用（默认） 1：启用	否

删除掉旧的不需要的@mount[0]节点（如果有的话，多次执行）：

```
DF1A:$ uci delete fstab.@mount[0]
DF1A:$ uci delete fstab.@mount[0]
uci: Entry not found
```

启动时自动挂载（记得用mmcblk0p1的UUID）：

```
#添加的匿名节点号请记住,需要使用该编号
DF1A:$ uci add fstab mount
cfg044d78
#使用该匿名节点编号编写自动挂载语法
DF1A:$ uci set fstab.cfg044d78.target=/overlay
DF1A:$ uci set fstab.cfg044d78.fstype=ext4
DF1A:$ uci set fstab.cfg044d78.options='rw,sync'
DF1A:$ uci set fstab.cfg044d78.enabled=1
DF1A:$ uci set fstab.cfg044d78.enabled_fsck=0
#请注意,uuid使用查看到的UUID,每个人操作获得到的UUID不同
DF1A:$ uci set fstab.cfg044d78.uuid='9092ad2e-27fd-4b44-9916-32556ce705a8'
DF1A:$ uci commit fstab
```

这里需要重新启动系统使设置生效：

```
DF1A:$ reboot
```

检测是否挂载成功：

```
#使用df命令测试
DF1A:$ df |grep mmcblk0p1
/dev/mmcblk0p1            6597544    15600    6223764    0% /overlay
#使用mount命令测试
DF1A:$ mount|grep mmcblk0p1
/dev/mmcblk0p1 on /overlay type ext4 (rw,relatime,data=ordered)
```

7.2.2 挂载Swap内存交换分区

存储区挂载完成后，/etc/config/fstab这个文件就是存储器上的版本，也就是说看起来好像没修改过。那就对了，挂载存储区时/overlay是rootfs_data，现在是在外部存储器上了，这个文件是在之前复制的，所以没变化，不过不会有影响。我们现在就用这个配置文件实现Swap内存交换分区的挂载。

查找global的匿名节点号：

```
DF1A:$ uci show -X fstab.@global[0]
fstab.cfg023fd6=global
fstab.cfg023fd6.anon_swap='1'
fstab.cfg023fd6.auto_swap='1'
fstab.cfg023fd6.auto_mount='0'
fstab.cfg023fd6.delay_root='5'
fstab.cfg023fd6.check_fs='0'
fstab.cfg023fd6.anon_mount='0'
```

打开自动匿名和自动挂载Swap内存交换分区的选项：

```
DF1A:$ uci set fstab.cfg023fd6.anon_swap=1
DF1A:$ uci set fstab.cfg023fd6.auto_swap=1
DF1A:$ uci commit fstab
```

重置fstab使设置生效：

```
DF1A:$ /etc/init.d/fstab reload
```

检测Swap内存交换分区是否完成挂载：

```
DF1A:$ cat /proc/meminfo |grep Swap
SwapCached:            0 kB
SwapTotal:        925948 kB
SwapFree:         925948 kB
```

7.2.3 挂载/mnt/extdisk

挂载/mnt/extdisk主要仅用于进行大容量的文件共享和管理，不适合使用外部存储器作为Overlay分区。

关闭Overlay分区挂载

如读者在前面章节做过/overlay的挂载，请关闭Overlay分区挂载，操作方法如下。如果读者没有操作过前面章节，可以跳过这一步骤。

设备断电后拔出TF卡，重新上电（不要插TF卡），确认当前Overlay分区使用rootfs_data：

```
DF1A:$ df|grep /overlay
/dev/mtdblock7        6336       556      5780    9% /overlay
overlayfs:/overlay    6336       556      5780    9% /
```

删除掉fstab中的自动挂载配置：

```
DF1A:$ uci show -X fstab.@mount[0]
fstab.cfg044d78=mount
fstab.cfg044d78.target='/overlay'
fstab.cfg044d78.fstype='ext4'
fstab.cfg044d78.options='rw,sync'
fstab.cfg044d78.enabled='1'
fstab.cfg044d78.enabled_fsck='0'
fstab.cfg044d78.uuid='9092ad2e-27fd-4b44-9916-32556ce705a8'
DF1A:$ uci delete fstab.cfg044d78
DF1A:$ uci commit fstab
```

重置fstab使其立即生效：

```
DF1A:$ /etc/init.d/fstab reload
```

插入TF卡，测试是否有影响：

```
DF1A:$ [   44.062551] mmc_host mmc0: Bus speed (slot 0) = 150000000Hz (slot req
400000Hz, actual 398936HZ div = 188)
[   44.345997] mmc_host mmc0: Bus speed (slot 0) = 150000000Hz (slot req
25000000Hz, actual 25000000HZ div = 3)
[   44.356030] mmc0: new high speed SDHC card at address aaaa
[   44.362725] mmcblk0: mmc0:aaaa SS08G 7.40 GiB
[   44.371023]  mmcblk0: p1 p2
DF1A:$ df|grep mmcblk0p1
```

启动时将TF卡自动挂载为/mnt/extdisk（记得用mmcblk0p1的UUID）：

```
#创建目录
DF1A:$ mkdir -p /mnt/extdisk
#添加的匿名节点号请记住，需要使用该编号
DF1A:$ uci add fstab mount
```

```
cfg044d78
#使用该匿名节点编号编写自动挂载语法
DF1A:$ uci set fstab.cfg044d78.target=/mnt/extdisk
DF1A:$ uci set fstab.cfg044d78.fstype=ext4
DF1A:$ uci set fstab.cfg044d78.options='rw,sync'
DF1A:$ uci set fstab.cfg044d78.enabled=1
DF1A:$ uci set fstab.cfg044d78.enabled_fsck=0
#注意将uuid的值替换为实际的UUID值
DF1A:$ uci set fstab.cfg044d78.uuid='9092ad2e-27fd-4b44-9916-32556ce705a8'
DF1A:$ uci commit fstab
```

重置fstab使设置生效：

```
DF1A:$ /etc/init.d/fstab reload
```

检测是否挂载成功：

```
DF1A:$ df |grep mmcblk0p1
/dev/mmcblk0p1           6597544        15600      6223764     0% /mnt/extdisk
```

7.2.4 挂载的注意事项

去掉扩展的存储器会如何：如果拔掉存储器，系统会默认返回使用Flash上的rootfs_data来实现Overlay分区，启动可以正常完成，但是之后安装的东西都不见了，系统回到原点。不过别怕，再插上之前的那个磁盘，又可以恢复，除非那个外部存储器坏掉了，这时候就要重复本章前面的内容重建了。

插入其他设备会报错：可能是供电异常或磁盘顺序变化，建议重启再看看。

检查/etc/config/fstab配置文件：当系统启动后，我们检查这个文件发现怎么还是旧的？这就对了，因为你在挂载之前修改的是rootfs_data下的/etc/config/fstab，而挂载之后看到的/etc/config/fstab是外部存储器上的，这个还没改过呢！不过没关系，不影响使用。

7.3 Windows文件共享

Windows文件共享是Windows系列操作系统十分重要的一项功能，可以让其他计算机像访问本地磁盘一样访问网络共享文件夹，也可以用于将文件共享给机顶盒或手机使用。本节介绍使用7.2节中的存储器实现简易的家庭存储解决方案。

7.3.1 安装软件包

准备共享文件夹：

```
DF1A:$ mkdir -p /mnt/extdisk/pub
#给予该文件夹777权限,防止演示过程中出现文件权限错误
DF1A:$ chmod 777 /mnt/extdisk/pub
```

安装软件包：实现Windows与Linux之间共享数据的软件叫Samba，该软件可以在Linux下实现网上邻居，并且可以被Windows探测发现。

```
DF1A:$ opkg update
DF1A:$ opkg install samba36-server
```

7.3.2 配置samba服务

UCI配置文件samba的内容如下:
```
config samba
    option name 'siwifi'
    option workgroup 'WORKGROUP'
    option description 'OpenWrt'
    option homes '1'
config sambashare
    option name 'USB'
    option path '/mnt/sda1/'
    option read_only 'no'
    option guest_ok 'yes'
    option create_mask '0777'
    option dir_mask '0777'
```

samba.@samba匿名节点选项参数说明如表7-3所示。samba.@sambashare匿名节点选项参数说明如表7-4所示。

表7-3 samba.@samba 匿名节点选项参数说明

选项	说明	可选值及说明	必填
charset	文件名编码格式	UTF-8、cp936，默认为UTF-8	否
name	服务的名称	字符串，默认为主机名称	否
workgroup	网上邻居工作组	字符串，默认为WORKGROUP	否
description	对设备的描述信息	字符串	否
homes	是否启用用户文件夹共享模式	0: 禁用（默认） 1: 启用	否

表7-4 samba.@sambashare 匿名节点选项参数说明

选项	说明	可选值及说明	必填
name	共享的名称	字符串，默认为空	是
path	共享所对应的设备上的路径	字符串，共享路径，默认为空	是
users	指定什么用户可以访问这个共享	字符串，有权限的系统用户，默认为空	否
read_only	共享是否为只读模式	no: 否（默认） yes: 是	否
guest_ok	是否支持访客模式	no: 否（默认） yes: 是	否
create_mask	共享所创建的文件权限	数字，4位权限编码，默认为0744	否
dir_mask	共享所创建的目录权限	数字，4位权限编码，默认为0755	否

配置共享目录

清理掉旧的共享信息设定：
```
DF1A:$ uci delete samba.@sambashare[0]
DF1A:$ uci delete samba.@sambashare[0]
uci: Entry not found
DF1A:$ uci commit samba
```

设置共享文件夹参数：
```
#增加一个新的sambashare节点
```

```
DF1A:$ uci add samba sambashare
cfg04e23c
DF1A:$ uci set samba.cfg04e23c.name='shares'
DF1A:$ uci set samba.cfg04e23c.path='/mnt/extdisk/pub'
DF1A:$ uci set samba.cfg04e23c.guest_ok='yes'
DF1A:$ uci set samba.cfg04e23c.create_mask='0777'
DF1A:$ uci set samba.cfg04e23c.dir_mask='0777'
DF1A:$ uci set samba.cfg04e23c.read_only='no'
DF1A:$ uci commit samba
```

启动samba服务：

```
DF1A:$ /etc/init.d/samba restart
```

设置samba服务开机自动启动：

```
DF1A:$ /etc/init.d/samba enable
```

7.3.3 访问共享文件夹

按照以下步骤可以进行测试。

（1）在Windows系统屏幕左下方输入框中输入"\\192.168.4.1\"后将弹出可选文件夹（见图7-1），也可以使用Win+R组合键弹出运行框输入该地址。该地址即开发板的演示地址，请根据你的开发板LAN口的IP地址来填写。

图7-1 在Windows中输入"\\192.168.4.1\"后将检索到共享文件夹

（2）用鼠标单击弹出的"\\192.168.4.1\shares"即可打开共享的文件夹（见图7-2）。

图7-2 打开共享的文件夹

（3）此时你可以像使用本地磁盘一样对该文件夹进行操作，如复制一些文件进去。
（4）在开发板中查看复制进来的文件。

```
DF1A:$ ls -l /mnt/extdisk/pub/
-rwxrw-rw-    1 nobody   nogroup        80873 Sep  5 12:12 2.6.5.docx
-rwxrw-rw-    1 nobody   nogroup        70895 Sep  5 09:41 2.7.0.docx
-rwxrw-rw-    1 nobody   nogroup        72823 Sep  5 12:15 2.7.1.docx
-rwxrw-rw-    1 nobody   nogroup        76220 Sep  5 12:40 2.8.0.docx
-rwxrw-rw-    1 nobody   nogroup       162486 Sep  3 13:34 2.9.2.docx
```

7.4 FTP文件共享

FTP（File Transfer Protocol，文件传输协议）是目前广泛支持的一种网络文件传输技术。在OpenWrt下我们使用vsftpd，它是一个快速、高效、轻便、安全的FTP服务器。

7.4.1 安装软件包

安装软件包：

```
DF1A:$ opkg update
DF1A:$ opkg install vsftpd vsftpd-tls
```

7.4.2 配置vsftpd服务

开发板中的vsftpd软件目前不支持UCI配置，因此我们使用Vim编辑器完成配置文件设定，文件名称为/etc/vsftpd.conf：

```
#后台运行
background=YES
#监听端口
listen=YES
#是否允许匿名访问FTP
anonymous_enable=NO
#本地允许,并且可写
local_enable=YES
write_enable=YES
#用户上传文件的文件权限
```

```
local_umask=022
#是否检测Shell
check_shell=NO
#开启这个选项表示每个目录下如果有message_file将显示信息
#dirmessage_enable=YES
#开启FTP欢迎语,默认可以注释掉,不使用
#ftpd_banner=Welcome to blah FTP service.
#保留会话,请保留
session_support=NO
#是否记录操作日志
#syslog_enable=YES
```

启动vsftpd服务:

```
DF1A:$ /etc/init.d/vsftpd restart
```

设置vsftpd服务开机自动启动:

```
DF1A:$ /etc/init.d/vsftpd enable
```

7.4.3 访问FTP服务

WinSCP是一个强大的客户端软件,它支持SFTP、FTP、SCP三种模式,这里我们仅使用FTP功能进行测试。

(1)在WinSCP登录界面(见图7-3)中单击"新建",在新窗口中填写FTP连接信息。"主机名"填写开发板IP地址,"端口号"填写21,"用户名"与"密码"填写SSH登录的账号和密码,比如admin/admin。

图7-3 WinSCP登录配置界面

(2)单击"保存",在新对话框选择"保存密码",单击"确定",返回登录对话框。单击刚才创建的连接信息,单击"登录"。

(3)登录后将显示出文件传输界面,界面操作与SCP相同(见图7-4)。

图7-4 WinSCP主界面

7.5 BT远程下载

BT是一种互联网上的P2P传输协议，全名叫"BitTorrent"，中文全称为"比特流"，最初的创造者是布拉姆·科恩，现在则独立发展成一个有广大开发者群体的开放式传输协议。我们使用Aria2软件可实现通过BT、FTP、HTTP等协议下载。

7.5.1 安装软件包

Aria2一共包含两部分，分别为Aria2服务端软件包与Aria2客户端软件包，使用jsonrpc方式通信，其中服务端安装于开发板中，客户端有对应Windows、macOS、iOS、Android等多种操作系统的版本。下面我们首先安装开发板中的Aria2服务端软件包。

安装软件包：

```
DF1A:$ opkg update
DF1A:$ opkg install libxml2 aria2
```

准备文件：

```
#在外部存储器中创建目录
DF1A:$ mkdir -p /mnt/extdisk/aria2/ /mnt/extdisk/pub/downloads/
#准备默认文件
DF1A:$ touch /mnt/extdisk/aria2/aria2.session
#设置权限
DF1A:$ chmod 666 /mnt/extdisk/aria2/aria2.session
DF1A:$ chmod 777 /mnt/extdisk/aria2/ /mnt/extdisk/pub/downloads/
```

7.5.2 配置Aria2服务

开发板中的Aria2软件目前不支持UCI配置，因此我们使用Vim编辑器完成配置文件设定，文件名称为/mnt/extdisk/aria2/aria2.conf。

aria2.conf选项参数说明如表7-5所示。

表 7-5 aria2.conf 匿名节点选项参数说明

选项	说明	可选值及说明
dir	下载文件的保存路径	路径
disk-cache	启用磁盘缓存，0为禁用缓存，需1.16以上版本	以M结尾表示MB，默认为16M
file-allocation	文件预分配方式，能有效减少磁盘碎片，falloc和trunc则需要文件系统和内核支持。NTFS建议使用falloc，EXT3/4建议trunc，Mac下需要注释此项	可选none、falloc、trunc、prealloc，默认为prealloc
continue	断点续传	true、false
log-level	日志等级	debug、info、notice、warn、error
log	记录日志的文件，如果不指定这个参数同时没有设定为服务进程，则log输出到stdout	文件名或为空
max-concurrent-downloads	最大同时下载任务数，运行时可修改	默认为5
max-connection-per-server	同一服务器连接数，添加时可指定	默认为1
min-split-size	最小文件分片大小	取值范围为1~1024M，默认为20M
split	单个任务最大分片数	默认为5
max-overall-download-limit	整体下载速度限制	默认为0（表示不限速），可填写单位K、M
max-download-limit	单个任务下载速度限制	默认为0（表示不限速），可填写单位K、M
max-overall-upload-limit	整体上传速度限制	默认为0（表示不限速），可填写单位K、M
max-upload-limit	单个任务上传速度限制	默认为0（表示不限速），可填写单位K、M
disable-ipv6	禁用IPv6	true、false
input-file	从会话文件中读取下载任务	文件名
save-session	在Aria2退出时将"错误/未完成"的下载任务保存到会话文件中	文件名，与input-file相同即可
save-session-interval	定时保存会话，需1.16.1以上版本	默认为0（退出时才保存），单位为秒
enable-rpc	启用RPC接口	true、false（默认）
rpc-allow-origin-all	允许所有来源访问RPC接口	true、false（默认）
rpc-listen-all	允许非外部访问RPC接口	true、false（默认）
event-poll	事件轮询方式	epoll、kqueue、port、poll、select，不同系统默认值不同
rpc-listen-port	RPC监听端口	默认为6800
rpc-secret	RPC授权令牌	专用格式字符串
follow-torrent	当下载的是一个种子（以.torrent结尾）时，自动开始BT任务	true（默认）、false
listen-port	BT监听端口	默认为6881-6999
bt-max-peers	单个种子最大连接数	默认为55
enable-dht	打开IPv4 DHT功能	true（默认）、false
enable-dht6	打开IPv6 DHT功能	true（默认）、false
dht-listen-port	IPv4 DHT网络监听端口	默认为6881-6999
dht-file-path	IPv4 DTH数据文件	文件名
dht-file-path6	IPv6 DTH 数据文件	文件名
bt-enable-lpd	本地节点查找	true、false（默认）

续表

选项	说明	可选值及说明
enable-peer-exchange	开启种子交换	true（默认）、false
bt-request-peer-speed-limit	每个种子限速	默认为50K，可填写单位K、M
peer-id-prefix	指定对等体ID的前缀	例如：-TR2770-
user-agent	Aria2客户端伪装	例如：Transmission/2.77
seed-ratio	当种子的分享率达到这个数时，自动停止作种	0为一直作种，默认为1.0
force-save	强制保存会话，即使任务已经完成	true、false（默认）
bt-hash-check-seed	BT校验相关	true（默认）、false
bt-seed-unverified	继续之前的BT任务时，无须再次校验	true、false（默认）
bt-save-metadata	保存磁力链接元数据为种子文件（.torrent文件）	true、false（默认）
bt-tracker	设置BT的Tracker地址	多个地址用逗号分隔

关于bt-tracker

在线下载最新的bt-tracker，可以提高BT的下载速度，原始文件为换行格式，需要手工修改为配置参数格式。

配置文件/mnt/extdisk/aria2/aria2.conf示例：

```
dir=/mnt/extdisk/pub/downloads
disk-cache=32M
file-allocation=trunc
continue=true
#downloads
max-concurrent-downloads=10
max-connection-per-server=5
min-split-size=10M
split=5
max-overall-download-limit=0
max-download-limit=0
max-overall-upload-limit=2M
max-upload-limit=0
disable-ipv6=true
#sesisons
input-file=/mnt/extdisk/aria2/aria2.session
save-session=/mnt/extdisk/aria2/aria2.session
save-session-interval=60
#rpc
enable-rpc=true
rpc-allow-origin-all=true
rpc-listen-all=true
rpc-listen-port=6800
#bt
follow-torrent=true
listen-port=6881-6999
enable-dht=true
#enable-dht6=false
dht-listen-port=6881-6999
dht-file-path=/mnt/extdisk/aria2/dht.dat
dht-file-path6=/mnt/extdisk/aria2/dht6.dat
```

```
bt-enable-lpd=false
enable-peer-exchange=false
peer-id-prefix=-TR2770-
user-agent=Transmission/2.77
seed-ratio=0
bt-seed-unverified=true
bt-save-metadata=true
#tracker
bt-tracker=udp://××.138.0.158:6969/announce,udp://××.80.120.114:2710/
announce,udp://××.80.120.113:2710/announce,udp://××.158.213.92:1337/
announce,udp://××.225.17.100:1337/announce,udp://××.19.107.254:80/ann
ounce,udp://××.241.58.209:6969/announce,udp://××.83.20.20:6969/annou
nce,udp://××.206.19.247:6969/announce,udp://××.44.243.4:1337/announc
e,udp://××.235.174.46:2710/announce,udp://××.154.52.99:80/announce,u
dp://××.211.168.204:2710/announce,udp://××.143.148.21:2710/announce,
udp://××.234.156.205:451/announce,udp://××.37.235.149:6969/announce,
udp://××.100.245.181:6969/announce,udp://××.56.74.11:6969/announce,u
dp://××.15.226.113:6969/announce,udp://××.105.151.164:6969/announce
```

7.5.3 启动Aria2服务

测试启动：

```
DF1A:$ aria2c -c --conf-path=/mnt/extdisk/aria2/aria2.conf --log-level=debug
```

- 如果启动完成后没有退出，则表示配置有效。
- 用组合键Ctrl+C退出当前启动。

设置开机自动启动：

```
#编辑/etc/rc.local文件，在exit 0前增加启动语法，例如：
# Put your custom commands here that should be executed once
# the system init finished. By default this file does nothing.
aria2c -c --conf-path=/mnt/extdisk/aria2/aria2.conf -D
exit 0
```

这里需要重新启动系统使设置生效：

```
DF1A:$ reboot
```

测试Aria2是否能运行：

```
DF1A:$ netstat -lnp|grep aria2c
tcp        0      0 0.0.0.0:6800            0.0.0.0:*               LISTEN      1989/aria2c
```

7.5.4 Aria2客户端

Aria2有多种客户端软件，我们以Windows版（该软件也提供macOS版）的AriaNg-Native为例进行客户端介绍。要使用Android客户端，请在手机商店中搜索Aria2App。

配置AriaNg-Native

（1）下载AriaNg-Native软件，并且安装。

（2）启动软件主界面，单击左侧的"AriaNg设置"，然后单击右侧选项卡"RPC(localhost

名称)",在"Aria2 RPC地址"中填写开发板IP地址(见图7-5)。

图7-5　AriaNg设置

(3)单击右下角的"Reload Page",然后单击"Aria2状态"查看是否已连接(见图7-6)。

图7-6　查看Aria2状态

这里我们以Ubuntu操作系统为例演示下载。

(1)单击"正在下载",单击"新建"。

(2)粘贴下载地址,下载BT种子(见图7-7)。

图7-7　下载BT种子

(3)单击"立即下载",然后就耐心等待吧,种子下载完毕后会自动下载文件(见图7-8)。

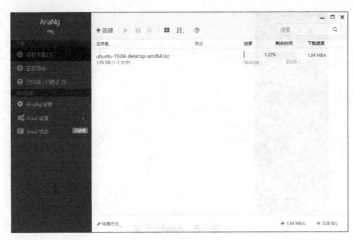

图7-8 正在下载文件

查看下载的文件

整个下载过程是由开发板完成的,因此在下载过程中可以随时关闭客户端,不会受到计算机关闭AriaNg-Native的影响。要查看下载的文件,大家可以选择使用前文介绍过的FTP或Windows文件共享。AriaNg_Native的使用方法类似迅雷软件,具体使用方法可以多多摸索。

7.6 PPTP客户端

点对点隧道协议(PPTP)是一种支持多协议虚拟专用网络(VPN)的网络技术。我们在OpenWrt系统中实现PPTP客户端,可以允许计算机、手机等设备通过OpenWrt访问远程局域网。

7.6.1 安装软件包

安装软件包:

```
DF1A:$ opkg update
DF1A:$ opkg install ppp-mod-pptp
```

安装完成后要重启系统:

```
DF1A:$ reboot
```

7.6.2 配置PPTP客户端

UCI配置文件network对PPTP节点进行支持:

```
config 'interface' 'vpn'
        option 'ifname'    'pptp-vpn'
        option 'proto'     'pptp'
        option 'username'  '账号'
        option 'password'  '密码'
        option 'server'    '服务器地址'
        option 'buffering' '1'
```

network.vpn选项参数说明如表7-6所示。

表7-6 network.vpn 选项参数说明

选项	说明	可选值及说明	必填
ifname	VPN服务的设备名称	pptp-vpn	是
proto	所采用协议	pptp	是
username	账号	字符串，PPTP的账号	是
password	密码	字符串，PPTP的密码	是
server	服务器地址	字符串，IP地址	
buffering	缓冲	0：禁用 1：启用	

配置举例：

```
network.vpn=interface
network.vpn.ifname=pptp-vpn
network.vpn.proto=pptp
network.vpn.username=hoowa
network.vpn.password=123456
network.vpn.server=60.xx.xx.xx
network.vpn.buffering=1
```

可以使用计算机确认该PPTP服务是否有效。

账号、密码、服务器地址请根据实际申请到的填写，以上只是举例。

7.6.3 启动PPTP客户端

PPTP客户端作为network配置的一个interface节点，在系统重新启动后会自动尝试连接。如果要手动测试连接，可以使用下面这个语法：

```
DF1A:$ /etc/init.d/network reload
```

检测是否连接成功（以查看到pptp-vpn网络接口为准）：

```
DF1A:$ ifconfig pptp-vpn
pptp-vpn  Link encap:Point-to-Point Protocol
          inet addr:10.10.0.11  P-t-P:10.0.0.254  Mask:255.255.255.255
          inet6 addr: fe80::f18c:5285:164d:9e4a/10 Scope:Link
          UP POINTOPOINT RUNNING NOARP MULTICAST  MTU:1450  Metric:1
          RX packets:36 errors:0 dropped:0 overruns:0 frame:0
          TX packets:50 errors:0 dropped:0 overruns:0 carrier:0
          collisions:0 txqueuelen:3
          RX bytes:2083 (2.0 KiB)  TX bytes:3069 (2.9 KiB)
```

查看日志了解连接过程：

```
DF1A:$ logread|grep pptp
```

7.6.4 终端通过PPTP连接外网

完成默认的VPN拨号后，本地系统可以通过VPN连接外网，但局域网下的计算机、手机等设备无法使用外网，这时要修改防火墙配置，将VPN增加到WAN区域后。

修改WAN域范围，增加VPN：

```
#找到WAN口的zone匿名节点编号
DF1A:$ uci show -X firewall|grep "name='wan'"
firewall.cfg0cdc81.name='wan'
#增加VPN域
DF1A:$ uci add_list firewall.cfg0cdc81.network='vpn'
DF1A:$ uci commit firewall
#使修改立即生效
DF1A:$ /etc/init.d/firewall reload
```

8 SF16A18的LuCI界面

LuCI是OpenWrt默认的Web管理界面，项目开始于2008年。LuCI使用Lua编程语言开发，非常适合嵌入式系统，具有模块化设计，易于扩展、维护，非常小巧，具有很好的性能和更快的运行时间。LuCI最初采用MVC（模型-视图-控制器模式）Web框架，逐渐发展为面向对象的库和模板，并将接口拆分为模型和视图。

SF16A18-LuCI配置界面是SiFlower基于LuCI框架进行二次开发的，非常简洁和易于配置，并且提供了修改页面风格和样式的UCI配置文件，不用修改任何代码就可以实现页面样式的快速修改，非常适合生产环境。

LuCI默认使用uhttpd作为Web服务器。uhttpd提供Lua插件和ubus支持。uhttpd配置文件中会配置网站的htdocs目录(/www)和CGI(www/cgi-bin)目录。

8.1 SF16A18-LuCI目录结构

8.1.1 SF16A18-LuCI源代码目录

LuCI源代码需要通过源代码工程下的scripts/feeds update命令进行更新，获取的LuCI的源代码存放在feeds/luci目录，而SF16A18-LuCI源代码随工程自带，用户不用手动更新获取，源代码默认存放在package/siflower/luci-siflower目录。SF16A18-LuCI源代码目录结构与LuCI源代码目录结构保持一致，如图8-1所示。

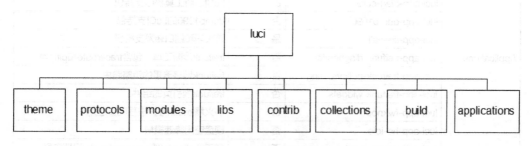

图8-1 LuCI源代码目录结构

其中各目录的含义如下。

- collections：包含LuCI SSL、最小主题包和默认主题。
- applications：单个模块应用或插件。每个应用包含model、view、controller、页面的i18n多语言包、工具和配置等。
- build：i18n多语言生成工具、文档生成工具等。
- contrib：贡献包。LuCI之外的贡献包放在这个目录，如uhttpd的Lua支持包等。
- theme：前端主题。不同主题使用不同的目录。
- modules：存放应用的集合。
- protocols：网络协议支持，支持relay、ppp、IPv6、3G、openconnect等协议。
- libs：通用库文件，包含JSON库、IP和网段处理库、nixio库、httpclient库、JSON-C库等。

SF16A18-LuCI的软件包

SF16A18-LuCI默认支持的软件包如表8-1所示。

表8-1 SF16A18-LuCI 默认支持的软件包

软件包类别	软件包名称	默认选择	说明
Collections	luci	是	默认的LuCI包，必选
	luci-ssl	是	使LuCI支持HTTPS
Modules	luci-base	是	LuCI核心库
	translations	是	默认支持English和Chinese
	luci-mod-admin-86v	否	SiFlower 86面板产品的LuCI模块
	luci-mod-admin-flash	否	LuCI Flash。小存储设备使用
	luci-mod-admin-full	是	LuCI默认的全功能控制管理模块
	luci-mod-failsafe	否	故障保护系统升级模块
	luci-mod-freifunk	否	LuCI freifunk 模块
	luci-mod-freifunk-community	否	freifunk Community元包
	luci-mod-rpc	否	LuCI JSON API
Applications	luci-app-adblock	否	广告屏蔽软件adblock的LuCI支持包
	luci-app-ahcp	否	Ad-Hoc的LuCI支持包
	luci-app-asterisk	否	Asterisk的LuCI支持包
	luci-app-commands	否	Shell命令的LuCI支持包
	luci-app-ddns	否	动态DNS客户端的LuCI支持包
	luci-app-diag-core	否	LuCI诊断工具核心支持包
	luci-app-dump1090	否	dump1090的LuCI支持包
	luci-app-firewall	是	防火墙和端口转发支持包
	luci-app-freifunk-diagnostics	否	freifunk诊断工具，包含traceroute和ping等
	luci-app-freifunk-policyrouting	否	freifunk网格流量的策略路由
	luci-app-freifunk-widgets	否	freifunk索引页的组件
	luci-app-fwknopd	否	防火墙knock守护进程配置应用
	luci-app-hd-idle	否	硬盘空闲降速模块
	luci-app-meshwizard	否	基于shellscript的mesh networks设置向导
	luci-app-minidlna	否	miniDLNA的LuCI支持包
	luci-app-mjpg-streamer	否	MJPG-Streamer服务配置模块
	luci-app-mmc-over-gpio	否	MMC-over-GPIO配置模块

续表

软件包类别	软件包名称	默认选择	说明
Applications	luci-app-multiwan	否	Multiwan代理的LuCI支持
	luci-app-ntpc	否	NTP时间同步配置模块
	luci-app-ocserv	否	OpenConnect VPN的LuCI支持
	luci-app-olsr	否	OLSR配置和状态模块
	luci-app-olsr-services	否	OLSR服务支持模块
	luci-app-olsr-viz	否	OLSR可视化
	luci-app-openvpn	否	OpenVPN支持
	luci-app-p2pblock	否	freifunk P2P-Block插件支持
	luci-app-p910nd	否	p910nd打印机服务模块支持
	luci-app-pbx	否	PBX管理
	luci-app-pbx-voicemail	否	PBX管理语音邮件支持
	luci-app-polipo	否	Polipo代理
	luci-app-privoxy	否	Privoxy Web代理
	luci-app-qos	否	服务质量配置模块
	luci-app-radicale	否	radicale的carddav/caldav支持
	luci-app-radvd	否	Radvd支持
	luci-app-samba	是	网络共享Samba SMB/CIFS模块支持
	luci-app-shairplay	否	Shairplay支持
	luci-app-shairport	否	Shairport支持
	luci-app-siflower-ap	否	SiFlower定制的AP
	luci-app-siitwizard	否	siit ipv4-over-ipv6配置向导
	luci-app-splash	否	freifunk DHCP splash应用程序
	luci-app-statistics	否	统计应用支持
	luci-app-tinyproxy	否	tinyproxy-HTTP（S）代理配置
	luci-app-transmission	否	Transmission的LuCI支持
	luci-app-udpxy	否	udpxy的LuCI支持
	luci-app-upnp	是	通用即插即用配置模块
	luci-app-vstat	否	VnStat支持模块
	luci-app-voice-core	否	语音软件（核心）的LuCI支持
	luci-app-watchcat	否	Watchcat的LuCI支持
	luci-app-wol	否	LuCI对wake-on-lan的支持
	luci-app-wshaper	否	LuCI对wshaper的支持
Themes	luci-theme-86v	否	SiFlower 86面板主题
	luci-theme-bootstrap	否	OpenWrt官方默认的Bootstrap主题
	luci-theme-freifunk-bno	否	freifunk Berlin Nordost主题
	luci-theme-freifunk-generic	否	freifunk通用主题
	luci-theme-material	是	Material主题
	luci-theme-openwrt	否	OpenWrt主题
	luci-theme-siwifi	否	Siwifi主题
Protocols	luci-proto-3g	否	3G支持
	luci-proto-ipv6	否	支持dhcpv6、6in4、6to4、6rd、ds-lite、aiccu
	luci-proto-openconnect	否	OpenConnect VPN支持
	luci-proto-ppp	是	PPP、PPPoE、PPPoA、PPtP支持
	luci-proto-relay	是	中继伪桥支持

续表

软件包类别	软件包名称	默认选择	说明
Libraries	luci-lib-httpclient	否	HTTP（S）客户端库
	luci-lib-ip	是	用于IP计算和路由信息的Lua库
	luci-lib-json	是	LuCI JSON库
	luci-lib-jsonc	否	JSON-C的Lua绑定库
	luci-lib-luaneightbl	否	用于IPv6邻居的neightbl-Lua库
	luci-lib-nixio	是	nixio posix库
	luci-lib-px5g	否	RSA/X.509密钥生成器（Lucid SSL支持所需）

LuCI本身包含众多软件包，即使什么也不选，系统也会默认选择两个软件包：luci-base和luci-mod-admin-full。这两个软件包在moduels目录下，是LuCI最基本的软件包，SF16A18-LuCI默认也选中这两个软件包。

1. luci-base

luci-base是LuCI最核心的软件包，它包含：

- LuCI Web核心框架，如dispatcher、http、ccache等。
- 核心库，如cbi、uci、ipkg、network、i18n、cgi、sys、template、degbug等。
    - cgi.lua：cgi的主要处理程序。
    - template：模板解析器，用于将HTML文件解析到Lua文件中，并保存预编译的模板文件。
    - http.lua：Web框架上层HTTP的功能。
    - i18n.lua：LuCI的语言翻译库。
    - ip.lua：IP计算功能。
    - sys.lua：Linux与POSIX系统实用程序。
    - util.lua：基本函数库。
    - dispatcher.lua：LuCI页面请求分发和调度器。
    - uci.lua：Lua使用UCI配置文件支持库。
- 基本view页面，如cbi组件、footer.htm、header.htm、syauth.htm、error404.htm等。
- 多语言po文件，包含i18n的base.po文件。
- UCI配置文件，包含主题、多语言、认证、缓存等配置的LuCI文件和ucitrack。
- HTTP请求的主页面index.htm、HTTP请求默认的cgi文件luci、存放静态资源文件的luci-static（包含CSS、JS及网页图片等文件及主题目录）。
- 其他文件，如luci-reload等。

2. luci-mod-admin-full

luci-mod-admin-full是LuCI默认的全功能管理模块，包含静态资源文件、管理者后台controller文件、model文件、cbi文件、view静态界面。

8.1.2 LuCI MVC模型

LuCI是基于MVC框架的模型，主要分为3个部分。

- Model（模型）：模型是数据和业务的处理规则。
- View（视图）：视图是用户看到并与之交互的界面。

● Controller（控制器）：控制器接收用户的输入并调用Model和View实现用户的需求。

除了以上提到的3个模块，LuCI还有自己封装的一些函数库及提供系统功能的一些程序，这些程序包含在公共模块（LuCI的系统文件及用到的各个库）中。

默认安装luci-base和luci-mod-admin-full软件包后，LuCI包含的Model、View、Controller信息如下。

Model（模型）

Model对应的安装目录为/usr/lib/lua/luci/model。Model通过cbi模块和UCI进行交互，将页面和配置关联起来，绑定UCI配置文件，通过前端View呈现配置数据，并将页面的设置写到路由器当中。

- cbi：界面元素的组合，用Lua文件描述UCI配置文件的结构，并将其转化为HTML呈现给前端。
- admin_network：管理者的network相关协议处理，会包含network.lua文件。
- uci.lua：UCI的接口库，需要向配置文件写数据时会调用其中的函数。
- network.lua：network UCI配置文件对应的model绑定处理。
- firewall.lua：firewall UCI配置文件对应的model绑定处理。
- ipkg.lua：opkg软件包对应的model绑定处理。

View（视图）

View对应的安装目录为/usr/lib/lua/luci/view。该目录存放的都是HTML文件，用于定义页面的模板和cbi控件模板。

- theme：主题目录下包含该主题通用的footer.htm和header.htm。
- cbi：当使用cbi方式调用时，用到的所有控件模板都定义在此。
- admin_status：管理者界面的status标题下的页面模板。
- admin_system：管理者界面的system标题下的页面模板。
- admin_network：管理者界面的网络配置的模板，网络配置的相关界面都在此目录中。
- admin_uci：该目录对应有关页面提交数据按钮控件的模板。
- sysauth.htm：授权页面。
- footer.htm：引用主题模板下的footer.htm，包含"保存应用""恢复""确定"的处理。
- header.htm：引用主题模板下的header.htm。

Controller（控制器）

Controller对应的安装目录为/usr/lib/lua/luci/controller。

- controller下的目录通常以用户名命名，如admin。
- controller/admin表示管理者目录。admin下的每一个文件分别对应LuCI管理者页面上的一级标题。每一个.lua文件都定义了index()函数作为菜单入口，在该函数中调用entry()函数，用于注册菜单项及向LuCI调度树注册页面菜单入口。

8.1.3 SF16A18-LuCI页面的构建

LuCI根据不同场景需求，提供了多种构建页面的方式，如template、call、cbi、form等，这些方式灵活地把Model、View、Controller、UCI配置关联起来，实际开发中可以根据需要灵活选择和组合。SF16A18-LuCI使用template和call方式实现了简洁的Web页面管理，把菜单、页面、页面逻辑、UCI配置关联起来。

SF16A18-LuCI的主要目录结构如下（见图8-2）。

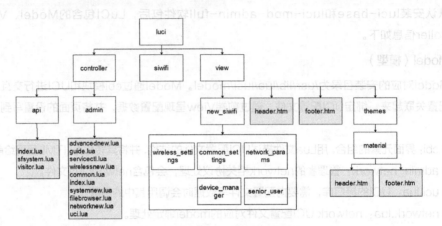

图8-2　SF16A18-LuCI主要目录结构

- api：负责提供API接口支持。
- siwifi：siwifi特有的库，提供云端/本地客户访问接口、获取设备信息等库。
- controller/admin/*.lua：目录下的.lua文件，提供前端页面获取系统信息的接口。这些文件会调用siwifi目录下的库。
- view/new_siwifi：页面HTML模板，包含5个子目录，按照名称和功能划分，通过controller的.lua文件指定页面加载的模板。
- 每个页面都包含view下的header.htm和footer.htm文件，而header.htm和footer.htm会加载themes/material目录下的header.htm和footer.htm文件。
- 其他没有列出的目录、文件与LuCI相同。

SF16A18-LuCI使用了Controller、View和Model uci.lua库来实现菜单、页面、配置三者之间的关联。图8-3列出了SF16A18-LuCI页面关联关系及引用的资源。

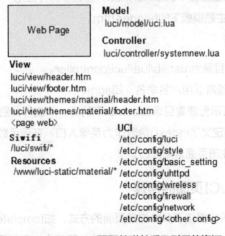

图8-3　SF16A18-LuCI页面关联关系及引用的资源

下面我们将结合SF16A18-LuCI的页面,来说明每个页面是如何实现这种关联的。

1. 登录页面

用浏览器打开Web配置服务器,接收用户连接请求时,首先打开/www/index.htm文件,在此文件中根据连接属性指定的路径执行cgi-bin/luci脚本。脚本会执行/usr/lib/lua/luci/sgi/cgi.lua中的run函数,这意味着LuCI正式启动。run()函数接收HTTP的请求,并创建协程httpdispatch。该协程调用dispatch()函数,用于调度以及初始化LuCI的各个部分(加载模板引擎、初始化多语言,构建菜单节点树等)。接下来会执行controller下的index.lua文件,该文件指定了登录页面的HTML文件为luci/view/sysauth.htm。

登录流程如图8-4所示。

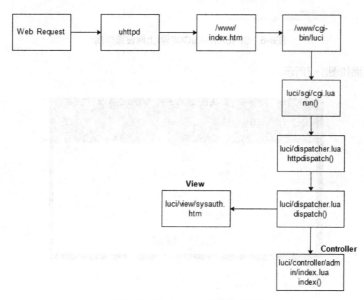

图8-4　SF16A18-LuCI登录流程

登录认证页面如图8-5所示。

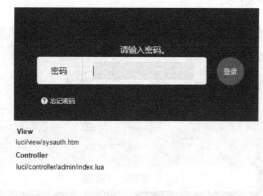

图8-5　SF16A18-LuCI登录认证页面

2. 向导页面

上网设置页面如图8-6所示。

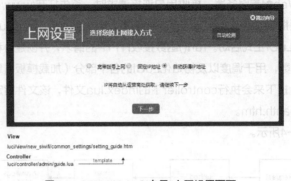

图8-6 SF16A18-LuCI向导-上网设置页面

无线设置页面如图8-7所示。

图8-7 SF16A18-LuCI向导-无线设置页面

设置完成页面如图8-8所示。

图8-8 SF16A18-LuCI向导-设置完成页面

3. 常用设置

"常用设置"中包含"连接设备管理""上网设置""无线设置"3个设置,这3个设置在同一个页面,根据选择的项目来隐藏其他项。

连接设备管理页面如图8-9所示。

图8-9　SF16A18-LuCI常用设置-连接设备管理页面

上网设置页面如图8-10所示。

图8-10　SF16A18-LuCI常用设置-上网设置页面

无线设置页面如图8-11所示。

图8-11 SF16A18-LuCI常用设置-无线设置页面

4. 高级设置

高级设置的所有功能菜单如图8-12所示。

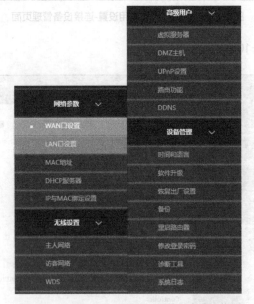

图8-12 SF16A18-LuCI高级设置的所有功能菜单

高级设置页面较多，我们不贴出具体页面，用表格列出页面的关联关系。高级设置功能页面的关联关系如表8-2~表8-5所示。

表 8-2 网络参数功能页面的关联关系

功能	构建方式	Controller	View	调用接口及函数	
WAN口设置	template	controller/admin/ networknew.lua	view/new_siwifi/network_ params/wan.html	controller/admin/ networknew.lua	get_wan
					set_wan
					set_pppoe_ advanced
LAN口设置	template	controller/admin/ networknew.lua	view/new_siwifi/network_ params/lan.htm	controller/admin/ networknew.lua	get_lan
					set_lan
MAC地址	template	controller/admin/ networknew.lua	view/new_siwifi/network_ params/mac.htm	controller/admin/ networknew.lua	get_mac
					set_lan
DHCP服务器	template	controller/admin/ networknew.lua	view/new_siwifi/network_ params/dhcp.htm	controller/admin/ networknew.lua	get_dhcp
					get_dhcp_ devices
					set_dhcp
IP与MAC绑定设置	template	controller/admin/ networknew.lua	view/new_siwifi/network_ params/ip_mac.htm	controller/admin/ networknew.lua	get_ip_mac_ bind_table
					get_ip_mac_ online_table

表 8-3 无线设置功能页面的关联关系

功能	构建方式	Controller	View	调用接口及函数	
主人网络	template	controller/admin/ wirelessnew.lua	view/new_siwifi/network_ params/owner.htm	controller/admin/ wirelessnew.lua	get_wifi_iface
					get_freq_ intergration
					set_wifi_iface
					set_freq_ intergration
访客网络	template	controller/admin/ wirelessnew.lua	view/new_siwifi/network_ params/visitor.htm	controller/admin/ wirelessnew.lua	get_customer_ wifi_iface
					set_customer_ wifi_iface
WDS	template	controller/admin/ wirelessnew.lua	view/new_siwifi/network_ params/wds.htm	controller/admin/ wirelessnew.lua	get_wds_info
					wds_disable
					wds_enable
					wds_getrelip
					wds
					wifi_scan
					wds_sta_is_ disconnected
					wifi_connect

表 8-4 高级用户功能页面的关联关系

功能	构建方式	Controller	View	调用接口及函数	
虚拟服务器	template	controller/admin/advancednew.lua	view/new_siwifi/senior_user/virtual_server.htm	controller/admin/advancednew.lua	get_virutal_server
					set_virutal_server
DMZ主机	template	controller/admin/advancednew.lua	view/new_siwifi/senior_user/dnz.htm	controller/admin/advancednew.lua	get_dmz_host
					set_dmz_host
UPnP设置	template	controller/admin/advancednew.lua	view/new_siwifi/senior_user/upnp.htm	controller/admin/advancednew.lua	get_UPnP
					set_UPnP
路由功能	template	controller/admin/advancednew.lua	view/new_siwifi/senior_user/router.htm	controller/admin/advancednew.lua	get_static_routing
					get_routing_table
					set_static_routing
DDNS	template	controller/admin/advancednew.lua	view/new_siwifi/senior_user/ddns.htm	controller/admin/advancednew.lua	get_ddns
					set_ddns

表 8-5 设备管理功能页面的关联关系

功能	构建方式	Controller	View	调用接口及函数	
时间和语言	template	controller/admin/systemnew.lua	view/new_siwifi/device_manager/time.htm	controller/admin/systemnew.lua	get_date
					set_date
					set_lang
					get_lang
软件升级	template	controller/admin/systemnew.lua	view/new_siwifi/device_manager/software_upgrade.htm	controller/admin/systemnew.lua	get_version
					upgrade
					ac_ota_upgrade
					ota_upgrade
					ap_upgrade
					ap_upgrade_check
恢复出厂设置	template	controller/admin/systemnew.lua	view/new_siwifi/device_manager/reset.htm	controller/admin/systemnew.lua	reset
备份	template	controller/admin/systemnew.lua	view/new_siwifi/device_manager/import_backup.htm	controller/admin/systemnew.lua	import_config
重启路由器	template	controller/admin/systemnew.lua	view/new_siwifi/device_manager/reboot.htm	controller/admin/systemnew.lua	restart
修改登录密码	template	controller/admin/systemnew.lua	view/new_siwifi/device_manager/modify_password.htm	controller/admin/systemnew.lua	set_password
诊断工具	template	controller/admin/systemnew.lua	view/new_siwifi/device_manager/debug.htm	controller/admin/systemnew.lua	start_diagnostic_tool
					get_diagnostic_result
					Stop_diagnostic_tool
系统日志	template	controller/admin/systemnew.lua	view/new_siwifi/device_manager/syslog.htm	controller/admin/systemnew.lua	get_log
					clean_log

8.2 界面的简易定制

开发板的路由器Web管理页面可以定制。我们可以根据用户需求自定义和修改路由器管理页面内部风格，包含管理页面的背景颜色、字体颜色和所有用到的图标，为用户量身打造一款属于自己的路由器管理页面。

8.2.1 修改图标

目前支持定制化的图标包含页面刷新加载图标（coverLoading.gif）、保存加载图标（saveLoading.gif）、企业logo图标（logo.png）以及其他页面中用到的小图标（sibasic.png）。这些文件放在路由器的/www/luci-static/material/images目录下。material是LuCI主题目录。

页面刷新加载图标（coverLoading.gif）

图片格式是GIF动画，分辨率是80像素×80像素，位深度是8位，默认样式如图8-13所示。

图8-13　默认的页面刷新加载图标

如果想更新该图片，要求新设计的图片大小、分辨率、位深度必须和该图片保持完全一致，重新设计的图片命名也必须和该图片完全一致。

保存加载图标（saveLoading.gif）

图片格式是GIF动画，分辨率是16像素×16像素，位深度是8位，默认样式如图8-14所示。

图8-14　保存加载图标

如果想更新该图片，要求新设计的图片大小、分辨率、位深度必须和该图片保持完全一致，重新设计的图片命名也必须和该图片完全一致。

企业logo图标（logo.png）

图片格式是PNG，分辨率是124像素×15像素，位深度是32位，默认样式如图8-15所示。

图8-15　企业logo图标

默认底色是透明色，字样为白色，为了易于分辨，把底色涂成黑色。如果想更新该图片，要求新设计的图片大小、分辨率、位深度必须和该图片保持完全一致，重新设计的图片命名也必须和该图片完全一致。

页面中用到的其他小图标（sibasic.png）

为了节省存储空间，其他小图标被放到一个PNG格式的图片文件中，分辨率是740像素×368像素，位深度是32位，默认种类、样式和排布如图8-16所示。

图8-16 其他小图标

该图标默认底色是透明色，里面有白色的图标，为了易于分辨，把底色涂成黑色。关于这些图标的使用位置，可以打开路由器的管理网页查看。用户如果想要更新其中部分或全部图标，需要在图8-16的基础上修改，新替换的图标在图片中的大小和位置必须保持完全一致，不然会显示异常。

8.2.2 修改字体和背景颜色

字体和背景颜色的修改是通过修改配置文件style实现的。配置文件路径为/etc/config/style。这个文件里面包含两部分，一部分是关于字体颜色的，另一部分是关于背景颜色的。

字体颜色修改说明

默认字体配置文件的内容如下。

```
#text color
config setting 'font'
#all text base
option base 'black'
#top nav text
option header 'white'
#bottom device info and technical support
option foot '#999'
#title
option title '#df0007'
#all button
option button 'white'
option input_disabled 'gray'
#help title
option help_and_wds 'white'
#help text
option help_text 'black'
#table top toolbar
option ToolBar '#8c000a'
#note tip notice
option note_tip '#fb6e52'
```

```
#common setting
#for red background , use white text
option com_base 'white'
#for dark background , use white text
option com_left 'white'
```

可以设置所有字体的颜色，内有注释，以#开头的均为注释。可以使用#ffffff代表白色，也可以使用#fff表示，或者用white、rgb（255,255,255）、rgba（255,255,255,1）等表示，具体规则与CSS的规则相同。字体选项说明如表8-6所示。

表8-6 字体选项说明

选项	说明
base	最基础的颜色，它几乎对所有字体生效，但会被覆盖，优先级最低
header	顶部导航栏
foot	底部的软件版本和技术支持热线的位置
title	高级设置中正文部分的标题
button	所有按钮，disabled表示无法使用的按钮
help	页面中的帮助窗口，有标题和正文
toolbar	表格上的工具栏，一般有删除、添加等
nav	导航栏，有顶部导航栏和左侧导航栏
notice	一些需要注意的文字
com_base	常用设置和设置向导页面中红色背景的文字，比如表头、错误提示
com_ left	常用设置中左侧暗色背景中的文字

背景颜色修改说明

默认配置文件的内容如下。

```
#background color
config setting 'background'
#bottom background
option html_and_body 'white'
#for all select and input
option select_input '#f4f4f4'
#for all disabled input
option input_disabled '#f5f5f5'
#for all button
option button '#df0007'
option button_disabled 'darkgray'
#top nav background
option header 'linear-gradient(red,#9b1b1b)'
#at bottom background, technical support
option foot '#dfdfdf'
#left nav background
option left_nav '#353535'
#help and wds div top background
option help_and_wds_Top '#df0007'
#help inner background
option Help 'white'
#common setting
```

```
#biggest background
option bigBg '#353535'
#right background at common setting
option rightBg 'white'
#for red background , use white text
option com_base '#df0007'
#login table background
option loginTable 'white'
```

我们可以设置所有背景色，并指出渐进色，内有注释。可以使用各种方法表示颜色，同字体颜色设置。背景选项说明如表8-7所示。

表 8-7 背景选项说明

选项	说明
html_and_body	设置页面最基本的背景色，仅对高级设置起作用
select_input	设置所有选择框和输入框的背景色
input_disable	设置无法使用的输入框的背景色
button	所有按钮背景色，disabled表示无法使用的按钮
header	顶部导航栏背景色
foot	底部的软件版本和技术支持热线的位置背景色
help	页面中的帮助窗口背景色
left_nav	设置高级设置中左侧导航栏的背景色
bigBg	常用设置和设置向导的背景色
rightBg	常用设置中右边主体的背景色
com_base	常用设置和设置向导页面中表头、错误提示等的背景色
login_table	登录页面的背景色

8.2.3 应用举例

image和style内容修改完成后，可以先通过SSH登录路由器后台，临时替换管理页面风格，查看效果是否是预期的。打开路由器管理页面后，通过SSH登录路由器，使用SCP工具将修改后的image目录和style文件传到路由器的相应位置，以替换原文件，然后按Ctrl+F5刷新网页来查看效果。文件位置如下。

图片路径：/www/luci-static/material/images/

配置文件路径：/etc/config/style

操作步骤：

1. 按照之前章节的说明，根据设计内容，创建images文件和修改style文件。
2. 通过WinSCP工具用设计的文件替换路由器后台对应的文件。

（1）打开WinSCP工具，单击"新建站点"，弹出对话框，"文件协议"选择SCP，"主机名"为192.168.4.1，"端口号"为22，"用户名"为admin，"密码"为admin，如图8-17所示。

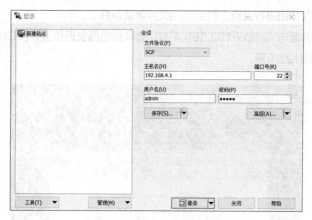

图8-17 使用WinSCP新建站点

（2）单击"登录"，登录路由器，进入SCP模式，进入本地修改好的images目录，同时在路由器后台目录上找到images目录所在的目录，将本地整个images目录拖到路由器的目录进行上传，如图8-18所示。

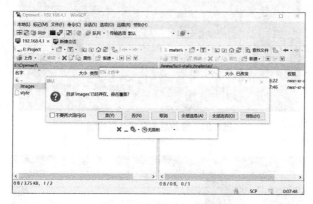

图8-18 上传images目录

（3）单击"是"按钮进行替换，然后images目录就替换好了。

（4）进入本地修改好的style文件所在的目录，同时在路由器后台目录找到style文件所在的目录，将本地style文件拖到路由器的目录进行上传，如图8-19所示。

图8-19 上传style文件

（5）单击"是"按钮进行替换，然后style文件就替换好了。

3. 用计算机浏览器访问http://192.168.4.1/，登录路由器管理页面，按Ctrl+F5查看修改后的效果，如图8-20~图8-22所示。

图8-20 修改后的上网设置页面

图8-21 修改后的登录认证页面

图8-22 修改后的连接设备管理页面

9 工具与命令

9.1 iPerf

网络开发中经常要测试两个网络节点之间的通信性能、服务质量（QoS）和带宽，这些参数包括带宽、时延、抖动和丢包率。能完成这类测试的工具有很多，测试方法也多种多样。在OpenWrt中，推荐使用iPerf来进行网络带宽测试。iPerf是一个开源并且跨平台的专业网络测试工具，它使用客户端-服务器模型，能够测量两个网络节点之间TCP和UDP带宽性能，还能提供带宽、网络延迟、丢包率等统计信息，可以用来测试一些网络设备，如路由器、防火墙、交换机等的性能。

iPerf目前有2个版本：iPerf2和iPerf3。iPerf3在iPerf2的基础上新增了一些功能，如设置拥塞控制算法（-C选项）、设置IPv6流标签（-L选项）、以JSON格式输出、进行磁盘读取/写入测试等。iPerf3不兼容iPerf2，iPerf3和iPerf2安装后所执行的命令名字也不一样，在使用时需要注意，确保测试两端的软件版本一致。

9.1.1 iPerf命令

开发板使用的iPerf版本为2.0.10。

```
DF1A:$ iperf -v
iperf version 2.0.10 (11 Aug 2017) pthreads
```

iPerf是基于客户端-服务器模式工作的，因此，要使用iPerf测试带宽，需要建立一个客户端和一个服务器。iPerf客户端和服务器，都使用同一个命令，只不过启动命令的选项不同而已。通过"iperf -h"可以查看支持的命令：

```
DF1A:$ iperf -h
Usage: iperf [-s|-c host] [options]
       iperf [-h|--help] [-v|--version]
Client/Server:
-b, --bandwidth #[kmgKMG | pps] bandwidth to send at in bits/sec or packets per second
-e, --enhancedreports   use enhanced reporting giving more tcp/udp and traffic information
-f, --format    [kmgKMG]   format to report: Kbits, Mbits, KBytes, MBytes
-i, --interval  #          seconds between periodic bandwidth reports
 -l, --len      #[kmKM]    length of buffer in bytes to read or write (Defaults: TCP=128K, v4 UDP=1470, v6 UDP=1450)
```

```
  -m, --print_mss               print TCP maximum segment size (MTU - TCP/IP header)
  -o, --output      <filename>  output the report or error message to this specified file
  -p, --port        #           server port to listen on/connect to
  -u, --udp                     use UDP rather than TCP
      --udp-counters-64bit      use 64 bit sequence numbers with UDP
  -w, --window      #[KM]       TCP window size (socket buffer size)
  -z, --realtime                request realtime scheduler
  -B, --bind        <host>      bind to <host>, an interface or multicast address
  -C, --compatibility           for use with older versions does not sent extra msgs
  -M, --mss         #           set TCP maximum segment size (MTU - 40 bytes)
  -N, --nodelay                 set TCP no delay, disabling Nagle's Algorithm
  -S, --tos         #           set the socket's IP_TOS (byte) field
Server specific:
  -s, --server                  run in server mode
  -t, --time        #           time in seconds to listen for new connections as well
as to receive traffic (default not set)
  -U, --single_udp              run in single threaded UDP mode
  -D, --daemon                  run the server as a daemon
  -V, --ipv6_domain             Enable IPv6 reception by setting the domain and socket to
AF_INET6 (Can receive on both IPv4 and IPv6)
Client specific:
  -c, --client      <host>      run in client mode, connecting to <host>
  -d, --dualtest                Do a bidirectional test simultaneously
  -n, --num         #[kmgKMG]   number of bytes to transmit (instead of -t)
  -r, --tradeoff                Do a bidirectional test individually
  -t, --time        #           time in seconds to transmit for (default 10 secs)
  -B, --bind [<ip> | <ip:port>] bind src addr(s) from which to originate traffic
  -F, --fileinput <name>        input the data to be transmitted from a file
  -I, --stdin                   input the data to be transmitted from stdin
  -L, --listenport  #           port to receive bidirectional tests back on
  -P, --parallel    #           number of parallel client threads to run
  -R, --reverse                 reverse the test (client receives, server sends)
  -T, --ttl         #           time-to-live, for multicast (default 1)
  -V, --ipv6_domain             Set the domain to IPv6 (send packets over IPv6)
  -X, --peer-detect             perform server version detection and version exchange
  -Z, --linux-congestion <algo> set TCP congestion control algorithm (Linux only)
Miscellaneous:
  -x, --reportexclude [CDMSV]   exclude C(connection) D(data) M(multicast)
S(settings) V(server) reports
  -y, --reportstyle C           report as a Comma-Separated Values
  -h, --help                    print this message and quit
  -v, --version                 print version information and quit
[kmgKMG] Indicates options that support a k,m,g,K,M or G suffix
Lowercase format characters are 10^3 based and uppercase are 2^n based
(e.g. 1k = 1000, 1K = 1024, 1m = 1,000,000 and 1M = 1,048,576)
The TCP window size option can be set by the environment variable
TCP_WINDOW_SIZE. Most other options can be set by an environment variable
IPERF_<long option name>, such as IPERF_BANDWIDTH.
```

iperf命令参数说明如表9-1所示。

表 9-1　iperf 命令参数说明

命令行选项	说明
客户端/服务器通用选项	
-b, --bandwidth #[kmgKMG | pps]	以bit/s或packet/s为单位发送带宽
-e, --enhancedreports	提供更多TCP/UDP和流量信息
-f, --format　[kmgKMG]	指定打印带宽数的格式。支持的格式包括以下几种 'k'：kbit/s 'K'：KB/s 'm'：Mbit/s 'M'：MB/s 'g'：Gbit/s 'G'：GB/s
-i, --interval #	设置报告之间的时间间隔，单位为秒。如果设置为非零值，就会按照此时间间隔输出测试报告。默认值为0
-l, --len　#[kmKM]	设置读写缓冲区的长度。TCP默认值为128KB，v4 UDP默认为1470字节，v6 UDP默认为1450字节
-m, --print_mss	输出TCP MSS值（通过TCP_MAXSEG支持）。MSS值一般比MTU值小40字节
-o, --output　<filename>	将报告或错误信息输出到指定的文件中
-p, --port　#	要侦听/连接到的服务器端口，默认是5001端口
-u, --udp　--udp-counters-64bit	使用UDP方式而不是TCP方式，参看-b选项。UDP中使用64位序列号
-w, --window　#[KM]	设置套接字缓冲区为指定大小，设置TCP窗口大小
-z, --realtime	请求实时调度
-B, --bind　<host>	绑定到主机地址或接口、组播地址（当主机有多个地址或接口时使用该参数）
-C, --compatibility	与低版本的iPerf一起使用时，可以使用兼容模式
-M, --mss　#	通过TCP_MAXSEG选项尝试设置TCP最大信息段的值。MSS值的大小通常是TCP/IP头减去40字节
-N, --nodelay	设置TCP无延迟选项，禁用Nagle's运算法则
-S, --tos　#	设置套接字的IP_TOS（字节）字段
服务器专用参数	
-s, --server	以服务器模式运行
-t, --time　#	侦听新连接和接收流量的时间，单位为秒，默认值未设置
-U, --single_udp	以单线程UDP模式运行
-D, --daemon	以daemon形式运行服务器
-V, --ipv6_domain	通过将域和套接字设置为AF_INET6来启用IPv6接收（可以在IPv4和IPv6上接收）
客户端专用参数	
-c, --client　<host>	以客户端模式运行，连接到指定host服务器的地址
-d, --dualtest	运行双测试模式。这将使服务器端反向连接到客户端，使用-L参数中指定的端口（或默认使用客户端连接到服务器端的端口）
-n, --num　#[kmgKMG]	传送的字节数
-B, --bind [<ip> | <ip:port>]	绑定源流量的src地址
-F, --fileinput <name>	从文件中输入传输的数据
-I, --stdin	从stdin中输入传输的数据
-L, --listenport #	指定服务端反向连接到客户端时使用的端口
-P, --parallel #	线程数，指定客户端与服务端之间使用的线程数，默认只运行一个线程
-R, --reverse	反向测试（客户端接收，服务器发送）

续表

命令行选项	说明
-T, --ttl #	多播生存时间,默认值为1
-V, --ipv6_domain	将域设置为IPv6(通过IPv6发送数据包)
-X, --peer-detect	执行服务器版本检测和版本交换
-Z, --linux-congestion <algo>	设置TCP拥塞控制算法(仅限Linux)
其他参数	
-x, --reportexclude [CDMSV]	报告中不包含C(connection)、D(data)、M(multicast)、S(settings)、V(server)
-y, --reportstyle C	以逗号分隔的值报告
-h, --help	帮助
-v, --version	查看iPerf版本

9.1.2 调整TCP连接

使用iPerf测试TCP连接时,需要根据TCP连接状况,调整TCP窗口大小(TCP window size),TCP窗口太小或太大,都会影响TCP的性能和测试结果。所以测试TCP连接时,首先要计算TCP窗口的大小。在给出TCP窗口大小的计算公式之前,我们先来了解一些通信术语。

● 往返路程时间(round-trip time,RTT):指在双方通信中,发送方的信号传播到接收方的时间,加上接收方回传消息到发送方的时间。

● 带宽时延乘积(bandwidth-delay product,BDP):这是一个数据链路的传输能力(bit/s)与往返路程时间(单位为秒)的乘积,即 $BDP=BW \times RTT$。

● TCP窗口大小(TCP window size,WIN SIZE):用于网络数据传输时的流量控制,以避免拥塞的发生,属于TCP协议的内容。对于接收端,它表示接收端当前允许发送端发送的字节数,通常也称为接收端窗口大小,简称RWIN;对于发送端,它表示发送端当前允许发送的字节数,通常也称为发送端窗口大小,简称TWIN。

● 带宽(bandwidth,BW):网络的吞吐量,端到端之间可以传输的最大速率。如百兆以太网理论带宽为100Mbit/s,千兆以太网理论带宽为1Gbit/s。带宽通常用吞吐量来表示,如1秒内传输的bit的数量。

● 丢包(loss):如果线路带宽占用过高,数据包从一端传输到另外一端的途中,会产生丢失。丢包率是网络的一个重要指标,一旦产生丢包,说明网络带宽不足,系统会自动重传,导致继续劣化,所以丢包率超过10%以后,网络质量会迅速劣化。

● 抖动(jitter):如果延时不稳定,忽快忽慢,网络就还存在抖动。在VoIP、视频会议等场合,抖动是一个非常重要的指标。

带宽时延乘积及带宽、TCP窗口理论计算公式如下。

(1)带宽时延乘积: $BDP=BW \times RTT$

其中,BDP为带宽时延乘积,BW为带宽,RTT为往返路程时间。

(2)带宽(TCP吞吐量): $BW \leq WIN\ Size/RTT$

其中,$WIN\ Size$为TCP窗口大小。

(3)TCP窗口大小: $WIN\ Size \geq BW \times RTT$

在实际测试中，可以以公式计算得到的TCP窗口大小为基准，增大或者减小TCP窗口大小，得到性能的提升。

下面我们在开发板中，查看和调整TCP窗口大小。

1. 查看TCP窗口大小

在OpenWrt中，以服务器方式运行iPerf。查看到默认TCP窗口大小为85.3KB（与使用的操作系统有关）。

```
DF1A:$ iperf -s
------------------------------------------------------------
Server listening on TCP port 5001
TCP window size: 85.3 KByte (default)
------------------------------------------------------------
```

2. 调整TCP窗口大小

调整TCP窗口大小为100KB。但是iPerf将TCP窗口大小设置为200KB，是要求的2倍。

```
DF1A:$ iperf -s -w 100K
------------------------------------------------------------
Server listening on TCP port 5001
TCP window size:  200 KByte (WARNING: requested  100 KByte)
------------------------------------------------------------
```

调整TCP窗口大小为0KB。iPerf对设置的参数提出了警告，然后保持了默认的85.3KB。

```
DF1A:$ iperf -s -w 0K
WARNING: TCP window size set to 0 bytes. A small window size
will give poor performance. See the Iperf documentation.
------------------------------------------------------------
Server listening on TCP port 5001
TCP window size: 85.3 KByte (default)
------------------------------------------------------------
```

调整TCP窗口大小为150KB、160KB和1MB。我们发现只要设置超过160KB，TCP窗口大小都不会超过320KB，似乎320KB是iPerf所在系统的最大值。

```
DF1A:$ iperf -s -w 150K
------------------------------------------------------------
Server listening on TCP port 5001
TCP window size:  300 KByte (WARNING: requested  150 KByte)
------------------------------------------------------------
DF1A:$ iperf -s -w 160K
------------------------------------------------------------
Server listening on TCP port 5001
TCP window size:  320 KByte (WARNING: requested  160 KByte)
------------------------------------------------------------
DF1A:$ iperf -s -w 1M
------------------------------------------------------------
Server listening on TCP port 5001
TCP window size:  320 KByte (WARNING: requested 1.00 MByte)
------------------------------------------------------------
```

从上面的测试实验可以看出，通过iPerf设置TCP窗口大小，iPerf只是把这个参数作为建议传递给操作系统，而操作系统会根据当前系统限制给出实际的TCP窗口大小的调整值。

通常来说，操作系统对TCP窗口大小有范围限制。对于目前使用的OpenWrt系统，操作系统允许的TCP窗口大小范围为[2KB，320KB]。

- 对于没有超过系统限制的情况，通常操作系统会把TCP窗口大小设定为指定参数的2倍，如参数为100KB，实际设置为200KB。
- 对于超过最大系统限制的情况，无论参数设置为多少，系统都会将TCP窗口大小设置为系统限制的最大值，如320KB。
- 对于设置值小于系统限制的情况，系统会根据参数将TCP窗口大小调整到系统允许范围内，如参数为1KB，实际设置为2.18KB。

修改TCP/IP协议栈参数

如果要修改OpenWrt TCP/IP协议栈的参数，就需要对内核网络参数进行修改。具体可以参考Linux内核文档中的Documentation/networking/目录。在OpenWrt系统中，通过修改/proc/sys/net/ipv4/、/proc/sys/net/core/、/etc/sysctl.conf参数，可以达到修改内核TCP/IP协议栈参数的目的。

常用可修改内核TCP/IP协议栈的参数如表9-2所示。

表9-2 常用可修改内核TCP/IP协议栈的参数

参数	描述	/etc/sysctl.conf 对应值	默认值（字节）
/proc/sys/net/core/rmem_default	默认的TCP数据接收窗口大小	net.core.rmem_default	163840
/proc/sys/net/core/rmem_max	最大的TCP数据接收窗口大小	net.core.rmem_max	163840
/proc/sys/net/core/wmem_default	默认的TCP数据发送窗口大小	net.core.wmem_default	163840
/proc/sys/net/core/wmem_max	最大的TCP数据发送窗口大小	net.core.wmem_max	163840
/proc/sys/net/ipv4/tcp_rmem	TCP套接字使用的接收缓冲区的设置，包含3个值，分别为最小接收缓冲区大小、接收缓冲区的初始大小、允许的最大接收缓冲区大小	net.ipv4.tcp_rmem	4096、87380、2018944
/proc/sys/net/ipv4/tcp_wmem	TCP套接字使用的写入缓冲区的设置，包含3个值，分别为最小写入缓冲区大小、写入缓冲区的初始大小、允许的最大写入缓冲区大小	net.ipv4.tcp_wmem	4096、16384、2018944
/proc/sys/net/ipv4/tcp_window_scaling	启用RFC 1323定义的Window scaling（窗口缩放）。要支持超过64KB的窗口，必须启用该值。16bit最大支持到 2^{16}=65 535bit（64KB），如果要支持超过上限，必须将此参数设置为1	net.ipv4.tcp_window_scaling	1

修改参数的方法，主要有两种。

- 通过"echo value"追加到文件里，如echo "1" > /proc/sys/net/ipv4/tcp_window_scaling，但使用这种方法，设备重启后，参数又会恢复为默认值。
- 把参数添加到/etc/sysctl.conf中，然后执行"sysctl -p"使参数永久生效。

修改/proc/sys/net/core/rmem_max的值可以扩大TCP接收窗口大小（作为服务器），将其修改为2MB。修改/proc/sys/net/core/wmem_max的值可以扩大TCP发送窗口大小（作为客户端）。下面以修改TCP接收窗口大小为例进行演示：

```
DF1A:$ echo 2097152 > /proc/sys/net/core/rmem_max
DF1A:$ iperf -s -w 2.1M
------------------------------------------------------------
Server listening on TCP port 5001
TCP window size: 4.00 MByte (WARNING: requested 2.10 MByte)
------------------------------------------------------------
DF1A:$ iperf -s -w 2M
------------------------------------------------------------
Server listening on TCP port 5001
TCP window size: 4.00 MByte (WARNING: requested 2.00 MByte)
------------------------------------------------------------
```

9.1.3 iPerf的安装

在OpenWrt下安装

在OpenWrt下,直接通过opkg命令安装iPerf。

```
DF1A:$ opkg install iperf
Package iperf (2.0.10-1) installed in root is up to date.
```

在Ubuntu下安装

为了保证版本和OpenWrt上的版本一致,需要安装iPerf 2.0.5版本。

```
HOST:~$ sudo apt-get install iperf
...
HOST:~$ iperf -v
iperf version 2.0.5 (2 June 2018) pthreads
```

在CentOS下安装

安装iPerf 2.0.5版本。

```
HOST:~$ sudo yum install iperf.x86_64
HOST:~$ iperf -v
iperf version 2.0.5 (08 Jul 2010) pthreads
```

在Windows下安装

我们下载iperf-2.0.9-win64.zip,下载后无须安装,直接解压即可(见图9-1)。

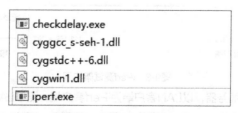

图9-1 在Windows下可直接解压使用iPerf

通过Windows CMD(命令提示符)启动iperf.exe就可以使用(见图9-2)。

```
C:\Users\Administrator\Downloads\Compressed\iperf-2.0.9-win64>iperf.exe -v
iperf version 2.0.9 (1 June 2016) pthreads

C:\Users\Administrator\Downloads\Compressed\iperf-2.0.9-win64>iperf.exe -h
Usage: iperf [-s|-c host] [options]
       iperf [-h|--help] [-v|--version]

Client/Server:
  -b, --bandwidth #[KMG | pps]  bandwidth to send at in bits/sec or packets per second
  -e, --enhancedreports    use enhanced reporting giving more tcp/udp and traffic information
  -f, --format    [kmKM]   format to report: Kbits, Mbits, KBytes, MBytes
  -i, --interval  #        seconds between periodic bandwidth reports
  -l, --len       #[KM]    length of buffer to read or write (default 8 KB)
  -m, --print_mss          print TCP maximum segment size (MTU - TCP/IP header)
  -o, --output    <filename> output the report or error message to this specified file
  -p, --port      #        server port to listen on/connect to
  -u, --udp                use UDP rather than TCP
  -w, --window    #[KM]    TCP window size (socket buffer size)
  -z, --realtime           request realtime scheduler
  -B, --bind      <host>   bind to <host>, an interface or multicast address
  -C, --compatibility      for use with older versions does not sent extra msgs
  -M, --mss       #        set TCP maximum segment size (MTU - 40 bytes)
  -N, --nodelay            set TCP no delay, disabling Nagle's Algorithm
  -V, --ipv6_domain        Set the domain to IPv6
```

图9-2 通过Windows CMD运行iPerf

9.1.4 iPerf使用实例

下面以一个实例来说明iPerf的使用方法和常用的路由器网络测试方法。iPerf具有Window客户端，也有Linux客户端。我们这里使用Windows iPerf作为客户端。大家实际使用时，只要注意两端iPerf的版本匹配，可以自由选择系统平台。

在使用iPerf时，需要建立一个服务器用于丢弃流量，建立一个客户端用于产生流量。下面我们将使用iPerf的TCP和UDP的两种模式对链路进行测试。

1. 通过TCP模式可以测试吞吐量。以开发板（OpenWrt）和LAN客户端（有线客户端或无线客户端）互为服务器和客户端，进行有线、无线网络上行和下行流量测试（见图9-3）。

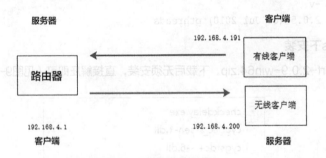

图9-3 iPerf测试架构图

- 以开发板端为iPerf服务器，以LAN客户端为iPerf客户端，测试LAN客户端的上行流量。
- 以开发板端为iPerf客户端，以LAN客户端为iPerf服务器，测试LAN客户端的下行流量。

2. 通过UDP模式可以测试数据包的吞吐量、丢包率和延迟指标。

测试有线TCP上行带宽

以开发板为iPerf服务器，以有线客户端为iPerf客户端，有线客户端通过LAN口（100Mbit/s）有线连接到开发板上，测试有线客户端的上行流量（见图9-4）。通过ping命令得出

RTT=1ms=0.001s（实际为0.5ms，为了计算方便，我们使用1ms），因为我们的LAN网卡网速为100Mbit/s，BW=100Mbit/s，$RWIN\ size=BW\times RTT/8$=100 000 000×0.001/8=12500（byte）=12.5（MB），所以将TCP窗口大小设置为12.5MB。

图9-4　iPerf测试TCP有线网络上行流量

（1）在开发板端，以服务器模式启动iPerf。修改服务器接收窗口大小为12.5MB，启动服务器，默认端口为5001，监听客户端连接（见图9-5）。

```
DF1A:$ iperf -s -w 6.25MB -i 3 -e
-s：以服务器方式运行。
-w：设置TCP窗口大小为6.25MB，系统实际会将其设置为12.5MB。
-i：设置每3s报告一次结果。
-e：提供更多流量信息。
```

```
admin@SiWiFi3d04:/# echo 13107200 > /proc/sys/net/core/rmem_max
admin@SiWiFi3d04:/# cat /proc/sys/net/core/rmem_max
13107200
admin@SiWiFi3d04:/# iperf -s -w 6.25MB -i 3 -e
------------------------------------------------------------
Server listening on TCP port 5001 with pid 10447
Read buffer size:  128 KByte
TCP window size: 12.5 MByte (WARNING: requested 6.25 MByte)
------------------------------------------------------------
```

图9-5　开发板端修改TCP/IP内核参数并启动iPerf服务器

（2）在有线客户端，以客户端模式启动iPerf（见图9-6）。

```
D:\iperf-2.0.9-win64>iperf.exe -c 192.168.4.1 -w 12.5M -i 3 n 500M -e
-c：连接服务器的IP地址。
-w：设置TCP窗口大小为12.5MB，客户端为Window 10系统，默认支持12.5MB TCP窗口大小，不用修改系统网络参数。
-i：设置每3s报告一次结果。
-n 500M：传输500MB数据。
-e：提供更多流量信息。
```

图9-6 在Windows有线客户端启动iPerf客户端

（3）随着客户端连接进来，服务器也进行了输出（见图9-7）。可以按Ctrl+C组合键中断输出。

图9-7 服务器输出测试结果

可以看到服务器每隔3s输出数据，一共传输500MB数据（见Transfer列），传输了44.29s（见Interval列），带宽为94.7Mbit/s（见Bandwidth列），符合百兆带宽标准。

测试有线TCP下行带宽

以有线客户端为服务器端，以路由器为客户端，进行有线客户端下行带宽测试（见图9-8）。

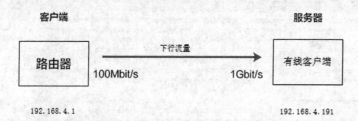

图9-8　iPerf测试TCP有线网络下行流量

（1）在有线客户端，以服务器模式启动iPerf（见图9-9）。

```
D:\iperf-2.0.9-win64>iperf.exe -s -w 12.5M -i 3 -e
```
-s：以服务器方式运行。
-w：设置TCP窗口大小为12.5MB。
-i：设置每3s报告一次结果。
-e：提供更多流量信息。

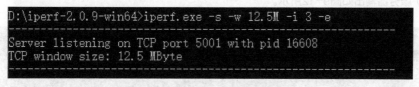

图9-9　在Windows下启动iPerf服务端

（2）在开发板端，以客户端模式启动iPerf。

```
DF1A:$ iperf -c 192.168.4.191 -w 6.25M -i 3 n 500M -e
```
-c：连接服务器的IP地址。
-w：设置TCP窗口大小为6.25MB，实际为12.5MB。
-i：设置每3s报告一次结果。
-n 500M：传输500MB数据。
-e：提供更多流量信息。

先更改wmem_max的大小（见图9-10）。

```
DF1A:~$ echo 13107200 > /proc/sys/net/core/wmem_max
DF1A:~$ cat /proc/sys/net/core/wmem_max
13107200
```

图9-10　在开发板端修改TCP/IP内核参数

然后启动iPerf客户端（见图9-11）。

```
DF1A:~$ iperf -c 192.168.4.191 -w 6.25M -i 3 -n 500M -e
------------------------------------------------------------
Client connecting to 192.168.4.191, TCP port 5001 with pid 8615
Write buffer size:  128 KByte
TCP window size: 12.5 MByte (WARNING: requested 6.25 MByte)
------------------------------------------------------------
[  3] local 192.168.4.1 port 33864 connected with 192.168.4.191 port 5001
[ ID] Interval       Transfer     Bandwidth       Write/Err  Rtry     Cwnd/RTT
[  3] 0.00-3.00 sec  41.3 MBytes  115 Mbits/sec   330/0        0      255K/8399 us
[ ID] Interval       Transfer     Bandwidth       Write/Err  Rtry     Cwnd/RTT
[  3] 3.00-6.00 sec  32.3 MBytes  90.2 Mbits/sec  258/0        0      255K/10189 us
[ ID] Interval       Transfer     Bandwidth       Write/Err  Rtry     Cwnd/RTT
[  3] 6.00-9.00 sec  34.0 MBytes  95.1 Mbits/sec  272/0        0      255K/10245 us
[ ID] Interval       Transfer     Bandwidth       Write/Err  Rtry     Cwnd/RTT
[  3] 9.00-12.00 sec 34.5 MBytes  96.5 Mbits/sec  276/0        0      255K/9470 us
[ ID] Interval       Transfer     Bandwidth       Write/Err  Rtry     Cwnd/RTT
[  3] 12.00-15.00 sec 34.5 MBytes 96.5 Mbits/sec  276/0        0      255K/9768 us
[ ID] Interval       Transfer     Bandwidth       Write/Err  Rtry     Cwnd/RTT
[  3] 15.00-18.00 sec 34.4 MBytes 96.1 Mbits/sec  275/0        0      255K/9741 us
[ ID] Interval       Transfer     Bandwidth       Write/Err  Rtry     Cwnd/RTT
[  3] 18.00-21.00 sec 34.3 MBytes 95.8 Mbits/sec  274/0        0      255K/9924 us
[ ID] Interval       Transfer     Bandwidth       Write/Err  Rtry     Cwnd/RTT
[  3] 21.00-24.00 sec 32.1 MBytes 89.8 Mbits/sec  257/0        0      255K/9692 us
[ ID] Interval       Transfer     Bandwidth       Write/Err  Rtry     Cwnd/RTT
[  3] 24.00-27.00 sec 35.8 MBytes 100 Mbits/sec   286/0        0      255K/9820 us
[ ID] Interval       Transfer     Bandwidth       Write/Err  Rtry     Cwnd/RTT
[  3] 27.00-30.00 sec 32.1 MBytes 89.8 Mbits/sec  257/0        0      255K/9313 us
```

图9-11　在开发板端启动iPerf客户端

（3）服务器输出如图9-12所示。

```
D:\iperf-2.0.9-win64>iperf.exe -s -w 12.5M -i 3 -e
------------------------------------------------------------
Server listening on TCP port 5001 with pid 15452
TCP window size: 12.5 MByte
------------------------------------------------------------
[  4] local 192.168.4.191 port 5001 connected with 192.168.4.1 port 44274
[ ID] Interval       Transfer     Bandwidth       Reads    Dist(bin=16.0K)
[  4] 0.00-3.00 sec  33.9 MBytes  94.9 Mbits/sec  8710     8710:0:0:0:0:0:0:0
[  4] 3.00-6.00 sec  33.9 MBytes  94.9 Mbits/sec  8777     8777:0:0:0:0:0:0:0
[  4] 6.00-9.00 sec  33.9 MBytes  94.9 Mbits/sec  8747     8747:0:0:0:0:0:0:0
[  4] 9.00-12.00 sec 34.0 MBytes  94.9 Mbits/sec  8725     8725:0:0:0:0:0:0:0
[  4] 12.00-15.00 sec 33.9 MBytes 94.9 Mbits/sec  8676     8676:0:0:0:0:0:0:0
[  4] 15.00-18.00 sec 33.9 MBytes 94.9 Mbits/sec  8761     8761:0:0:0:0:0:0:0
[  4] 18.00-21.00 sec 33.9 MBytes 94.9 Mbits/sec  8766     8766:0:0:0:0:0:0:0
[  4] 21.00-24.00 sec 33.9 MBytes 94.9 Mbits/sec  8687     8686:0:0:1:0:0:0:0
[  4] 24.00-27.00 sec 33.9 MBytes 94.9 Mbits/sec  8442     8441:0:0:1:0:0:0:0
[  4] 27.00-30.00 sec 34.0 MBytes 94.9 Mbits/sec  8730     8730:0:0:0:0:0:0:0
[  4] 30.00-33.00 sec 33.9 MBytes 94.9 Mbits/sec  8651     8651:0:0:0:0:0:0:0
[  4] 33.00-36.00 sec 33.9 MBytes 94.9 Mbits/sec  8640     8639:0:1:0:0:0:0:0
[  4] 36.00-39.00 sec 33.9 MBytes 94.9 Mbits/sec  8669     8666:2:0:1:0:0:0:0
[  4] 39.00-42.00 sec 34.0 MBytes 94.9 Mbits/sec  8543     8543:0:0:0:0:0:0:0
[  4] 0.00-44.19 sec 500 MBytes   94.9 Mbits/sec  127836   127829:3:1:2:1:0:0:0
```

图9-12　服务器输出测试结果

可以看到服务器每隔3s输出数据，一共传输500MB数据（见Transfer列），传输时间为44.19s（见Interval列），带宽为94.9Mbit/s（见Bandwidth列），符合百兆带宽标准。

测试有线UDP的丢包和延迟抖动

iPerf可以用于测试UDP数据包吞吐量、丢包率和延迟指标。UDP协议是一个非面向连接的轻量级传输协议，并且不提供可靠的数据传输服务，因此对UDP应用的关注点不是数据传输有多快，而是它的丢包率和延迟指标。

以开发板为iPerf服务器端，以有线客户端为iPerf客户端，通过iPerf的-u参数，服务器和客户端就可以使用UDP模式进行测试。

(1)在开发板端,以服务器模式启动iPerf(见图9-13)。

```
DF1A:$ iperf -s -u -w 6.25MB -i 3
```
-s:以服务器方式运行。
-u:UDP方式。
-w:设定buffer size,在进行UDP测试时,这个选项用来设置UDP Buffer Size(这里设置为12.5MB,默认为160KB)。
-i:设置每3s报告一次结果。

```
Server listening on UDP port 5001
Receiving 1470 byte datagrams
UDP buffer size: 12.5 MByte (WARNING: requested 6.25 MByte)
```

图9-13 开发板端作为UDP iPerf服务器

(2)有线客户端以客户端模式启动iPerf。从客户端到服务器之间的链路的理论带宽为100Mbit/s,用"-b 100M"参数进行测试(见图9-14)。

```
D:\iperf-2.0.9-win64>iperf.exe -c 192.168.4.1 -u -w 12.5M -b 100M -i 3 -t 20 -e
```
-c:连接服务器的IP地址。
-u:UDP方式。
-w:设置UDP Buffer Size为12.5MB。
-b:设置带宽(单位为bit/s),一般设置为测试的网络带宽的最大值。
-i:设置每3秒报告一次结果。
-t 20:一共测试20s。
-e:提供更多流量信息。

```
D:\iperf-2.0.9-win64>iperf.exe -c 192.168.4.1 -u -w 12.5M -b 100M -i 3 -t 20 -e
------------------------------------------------------------
Client connecting to 192.168.4.1, UDP port 5001 with pid 12280
Sending 1470 byte datagrams, IPG target: 117.60 us (kalman adjust)
UDP buffer size: 12.5 MByte
------------------------------------------------------------
[  3] local 192.168.4.191 port 59969 connected with 192.168.4.1 port 5001
[ ID] Interval       Transfer     Bandwidth        PPS
[  3]  0.00-3.00 sec  35.8 MBytes  100 Mbits/sec 8502 pps
[  3]  3.00-6.00 sec  35.8 MBytes  100 Mbits/sec 8503 pps
[  3]  6.00-9.00 sec  35.8 MBytes  100 Mbits/sec 8504 pps
[  3]  9.00-12.00 sec 35.8 MBytes  100 Mbits/sec 8503 pps
[  3] 12.00-15.00 sec 35.8 MBytes  100 Mbits/sec 8504 pps
[  3] 15.00-18.00 sec 35.8 MBytes  100 Mbits/sec 8503 pps
[  3]  0.00-20.00 sec  238 MBytes  100 Mbits/sec 8503 pps
[  3] Sent 170067 datagrams
```

图9-14 有线客户端以客户端模式启动iPerf

(3)服务器输出如图9-15所示。

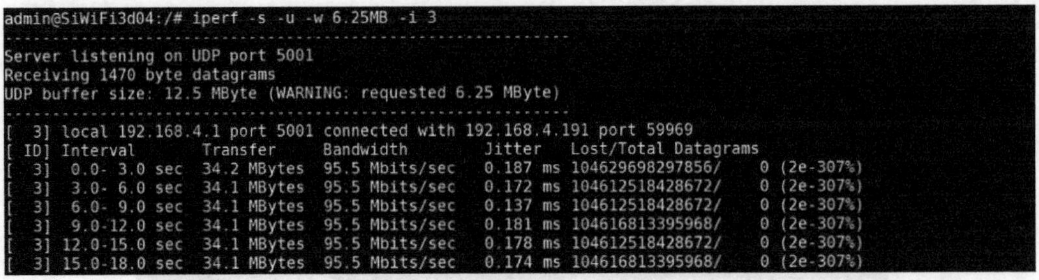

```
admin@SiWiFi3d04:/# iperf -s -u -w 6.25MB -i 3
------------------------------------------------------------
Server listening on UDP port 5001
Receiving 1470 byte datagrams
UDP buffer size: 12.5 MByte (WARNING: requested 6.25 MByte)
------------------------------------------------------------
[ 3] local 192.168.4.1 port 5001 connected with 192.168.4.191 port 59969
[ ID] Interval      Transfer     Bandwidth       Jitter   Lost/Total Datagrams
[ 3]  0.0- 3.0 sec  34.2 MBytes  95.5 Mbits/sec  0.187 ms 104629698297856/    0 (2e-307%)
[ 3]  3.0- 6.0 sec  34.1 MBytes  95.5 Mbits/sec  0.172 ms 104612518428672/    0 (2e-307%)
[ 3]  6.0- 9.0 sec  34.1 MBytes  95.5 Mbits/sec  0.137 ms 104612518428672/    0 (2e-307%)
[ 3]  9.0-12.0 sec  34.1 MBytes  95.5 Mbits/sec  0.181 ms 104616813395968/    0 (2e-307%)
[ 3] 12.0-15.0 sec  34.1 MBytes  95.5 Mbits/sec  0.178 ms 104612518428672/    0 (2e-307%)
[ 3] 15.0-18.0 sec  34.1 MBytes  95.5 Mbits/sec  0.174 ms 104616813395968/    0 (2e-307%)
```

图9-15 服务器输出测试结果

服务器输出带宽为95.5Mbit/s，抖动延迟见Jitter列，Lost列数据为丢包数量，Total Datagrams列数据为包数量，丢包率为0(2e-307%)。

iPerf有很多测试参数，大家可以根据实际情况进行设置。在上面的实例中，我们以开发板和LAN有线客户端为例进行测试，给大家留一个任务：通过LAN无线连接（2.4GHz或5GHz）到开发板，测试无线网络的吞吐量和丢包率。

9.2 网络测试工具

在网络环境中，连接到目标可能需要经过不同层级的路由器和交换机，我们经常要测试网络的连通性，并对出现的故障进行快速有效的定位和排除。值得庆幸的是，Linux下有很多优秀的网络测试工具，可以帮我们测试和分析网络问题。

开发板中默认内置了一些非常有用的网络测试工具，我们需要熟悉并掌握这些测试工具。

9.2.1 traceroute

traceroute是Linux下诊断网络问题时常用的工具，可以定位从源主机到目标主机之间经过了哪些路由器，以及到达各个路由器的耗时。我们通常使用traceroute命令来初步定位数据包丢失的节点。

使用traceroute进行测试时，默认对每一条路径上的每个设备都发送3次测试数据包，输出结果每一行中包括测试的节点序号、主机IP地址、3次数据包到达节点所耗费的时间（单位为毫秒）。

traceroute命令的格式如下。

```
traceroute [-FIldnrv] [-f 1ST_TTL] [-m MAXTTL] [-p PORT] [-q PROBES]
[-s SRC_IP] [-t TOS] [-w WAIT_SEC] [-g GATEWAY] [-i IFACE]
[-z PAUSE_MSEC] HOST [BYTES]
```

traceroute参数说明如表9-3所示。

表9-3 traceroute 参数说明

参数	描述
-F	设置不分段位
-I	使用ICMP而不是UDP数据报回应
-d	设置Socket的SO_DEBUG选项，启用Socket层级的排错功能
-n	直接使用IP地址而非主机名称
-r	绕过路由表，直接将数据包发送到远程主机上
-v	详细显示指令的执行过程
-m	设置检测数据包的最大存活数值TTL的大小（最大跳数）
-p	设置探测使用的UDP协议的通信端口号，默认为33434
-q	设置每个TTL的探测数，默认为3
-s	设置本地主机送出数据包的IP地址
-t	设置检测数据包的TOS数值，默认为0
-w	等待响应的时间，单位为秒，默认为3
-g	设置来源路由网关，最多可设置8个

traceroute命令示例

以默认参数探测：

```
DF1A:$ traceroute www.ptpress.com.cn
traceroute to www.ptpress.com.cn(39.96.127.170), 30 hops max, 38 byte packets
 1  172.16.10.1 (172.16.10.1)  0.625 ms  0.556 ms  0.447 ms
 2  192.168.2.1 (192.168.2.1)  2.000 ms  1.725 ms  1.414 ms
 3  172.27.0.1 (172.27.0.1)  4.672 ms  4.956 ms  4.224 ms
 4  10.60.253.213 (10.60.253.213)  5.067 ms  4.756 ms  4.615 ms
 5  211.137.46.229 (211.137.46.229)  9.553 ms  4.989 ms  11.291 ms
 6  211.137.36.153 (211.137.36.153)  11.504 ms  22.884 ms  11.984 ms
 7  221.180.241.10 (221.180.241.10)  12.701 ms  12.460 ms  22.360 ms
 8  * * *
 9  172.20.131.146 (172.20.131.146)  12.133 ms  11.941 ms  21.483 ms
10  39.96.127.170 (39.96.127.170)  10.862 ms  11.595 ms  12.147 ms
```

输出结果中，域名www.ptpress.com.cn对应的IP地址为39.96.127.170，从当前主机到目标主机，最多经过30跳（30 hops max），每次检测发送的包大小为38字节（38 byte packets）。接下来每一行包含3部分：节点序号；主机IP地址；向每个网关发送3个（用-q参数可更改）数据包后，网关响应返回的时间（单位为ms）。每次检测都同时发送3个数据包，如果某一个数据包超时没有返回，则时间显示为*，有可能是出问题，也有可能是防火墙屏蔽掉了ICMP的返回信息，需要具体分析。

设定参数探测：

-q：设置每个TTL的探测数，这里设为4。
-m：设置最多跳数，这里设为10跳。
-v：显示详细输出。
-n：直接使用IP。

```
DF1A:$ traceroute -q 4 -m 10 -v -n www.ptpress.com.cn
traceroute to www.ptpress.com.cn (39.96.127.170), 10 hops max, 38 byte packets
 1  172.16.10.1 46 bytes to (null)  0.719 ms  0.612 ms  0.555 ms  0.501 ms
 2  192.168.2.1 46 bytes to (null)  2.760 ms  1.950 ms  1.909 ms  2.043 ms
 3  172.27.0.1 36 bytes to (null)  6.283 ms  4.881 ms  4.812 ms  4.552 ms
```

```
4   *   *   *   *
5   211.137.46.225 36 bytes to (null)   7.313 ms   6.982 ms   7.390 ms   6.952 ms
6   211.137.47.45 36 bytes to (null)   13.230 ms   32.539 ms   14.128 ms   15.334 ms
7   *   *   *   *
8   *   *   *   *
9   172.20.131.146 54 bytes to (null)   13.584 ms   12.324 ms   13.981 ms   13.810 ms
10  39.96.127.170 46 bytes to (null)   12.995 ms   12.737 ms   13.935 ms   12.968 ms
```

使用基本的UDP端口8212进行探测：

```
-p 8212:使用UDP协议的端口8212。
DF1A:$ traceroute -p 8212 www.ptpress.com.cn
traceroute to www.ptpress.com.cn (39.96.127.170), 30 hops max, 38 byte packets
1   172.16.10.1 (172.16.10.1)   0.850 ms   1.037 ms   0.502 ms
2   192.168.2.1 (192.168.2.1)   3.670 ms   2.037 ms   1.875 ms
3   172.27.0.1 (172.27.0.1)   4.925 ms   4.868 ms   4.851 ms
4   10.60.253.213 (10.60.253.213)   5.182 ms   5.085 ms   5.393 ms
5   211.137.46.229 (211.137.46.229)   5.385 ms   *   5.872 ms
6   211.137.36.153 (211.137.36.153)   13.330 ms   13.036 ms   13.009 ms
7   *   221.180.241.10 (221.180.241.10)   13.984 ms   *
8   *   *   *
9   172.20.131.146 (172.20.131.146)   17.492 ms   12.312 ms   12.334 ms
10  39.96.127.170 (39.96.127.170)   12.044 ms !C   12.915 ms !C   12.782 ms !C
```

9.2.2 ping

ping命令是最为常用的网络命令，它通常用来测试与目标主机的连通性。Linux和Windows默认都支持此命令。

ping命令的格式如下。

```
ping [OPTIONS] HOST
```

ping命令参数说明如表9-4所示。

表9-4　ping命令参数说明

参数	描述
-4,-6	使用IPv4或IPv6
-c CNT	设置ping的次数后停止
-s SIZE	设置发送包大小，默认为56字节
-t TTL	设置TTL（Time To Live），指定IP包被路由器丢弃之前允许通过的最大网段数
-I IFACE/IP	使用的网络接口或IP地址
-W SEC	等待第一个响应包的时间，单位为秒
-w SEC	设置ping多少秒后停止，默认不停止
-q	不显示指令执行过程，开头和结尾的相关信息除外
-p	设置填满数据包的范本样式

ping命令示例

以默认参数执行：

```
DF1A:$ ping www.ptpress.com.cn
PING www.ptpress.com.cn (39.96.127.170): 56 data bytes
```

```
64 bytes from 39.96.127.170 seq=0 ttl=54 time=16.115 ms
64 bytes from 39.96.127.170 seq=1 ttl=54 time=16.595 ms
64 bytes from 39.96.127.170 seq=2 ttl=54 time=15.146 ms
64 bytes from 39.96.127.170 seq=3 ttl=54 time=18.047 ms
64 bytes from 39.96.127.170 seq=4 ttl=54 time=14.968 ms
^C
--- www.ptpress.com.cn ping statistics ---
5 packets transmitted, 5 packets received, 0% packet loss
round-trip min/avg/max = 14.968/16.174/18.047 ms
```

设置ping的次数：
-c：指定ping几次后停止ping，这里设为5次。
-s 65507：使用65507字节包，这里设为5次测试。注意包太大可能导致目标主机拒绝。Windows下该参数的最大值为65500，Linux下该参数的最大值为65507。
-I eth0.2：使用WAN口。

```
DF1A:$ ping -c 5 -s 65507 -I eth0.2 172.16.10.1
PING 172.16.10.1 (172.16.10.1): 65507 data bytes
65515 bytes from 172.16.10.1: seq=0 ttl=64 time=5.917 ms
65515 bytes from 172.16.10.1: seq=1 ttl=64 time=5.921 ms
65515 bytes from 172.16.10.1: seq=2 ttl=64 time=5.558 ms
65515 bytes from 172.16.10.1: seq=3 ttl=64 time=5.638 ms
65515 bytes from 172.16.10.1: seq=4 ttl=64 time=5.511 ms
--- 172.16.10.1 ping statistics ---
5 packets transmitted, 5 packets received, 0% packet loss
round-trip min/avg/max = 5.511/5.709/5.921 ms
```

9.2.3 tcpdump

tcpdump是一款强大的网络抓包工具，允许用户拦截、过滤和显示发送或接收到的数据包，是网络分析和问题排查的首选工具。tcpdump基于libcap库实现其功能。

tcpdump命令的格式如下。

```
tcpdump version 4.5.1
libpcap version 1.5.3
tcpdump [-aAbdDefhHIJKlLnNOpqRStuUvxX] [ -B size ] [ -c count ]
        [ -C file_size ] [ -E algo:secret ] [ -F file ] [ -G seconds ]
        [ -i interface ] [ -j tstamptype ] [ -M secret ]
        [ -P in|out|inout ]
        [ -r file ] [ -s snaplen ] [ -T type ] [ -V file ] [ -w file ]
        [ -W filecount ] [ -y datalinktype ] [ -z command ]
        [ -Z user ] [ expression ]
```

tcpdump部分参数说明如表9-5所示。

表 9-5 tcpdump 部分参数说明

参数	描述
-a	将网络地址和广播地址转变成名称
-A	以ASCII格式打印每个数据包，方便抓取网页数据包
-b	在数据-链路层上选择协议
-d	将编译后的数据包匹配代码以人类可读的形式转储到标准输出并停止
-dd	将匹配信息包的代码以C语言程序段的格式输出
-ddd	将匹配信息包的代码以十进制的形式输出
-D	打印出系统中所有可以用tcpdump截包的网络接口，可以将接口名称或编号提供给-i标志以指定要捕获的接口
-e	在输出行打印出数据链路层的头部信息
-f	用数字显示网络地址
-l	使标准输出变为缓冲行形式，可以把数据导出到文件
-L	列出网络接口的已知数据链路
-n	不要解析域名，直接显示IP
-nn	指定将每个监听到的数据包中的域名转换成IP，端口从应用名称转换成端口号后显示
-N	不输出主机名中的域名部分
-O	不运行分组匹配代码优化程序
-q	快速输出，只输出较少的协议信息
-S	将TCP的序列号以绝对值而不是相对值的形式输出
-t	输出行中不打印时间戳
-u	输出未解码的NFS句柄
v, -vv, -vvv	显示更多的详细信息
-X	同时用HEX和ASCII显示报文的内容
-XX	同-X但同时显示以太网头部
-B <size>	将操作系统捕获缓冲区大小设置为size，以KB为单位
-c <count>	在收到指定数量（count）的数据包后退出
-E<alg:secret>	使用alg:secret来解密IPsec ESP数据包，算法可以是des-cbc、3des-cbc、blowfish-cbc、rc3-cbc、cast128-cbc或none，默认为des-cbc，secret是ESP密钥的ASCII文本
-F <file>	将file作为过滤器表达式的输入
-i <interface>	指定监听的网络接口
-C <file_size>	在将原始数据包写入保存文件之前，请检查该文件当前是否大于file_size，如果是，请关闭当前文件并打开一个新文件。第一个savefile之后的savefiles将使用-w参数指定名称，后面带一个数字，从1开始并继续向上。file_size的单位是MB。-P不将网络接口设置成混杂模式
-r <file>	从指定的文件（file）中读取包
-s <snaplen>	从每个分组中读取最开始的snaplen个字节，而不是默认的68个字节
-T <type>	将监听到的包直接解释为指定类型（type）的报文，常见的类型有RPC（远程过程调用）和SNMP（简单网络管理协议）
-W <filecount>	将原始数据包写入文件而不是解析并打印出来，-W与-C选项一起使用时，将限制创建的文件数量达到指定的数量filecount，并从头开始覆盖文件，从而创建一个"旋转"缓冲区

基本抓包

指定抓包接口：

```
DF1A:$ tcpdump -i eth0.2
```

监听所有网络接口，显示IP地址：

```
DF1A:$ tcpdump -nS
```

显示更详细报文（包括tos、ttl、id、offset、proto等）：

```
DF1A:$ tcpdump -nnvvS
```

用HEX和ASCII两列对比输出：

```
DF1A:$ tcpdump -nnvvXS
```

报文过滤

tcpdump支持报文过滤，通常抓包网络报文数量异常多，很多时候我们只关心与具体问题有关的数据报，而这些数据只占到很小的一部分，可以使用tcpdump的报文过滤器来实现抓取指定报文。

1. tcpdump过滤器分为3类。

（1）type（报文的类型）：host（主机）、net（网络）、port（端口）、portrange（端口范围）。

（2）dir（报文的方向）：用来过滤报文的源地址和目的地址，可选值为src、dst、src or dst、src and dst、ra、ta、addr1、addr2、addr3、addr4。如果没有指定则默认为src or dst。ra、ta、addr1、addr2、addr3、addr4仅用于IEEE 802.11 Wireless LAN。

（3）proto（报文协议）：用于过滤报文协议，实际使用时可省略proto关键字，可选值为ether、fddi、tr、wlan、ip、ip6、arp、rarp、decnet、tcp and udp，支持包头过滤高级语法。

2. 过滤器支持表达式和操作符。

（1）and或&&：与。

（2）or或||：或。

（3）not或!：非。

（4）操作符包括>、<、>=（或greater）、<=（或less）、=、!=。

3. 过滤器还可以进行组合来表达复杂的过滤逻辑，如"src portrange <port1-port2>""dst net<net>"等。

4. 支持报文高级过滤。需要用户对TCP/IP协议非常了解，以字节级别精确匹配协议。

（1）proto[x:y]：过滤从proto协议x字节开始的y字节数，比如ip[2:2]过滤出3、4字节（从0开始）。

（2）proto[x:y] & z = 0：proto[x:y]和掩码z的与操作结果为0。

（3）proto[x:y] & z !=0：proto[x:y]和掩码z的与操作结果不为0。

（4）proto[x:y] & z = z：proto[x:y]和掩码z的与操作结果为z。

（5）proto[x:y] = z：proto[x:y]等于z。

设置抓包网卡为混杂模式：为支持抓取更多数据包，可以开启网卡混杂模式。所谓混杂模式，用最简单的语言形容就是让网卡抓取任何经过它的数据包。

```
DF1A:$ tcpdump -i eth0.2
[  604.117201] device eth0.2 entered promiscuous mode
```

过滤指定主机：

```
#抓取所有经过eth0.2、目的或源地址是172.16.10.10的网络数据
DF1A:$ tcpdump -i eth0.2 host 172.16.10.10
```

过滤某个网段的数据：

```
#抓取所有经过eth0.2、目的或源地址在172.16.10.0网段的网络数据
DF1A:$ tcpdump -i eth0.2 net 172.16.10.0/24
```

过滤指定端口的数据：

```
#抓取所有经过eth0.2、目的或源端口是80的网络数据
DF1A:$ tcpdump -i eth0.2 port 80
```

过滤指定端口范围的数据：

```
#抓取目的或源端口是21~80的网络数据
DF1A:$ tcpdump portrange 21-80
```

过滤源地址和目的地址：

```
#过滤来自172.16.10.10地址的数据
DF1A:$ tcpdump src 172.16.10.10
#过滤发送到172.16.10.10地址的数据
DF1A:$ tcpdump dst 172.16.10.10
```

过滤指定协议：

```
#过滤UDP协议
DF1A:$ tcpdump udp
#过滤TCP协议
DF1A:$ tcpdump tcp
```

过滤表达式：

```
#过滤端口号为8281且为TCP协议的数据包
DF1A:$ tcpdump src port 8281 and tcp
#抓取所有经过eth0.2、目的地址是172.16.10.119或172.16.10.200、端口是80的TCP数据
DF1A:$ tcpdump -i eth0.2 '((tcp) and (port 80) and ((dst host 172.16.10.119) or (dst host 172.16.10.200)))'
```

过滤数据报大小：

```
#过滤小于30字节的数据包
DF1A:$ tcpdump less 30
#过滤大于64字节的数据包
DF1A:$ tcpdump >=64
```

高级包过滤：

```
#只抓SYN包
DF1A:$ tcpdump -i eth0.2 'tcp[13] = 2'
#抓SYN、ACK包
DF1A:$ tcpdump -i eth0.2 'tcp[13] = 18'
#抓大于300字节的数据包
DF1A:$ tcpdump -i eth0.2 'ip[2:2] > 300'
```

输出/读取文件

使用tcpdump捕捉数据报文时，默认会将其输出到屏幕上，数据会比较多，不方便查看。这时可以使用tcpdump提供的保存到文件中的功能，这样方便使用其他图形工具（如wireshark、 snort）进行分析。

将捕获的数据报文输出到文件中：

```
#将TCP协议且端口为8281的所有数据捕获到capture_tcp_port_8281.pcap文件中
DF1A:$ tcpdump -w capture_tcp_port_8281.pcap port 8281 && proto tcp
```

从文件中读取数据报文，显示到屏幕上：

```
#从文件capture_tcp_port_8281.pcap中读取报文,显示到终端屏幕上
DF1A:$ tcpdump -nr capture_tcp_port_8281.pcap
reading from file capture_tcp_port_8281.pcap, link-type EN10MB (Ethernet)
15:29:59.102238 IP 192.168.4.191.57578 > xx.34.111.149.80: Flags [P.], seq
815045008:815045028,
 ack 3161469738, win 63876, length 20
15:29:59.128865 IP xx.34.111.149.80 >192.168.4.191.57578: Flags [.], ack 20, win 15544,
length 0
15:29:59.128954 IP xx.34.111.149.80 >192.168.4.191.57578: Flags [P.], seq 1:23, ack 20,
win 15544, length 22
15:29:59.170607 IP 192.168.4.191.57578 >xx.34.111.149.80: Flags [.], ack 23, win 63854,
length 0
```

9.3 Wi-Fi命令

无线网络是路由器的核心功能。开发板提供了查看无线设备、无线设备信息等命令，通过这些命令，我们可以查看当前无线网络的状态、信道、连接的设备信息，扫描其他无线设备等。下面我们将详细介绍开发板中与无线网络相关的命令。

9.3.1 iw

iw是Linux下基于nl80211的新型无线设备的命令行配置程序，用于替换iwconfig。

iw支持的命令参数比较多，开发板中iw命令的格式如下。

```
iw [options] command
选项：
       --debug
       --version
命令：
       help [command]
       event [-t] [-r] [-f]
       features
       phy
       list
       phy <phyname> info
       dev
       dev <devname> info
       dev <devname> del
       dev <devname> interface add <name> type <type> [mesh_id <meshid>] [4addr
on|off] [flags <flag>*] [addr <mac-addr>]
       phy <phyname> interface add <name> type <type> [mesh_id <meshid>] [4addr
on|off] [flags <flag>*] [addr <mac-addr>]
       dev <devname> ibss join <SSID> <freq in MHz> [HT20|HT40+|HT40-
```

```
|NOHT|5MHz|10MHz|80MHz] [fixed-freq] [<fixed bssid>] [beacon-interval <TU>]
 [basic-rates <rate in Mbps,rate2,...>] [mcast-rate <rate in Mbps>] [key d:0:abcde]
       dev <devname> ibss leave
       dev <devname> station dump
       dev <devname> station set <MAC address> mesh_power_mode <active|light|deep>
       dev <devname> station set <MAC address> vlan <ifindex>
       dev <devname> station set <MAC address> plink_action <open|block>
       dev <devname> station del <MAC address>
       dev <devname> station get <MAC address>
       dev <devname> survey dump
       dev <devname> ocb leave
       dev <devname> ocb join <freq in MHz> <5MHZ|10MHZ>
       dev <devname> mesh leave
       dev <devname> mesh join <mesh ID> [[freq <freq in MHz> <HT20|HT40+
|HT40-|NOHT|80MHz>] [basic-rates <rate in Mbps,rate2,...>]], [mcast-rate <rate in
Mbps>] [beacon-interval <time in TUs>] [dtim-period <value>] [vendor_sync
on|off] [<param>=<value>]*
       dev <devname> mpath dump
       dev <devname> mpath set <destination MAC address> next_hop <next hop
MAC address>
       dev <devname> mpath new <destination MAC address> next_hop <next hop
MAC address>
       dev <devname> mpath del <MAC address>
       dev <devname> mpath get <MAC address>
       dev <devname> mpp dump
       dev <devname> mpp get <MAC address>
       dev <devname> scan [-u] [freq <freq>*] [ies <hex as 00:11:..>] [meshid
<meshid>] [lowpri,flush,ap-force] [randomise[=<addr>/<mask>]] [ssid <ssid>*|passive]
       dev <devname> scan sched_stop
       dev <devname> scan sched_start interval <in_msecs> [delay <in_secs>] [freqs
<freq>+] [matches [ssid <ssid>]+]] [active [ssid <ssid>]+
|passive] [randomise[=<addr>/<mask>]]
       dev <devname> scan trigger [freq <freq>*] [ies <hex as 00:11:..>] [meshid
<meshid>] [lowpri,flush,ap-force] [randomise[=<addr>/<mask>]] [ssid <ssid>*|passive]
       dev <devname> scan dump [-u]
       phy <phyname> reg get
       reg get
       reg set <ISO/IEC 3166-1 alpha2>
       dev <devname> link
       dev <devname> offchannel <freq> <duration>
       dev <devname> cqm rssi <threshold|off> [<hysteresis>]
       dev <devname> vendor send <oui> <subcmd> <filename|-|hex data>
       phy <phyname> set antenna_gain <antenna gain in dBm>
       phy <phyname> set antenna <bitmap> | all | <tx bitmap> <rx bitmap>
       dev <devname> set txpower <auto|fixed|limit> [<tx power in mBm>]
       phy <phyname> set txpower <auto|fixed|limit> [<tx power in mBm>]
       phy <phyname> set distance <auto|distance>
       phy <phyname> set coverage <coverage class>
       phy <phyname> set netns { <pid> | name <nsname> }
       phy <phyname> set retry [short <limit>] [long <limit>]
```

```
    phy <phyname> set rts <rts threshold|off>
    phy <phyname> set frag <fragmentation threshold|off>
    dev <devname> set channel <channel> [NOHT|HT20|HT40+|HT40-|5MHz|10MHz|80MHz]
    phy <phyname> set channel <channel> [NOHT|HT20|HT40+|HT40-|5MHz|10MHz|80MHz]
    dev <devname> set freq <freq> [NOHT|HT20|HT40+|HT40-|5MHz|10MHz|80MHz]
    dev <devname> set freq <control freq> [5|10|20|40|80|80+80|160] [<center1_freq> [<center2_freq>]]
    phy <phyname> set freq <freq> [NOHT|HT20|HT40+|HT40-|5MHz|10MHz|80MHz]
    phy <phyname> set freq <control freq> [5|10|20|40|80|80+80|160] [<center1_freq>[<center2_freq>]]
    phy <phyname> set name <new name>
    dev <devname> set mcast_rate <rate in Mbps>
    dev <devname> set peer <MAC address>
    dev <devname> set noack_map <map>
    dev <devname> set 4addr <on|off>
    dev <devname> set type <type>
    dev <devname> set meshid <meshid>
    dev <devname> set monitor <flag>*
    dev <devname> set mesh_param <param>=<value> [<param>=<value>]*
    dev <devname> set power_save <on|off>
    dev <devname> set bitrates [legacy-<2.4|5> <legacy rate in Mbps>*] [ht-mcs-<2.4|5><MCS index>*] [vht-mcs-<2.4|5> <NSS:MCSx,MCSy... | NSS:MCSx-MCSy>*] [sgi-2.4|lgi-2.4][sgi-5|lgi-5]
    dev <devname> get mesh_param [<param>]
    dev <devname> get power_save <param>
    dev <devname> auth <SSID> <bssid> <type:open|shared> <freq in MHz> [key 0:abcde d:1:6162636465]
    dev <devname> connect [-w] <SSID> [<freq in MHz>] [<bssid>] [key 0:abcde d:1:6162636465]
    dev <devname> disconnect
```

iw命令示例

获得所有无线设备的性能(2.4GHz和5GHz):

```
DF1A:$ iw list
#iw phy命令输出的结果与iw list命令输出的结果一样
DF1A:$ iw phy
```

获取无线信息:

```
#获得2.4GHz无线信息
DF1A:$ iw dev wlan0 info
#获得5GHz无线信息
DF1A:$ iw dev wlan1 info
```

AP模式时获取连接的STA统计信息:

```
#获得2.4GHz无线连接的STA统计信息
DF1A:$ iw dev wlan0 station dump
Station 6c:b7:49:d6:8f:24 (on wlan0)
        inactive time:  7000 ms
        rx bytes:       92446
        rx packets:     682
        tx bytes:       95424
```

```
        tx packets:      347
        tx retries:      14
        tx failed:       0
        signal:          -40 dBm
        signal avg:      -37 dBm
        tx bitrate:      72.2 Mbit/s MCS 7 short GI
        rx bitrate:      6.0 Mbit/s
        expected throughput:    34.57Mbit/s
        authorized:      yes
        authenticated:   yes
        preamble:        short
        WMM/WME:         yes
        MFP:             no
        TDLS peer:       no
        connected time: 485 seconds
#获得5GHz无线连接的STA统计信息
DF1A:$ iw dev wlan1 station dump
```

AP模式时获取已连接的STA（指定MAC地址）的统计信息：

```
DF1A:$ iw dev wlan1 station get 64:b2:41:d2:8f:23
Station 64:b2:41:d2:8f:23 (on wlan1)
        inactive time:   5020 ms
        rx bytes:        54544
        rx packets:      488
        tx bytes:        192485
        tx packets:      335
        tx retries:      69
        tx failed:       2
        signal:          -42 dBm
        signal avg:      -41 dBm
        tx bitrate:      180.0 MBit/s VHT-MCS 8 40MHz short GI VHT-NSS 1
        rx bitrate:      6.0 Mbit/s
        expected throughput:    42.114Mbit/s
        authorized:      yes
        authenticated:   yes
        preamble:        short
        WMM/WME:         yes
        MFP:             no
        TDLS peer:       no
        connected time: 142 seconds
```

AP模式时删除连接的STA（使其上不了网）：

```
DF1A:$ iw dev wlan1 station del 64:b2:41:d2:8f:23
```

测量无线信道：

```
#测量2.4GHz无线信道
DF1A:$ iw dev wlan0 survey dump
#测量5GHz无线信道
DF1A:$ iw dev wlan1 survey dump
```

无线扫描：

#查看开发板的无线设备接口，开发板中的无线接口有wlan0(2.4GHz)和wlan1(5GHz)

```
DF1A:$ ifconfig
#通过wlan0(2.4GHz)对周围2.4GHz网络进行扫描
DF1A:$ iw dev wlan0 scan
#通过wlan1(5GHz)对周围5GHz网络进行扫描
DF1A:$ iw dev wlan1 scan
```

监听所有接口事件：

```
#监听事件
DF1A:$ iw event
#打印时间戳
DF1A:$ iw event -t
#打印完整帧信息,包含auth/assoc/deauth/disassoc帧
DF1A:$ iw event -f
```

关闭无线连接：

```
DF1A:$ iw dev wlan0-wwan disconnect
```

获取链路状态

如果无线AP关联到其他AP（如无线中继）上，可以通过下面的命令获得相关信息。关于无线中继的内容，请参考5.4节。

未关联AP时输出：

```
DF1A:$ iw dev wlan0 link
Not connected.
```

关联AP时输出：

```
#wlan0-wwan已无线中继到其他AP
DF1A:$ iw dev wlan0-wwan link
Connected to 1c:40:e8:15:59:27 (on wlan0-wwan)
        SSID: LELE
        freq: 2432
        RX: 528461 bytes (1612 packets)
        TX: 4497 bytes (39 packets)
        signal: -28 dBm
        tx bitrate: 1.0 Mbit/s
        bss flags:       short-preamble short-slot-time
        dtim period:     1
        beacon int:      100
```

虚拟接口

iw可以添加虚拟接口。支持的虚拟接口类型有managed、ibss、monitor、mesh、wds。flags仅在虚拟接口类型为monitor时被使用，flags可设置为none、fcsfail、control、otherbss、cook。虚拟接口可以用于调试，例如通过monitor虚拟接口，使内核mac80211的monitor接口向用户空间传递额外数据。

添加虚拟接口：

```
#在2.4GHz无线设备上添加一个类型为monitor、名称为moni0的接口
DF1A:$ iw phy phy0 interface add moni0 type monitor
#也可以使用iw dev来添加虚拟接口
DF1A:$ iw dev wlan0 interface add moni1 type monitor flags control
```

```
#添加类型为managed的虚拟接口
DF1A:$ iw phy phy0 interface add wlanmanage type managed
```

使能虚拟接口：

```
#使能moni0接口
DF1A:$ ifconfig moni0 up
```

删除虚拟接口：

```
DF1A:$ iw dev moni0 del
```

添加、使能虚拟接口，并通过tcpdump调试：

```
DF1A:$ iw dev wlan0 interface add test0 type monitor flags none
DF1A:$ ifconfig test0 up
DF1A:$ tcpdump -i test0 -s 65000 -p -U -w /tmp/test.dump
```

国家/地区代码

获取国家/地区代码：

```
DF1A:$ iw reg get
```

以大写字母参考ISO 3166-1 alpha-2设置国家/地区代码：

```
DF1A:$ iw reg set CN
```

设定参数

设定参数前先down掉对应接口：

```
#先ifconfig down接口
DF1A:$ ifconfig wlan0 down
```

指定信道，设置特定频率：

```
#设置信道1的频率为HT40+
DF1A:$ iw dev wlan0 set channel 1 HT40+
```

设定频率：

```
DF1A:$ iw dev wlan0 set freq 2412 HT40+
```

设定发射功率：

```
DF1A:$ iw dev wlan0 set txpower auto
```

设定传送比特率：

```
DF1A:$ iw dev wlan0 set bitrates legacy-2.4 12 18 24
```

9.3.2 iwinfo

iwinfo封装了nl80211、madwifi、qcawifi、wl_ops等驱动接口，然后将这些接口整合出一套统一的API层，iwinfo接口通常需要无线驱动程序开发者来适配。

iwinfo命令格式如下：

```
iwinfo <device> info
iwinfo <device> scan
iwinfo <device> txpowerlist
iwinfo <device> freqlist
iwinfo <device> assoclist
iwinfo <device> countrylist
```

```
iwinfo <device> htmodelist
iwinfo <backend> phyname <section>
```

iwinfo参数说明如表9-6所示。

表 9-6　iwinfo 参数说明

参数	描述
<device> info	查看设备信息
<device> scan	扫描周围SSID
<device> txpowerlist	查看设备发送功率及当前使用的功率
<device> freqlist	查看设备的信道以及当前使用的信道
<device> assoclist	查看关联的终端信息
<device> countrylist	查看国家/地区列表
<device> htmodelist	查看HT通道带宽列表

iwinfo命令示例

查看设备信息：

```
DF1A:$ iwinfo wlan0 info
wlan0     ESSID: "SiWiFi-3cfc-2.4G"
          Access Point: 10:16:88:14:3C:FF
          Mode: Master  Channel: 1 (2.412 GHz)
          Tx-Power: 20 dBm  Link Quality: unknown/70
          Signal: unknown  Noise: unknown
          Bit Rate: unknown
          Encryption: WPA2 PSK (CCMP)
          Type: nl80211  HW Mode(s): 802.11bgn
          Hardware: unknown [Generic MAC80211]
          TX power offset: unknown
          Frequency offset: unknown
          Supports VAPs: yes  PHY name: phy0
```

上面的信息中显示"unknown"的内容，是因为没完全实现对应接口。

通过2.4GHz无线设备扫描周围SSID：

```
DF1A:$ iwinfo wlan0 scan
Cell 01 - Address: 1C:40:E8:15:59:27
          ESSID: "TEEGO"
          Mode: Master  Channel: 5
          Signal: -31 dBm  Quality: 70/70
          Encryption: WPA2 PSK (CCMP)
Cell 02 - Address: 8A:25:93:57:1F:61
          ESSID: "iNet-mi"
          Mode: Master  Channel: 1
          Signal: -59 dBm  Quality: 51/70
          Encryption: mixed WPA/WPA2 PSK (TKIP, CCMP)
```

查看2.4GHz设备发射功率和当前设置的发射功率：

```
DF1A:$ iwinfo wlan0 txpowerlist
  0 dBm (    1 mW)
  1 dBm (    1 mW)
```

```
  2 dBm (    1 mW)
  3 dBm (    1 mW)
  4 dBm (    2 mW)
  5 dBm (    3 mW)
  6 dBm (    3 mW)
  7 dBm (    5 mW)
  8 dBm (    6 mW)
  9 dBm (    7 mW)
 10 dBm (   10 mW)
 11 dBm (   12 mW)
 12 dBm (   15 mW)
 13 dBm (   19 mW)
 14 dBm (   25 mW)
 15 dBm (   31 mW)
 16 dBm (   39 mW)
 17 dBm (   50 mW)
 18 dBm (   63 mW)
 19 dBm (   79 mW)
* 20 dBm (  100 mW)
```

查看2.4GHz设备的信道和当前使用的信道：

```
DF1A:$ iwinfo wlan0 freqlist
* 2.412 GHz (Channel 1)
  2.417 GHz (Channel 2)
  2.422 GHz (Channel 3)
  2.427 GHz (Channel 4)
  2.432 GHz (Channel 5)
  2.437 GHz (Channel 6)
  2.442 GHz (Channel 7)
  2.447 GHz (Channel 8)
  2.452 GHz (Channel 9)
  2.457 GHz (Channel 10)
  2.462 GHz (Channel 11)
  2.467 GHz (Channel 12)
  2.472 GHz (Channel 13)
```

查看关联的终端信息：

```
DF1A:$ iwinfo wlan1 assoclist
62:B3:44:D5:8F:54  -34 dBm / unknown (SNR -34)  3500 ms ago
    Connected time       26 Seconds.
        RX: 6.0 MBit/s, MCS 0, 20MHz                      185 Pkts.
        TX: 86.7 MBit/s, MCS 0, 20MHz, short GI           107 Pkts.
```

查看国家/地区列表：

```
DF1A:$ iwinfo wlan0 countrylist
```

查看HT通道带宽列表：

```
#2.4GHz
DF1A:$ iwinfo wlan0 htmodelist
HT20 HT40
#5GHz
DF1A:$ iwinfo wlan1 htmodelist
```

HT20 HT40 VHT20 VHT40 VHT80

9.3.3 wifi

wifi命令是一个Shell脚本，可以打开和关闭无线接口、生成无线配置文件、重新加载无线配置、获得无线状态。wifi命令本身是基于ubus、uci等命令来获得无线参数和对无线网络进行操作的。

wifi命令格式如下：

```
/sbin/wifi [up|down|detect|reload|status]
```

wifi命令参数说明如表9-7所示。

表 9-7 wifi 命令参数说明

参数	描述
up	打开所有无线接口，启动无线网络
down	关闭所有无线接口，关闭无线网络
detect	生成默认的无线网络配置
reload	重新加载无线网络配置（通过ubus）
status	获得无线网络状态
	无参数时，重新加载配置，启动无线网络

wifi命令示例

启动无线网络：

```
DF1A:$ wifi up
```

关闭无线网络：

```
DF1A:$ wifi down
```

生成无线网络配置：

```
DF1A:$ wifi detect > /etc/config/wireless
```

重新加载无线网络配置：

```
DF1A:$ wifi reload
```

获得无线网络状态：

```
DF1A:$ wifi status
```

重启无线网络：

```
DF1A:$ wifi
```

9.3.4 sfwifi

sfwifi是SiFlower特有的命令，这个Shell脚本用于重新加载整个Wi-Fi模块。

sfwifi命令格式如下：

```
/bin/sfwifi [reset|remove|reload] [lb/hb]
```

sfwifi命令参数说明如表9-8所示。

表 9-8　sfwifi 命令参数说明

参数	描述
reset	重新加载整个Wi-Fi驱动程序
remove	卸载整个Wi-Fi驱动程序
reload lb/hb	重新加载2.4GHz/5GHz无线网络驱动程序

sfwifi命令示例

重新加载整个Wi-Fi驱动程序：

```
DF1A:$ sfwifi reset
```

卸载整个Wi-Fi驱动程序：

```
DF1A:$ sfwifi remove
```

重新加载2.4GHz无线网络驱动程序：

```
DF1A:$ sfwifi reload lb
```

重新加载5GHz无线网络驱动程序：

```
DF1A:$ sfwifi reload hb
```

9.3.5 wpa_supplicant

wpa_supplicant是由wpad-mini软件包提供的命令。wpa_supplicant是一个实现了IEEE 802.11i协议（无线接入标准）的WPA/WPA2认证的客户端程序。WPA和WPA2（RSN）是无线安全标准中的两种密钥管理规范。WPA（或WPA2）无线安全接入又包括使用IEEE 802.1x协议认证的企业版和使用PSK（预共享密钥）认证的个人版本。supplicant是无线客户端上实现WPA/802.1x认证功能的组件。wpa_supplicant是无线客户端上实现密钥管理和认证的请求软件（服务器使用的软件为hostapd）。

wpa_supplicant支持WPA/IEEE 802.11i功能和EAP方法(IEEE 802.1x Supplicant)。

- 支持的WPA/IEEE 802.11i功能：
  - WPA-PSK（WPA个人版）
  - 带EAP的WPA（RADIUS认证服务器）（WPA企业版）
  - CCMP、TKIP、WEP104、WEP40的密钥管理
  - WPA和完整的IEEE 802.11i/RSN/WPA2
  - RSN:PMKSA缓存，预认证
  - IEEE 802.11r
  - IEEE 802.11w
  - Wi-Fi Protected Setup（WPS）
- 支持的EAP 方法（IEEE 802.1x Supplicant）：
  - EAP-TLS
  - EAP-PEAP/MSCHAPv2（PEAPv0和PEAPv1）
  - EAP-PEAP/TLS（PEAPv0和PEAPv1）
  - EAP-PEAP/GTC（PEAPv0和PEAPv1）
  - EAP-PEAP/OTP（PEAPv0和PEAPv1）

- EAP-PEAP/MD5-Challenge（PEAPv0和PEAPv1）
- EAP-TTLS/EAP-MD5-Challenge
- EAP-TTLS/EAP-GTC
- EAP-TTLS/EAP-OTP
- EAP-TTLS/EAP-MSCHAPv2
- EAP-TTLS/EAP-TLS
- EAP-TTLS/MSCHAPv2
- EAP-TTLS/MSCHAP
- EAP-TTLS/PAP
- EAP-TTLS/CHAP
- EAP-SIM
- EAP-AKA
- EAP-AKA
- EAP-PSK
- EAP-FAST
- EAP-PAX
- EAP-SAKE
- EAP-IKEv2
- EAP-GPSK
- LEAP（注意：需要驱动程序支持）

wpa_supplicant官方模块如图9-16所示。

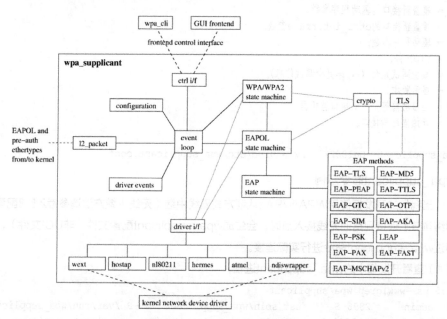

图9-16　wpa_supplicant官方模块

wpa_supplicant命令格式如下：

```
wpa_supplicant [-BddhKLqqtvW] [-P<pid file>] [-g<global ctrl>] \
        [-G<group>] \
        -i<ifname> -c<config file> [-C<ctrl>] [-D<driver>] [-H<hostapd path>] [-p<driver_param>] \
        [-b<br_ifname>] [-e<entropy file>] \
        [-o<override driver>] [-O<override ctrl>] \
        [-N -i<ifname> -c<conf> [-C<ctrl>] [-D<driver>] \
        [-p<driver_param>] [-b<br_ifname>] [-I<config file>] ...]
```

驱动程序:
  nl80211 = Linux nl80211/cfg80211
  wired = 有线以太网驱动程序

选项:
  -b = 可选网桥接口名称。
  -B = 在后台运行守护程序。
  -c = 配置文件。
  -C = ctrl_interface 参数（仅在不使用-c时使用）。
  -i = 接口名称。
  -I = 附加配置文件。
  -d = 增加调试冗余（-dd=更多调试信息）。
  -D = 驱动程序名称（可以是多个驱动程序: nl80211,wext）。
  -e = 熵文件。
  -g = 全局ctrl_interface。
  -G = 全局ctrl_interface组。
  -K = 在调试输出中包括密钥（密码等）。
  -t = 在调试消息中包含时间戳。
  -h = 帮助。
  -H = 连接到hostapd实例以管理状态更改。
  -L = 显示许可证（BSD）。
  -o = 覆盖新接口的驱动程序参数。
  -O = 覆盖新接口的ctrl_interface参数。
  -p = 驱动程序参数。
  -P = PID文件。
  -q = 减少调试冗余（-qq=更少调试信息）。
  -v = 显示版本。
  -W = 在启动前等待控制接口监视器。
  -N = 开始描述新接口。

示例:
  wpa_supplicant -Dnl80211 -iwlan0 -c/etc/wpa_supplicant.conf

wpa_supplicant命令示例

当我们的开发板使用WPA/WPA2认证方式无线中继（无线中继方法请参考2.5.3配置无线网络章节）连接到其他无线接入点时，会生成wpa_supplicant配置文件（非UCI文件），然后通过wpa_supplicant命令进行实际连接。

（1）查看开发板中的wpa_supplicant进程:

```
DF1A:$ ps -www|grep wpa_supplicant
 6580 admin     2896 S    /usr/sbin/wpa_supplicant -B -P /var/run/wpa_supplicant-wlan0-wwan.pid -D nl80211 -i wlan0-wwan -c /var/run/wpa_supplicant-wlan0-wwan.conf -C /var/run/wpa_supplicant
```

- -B：以守护进程方式运行wpa_supplicant。
- -P：指定pid文件为/var/run/wpa_supplicant-wlan0-wwan.pid。
- -D：指定使用nl80211驱动。
- -i：指定接口，这里的wlan0-wwan是无线UCI配置中的无线中继使用的接口。
- -c：wpa_supplicant配置文件/var/run/wpa_supplicant-wlan0-wwan.conf，由Netifd根据无线UCI配置生成。
- -C：如果设置了-c，此参数忽略。

（2）无线中继时生成的wpa_supplicant配置文件如下。

```
ctrl_interface=/var/run/wpa_supplicant
update_config=1
country=CN
network={
      scan_ssid=1
      ssid="LELE"
      key_mgmt=WPA-PSK
      psk="12345678"
      proto=WPA2
}
```

9.3.6 hostapd

hostapd是由wpad-mini软件包提供的命令。hostapd 是一个用户态的用于AP和认证服务器的守护进程，作为AP的认证服务器，负责控制管理STA的接入和认证。它实现了IEEE 802.11相关的接入管理（身份验证/关联）、IEEE 802.1x/WPA/WPA2/EAP认证、RADIUS客户端、EAP服务器和RADIUS认证服务器。hostapd也可以模拟AP功能，实现软AP（Soft AP）。

hostapd官方模块如图9-17所示。

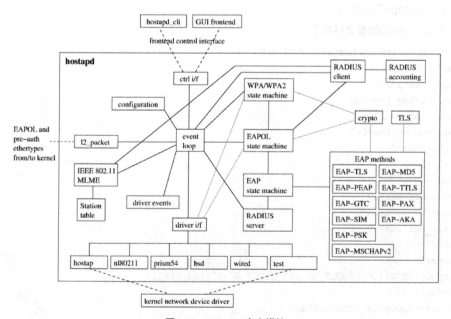

图9-17　hostapd官方模块

hostapd命令格式如下:

```
hostapd [-hdBKtv] [-P <PID file>] [-e <entropy file>] \
[-g <global ctrl_iface>] [-G <group>] \
<configuration file(s)>
```

无线设备可以设置表9-9所示的工作模式。

表9-9 无线工作模式

参数	描述
Master (AP)	变为无线接入点,提供无线接入服务
Managed (STA)	作为客户端连接其他无线接入点
Monitor	监听模式,监听周围无线信息
Ad-hoc	多台计算机直接相连

开发板中默认存在两个无线接口(分别对应2.4GHz和5.8GHz无线设备),以Master工作模式对外提供无线连接。而实际上这个工作是通过hostapd作为守护进程来实现的。

(1)查看开发板中hostapd进程:

```
DF1A:$ ps -www|grep hostapd
 1834 admin     2960 S    /usr/sbin/hostapd -P /var/run/wifi-phy1.pid -B
/var/run/hostapd-phy1.conf
 2446 admin     2964 S    /usr/sbin/hostapd -P /var/run/wifi-phy0.pid -B
/var/run/hostapd-phy0.conf
```

- -P:指定pid文件。
- -B:作为守护进程运行。
- /var/run/hostapd-phyX.conf:由Netifd根据无线UCI配置生成的配置文件。
- /var/run/hostapd-phy0.conf:2.4GHz无线AP认证配置文件。
- /var/run/hostapd-phy1.conf:5GHz无线AP认证配置文件。

(2)hostapd配置文件

hostapd-phy0配置文件如下。

```
DF1A:$ cat /var/run/hostapd-phy0.conf
driver=nl80211
logger_syslog=127
logger_syslog_level=2
logger_stdout=127
logger_stdout_level=2
country_code=CN
ieee80211d=1
hw_mode=g
channel=acs_survey
noscan=0
max_all_num_sta=40
ieee80211n=1
ht_coex=0
ht_capab=[LDPC][GF][SHORT-GI-20][SHORT-GI-40][RX-STBC1]
interface=wlan0
ctrl_interface=/var/run/hostapd
```

```
ap_isolate=1
disassoc_low_ack=1
preamble=1
wmm_enabled=1
ignore_broadcast_ssid=0
conditionally_ignore_bcast_ssid=0
uapsd_advertisement_enabled=1
wpa_passphrase=12345678
auth_algs=1
wpa=2
wpa_pairwise=CCMP
ssid=SiWiFi-3cfc-2.4G
wpa_key_mgmt=WPA-PSK
rsn_preauth=1
rsn_preauth_interfaces=br-lan
okc=1
bridge=br-lan
bssid=10:16:88:14:3c:ff
```

hostapd-phy1配置文件如下。

```
DF1A:$ cat /var/run/hostapd-phy1.conf
driver=nl80211
logger_syslog=127
logger_syslog_level=2
logger_stdout=127
logger_stdout_level=2
country_code=CN
ieee80211d=1
ieee80211h=1
hw_mode=a
channel=161
noscan=1
max_all_num_sta=40
ieee80211n=1
ht_coex=0
ht_capab=[HT40-][LDPC][GF][SHORT-GI-20][SHORT-GI-40][RX-STBC1]
vht_oper_chwidth=1
vht_oper_centr_freq_seg0_idx=155
ieee80211ac=1
vht_capab=[RXLDPC][SHORT-GI-80][RX-STBC-1][MAX-A-MPDU-LEN-EXP7]
interface=wlan1
ctrl_interface=/var/run/hostapd
ap_isolate=1
disassoc_low_ack=1
preamble=1
wmm_enabled=1
ignore_broadcast_ssid=0
conditionally_ignore_bcast_ssid=0
uapsd_advertisement_enabled=1
wpa_passphrase=12345678
```

```
auth_algs=1
wpa=2
wpa_pairwise=CCMP
ssid=SiWiFi-3d00
wpa_key_mgmt=WPA-PSK
rsn_preauth=1
rsn_preauth_interfaces=br-lan
okc=1
bridge=br-lan
bssid=10:16:88:14:3d:03
```

第三篇 深入浅出 OpenWrt 系统

10 交叉编译OpenWrt

10.1 安装VirtualBox虚拟机

进入第三篇,我们需要构建一个开发环境作为OpenWrt编译的宿主机环境。大家既可以直接使用主流的Linux发行版(如Ubuntu或CentOS)作为开发环境,也可以采用Windows下的虚拟机运行Linux系统。通常我们会采用后者,这样比较灵活和方便。

本书推荐使用VirtualBox这款虚拟机软件,它是一款开源、免费的虚拟机软件,学习和使用起来非常方便。

10.1.1 下载Windows版的VirtualBox软件

在Windows中安装VirtualBox需要下载虚拟机安装包和虚拟机扩展包。

● 下载虚拟机安装包。选择下载当前最新的Windows版(本书所用版本为VirtualBox 6.0.12 platform packages),单击"Windows hosts"进行下载(见图10-1)。下载后的文件为VirtualBox-6.0.12-133076-Win.exe。

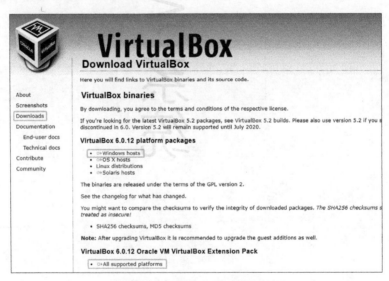

图10-1　VirtualBox官网下载界面

● 下载虚拟机扩展包。下载的虚拟机扩展包的版本要和安装的虚拟机的版本一致(本书所用为

VirtualBox 6.0.12 Oracle VM VirtualBox Extension Pack），单击"All supported platforms"进行下载。下载后的文件为Oracle_VM_VirtualBox_Extension_Pack-6.0.12.vbox-extpack。

10.1.2 安装VirtualBox虚拟机

1. 双击VirtualBox-6.0.12-133076-Win.exe打开安装程序，进入安装向导界面，单击"下一步"。

2. 可以修改VirtualBox的安装位置或按默认设置安装，单击"下一步"。

3. 快捷方式等设定按默认即可，单击"下一步"。

4. 单击"是"确认安装，再单击"安装"等待安装完毕。

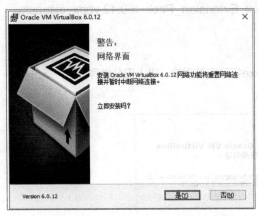 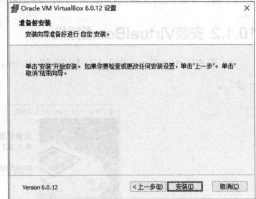

5. 安装过程中，会弹出安装通用串行总线控制器驱动程序的页面，单击"安装"即可。

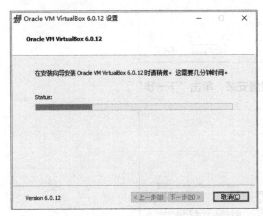 

6. 单击"完成"确认安装完成，将自动开启VirtualBox主界面。

10.1.3 安装VirtualBox虚拟机扩展包

虚拟机扩展包可以提供支持USB 2.0和USB 3.0设备VirtualBox RDP、磁盘加密、Intel卡的NVMe和PXE引导等功能。

1. 打开虚拟机软件，选择"管理"菜单下的"全局设定"功能菜单。

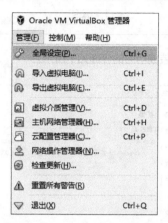

2. 在弹出的"全局设定"对话框中单击"扩展"选项。单击下图中标示的加号图标，以便加载扩展包。

3. 选择扩展包文件Oracle_VM_VirtualBox_Extension_Pack-6.0.12.vbox-extpack，弹出提示安装扩展包的对话框，单击"安装"。

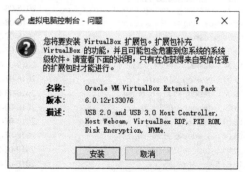

4. 阅读完许可协议后，单击"我同意"。默认会自动安装。

5. 提示安装成功后，单击"确定"即可。

6. 至此扩展包安装结束。安装成功后的界面如下所示。

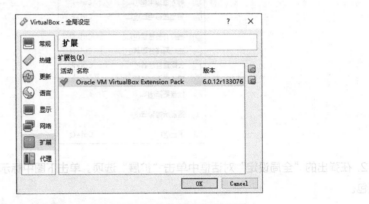

10.2 准备Ubuntu 16.04环境

安装完虚拟机软件VirtualBox后，我们需要创建一个虚拟机作为系统的编译环境，我们推荐使用Ubuntu作为开发用的虚拟机系统。

10.2.1 下载Ubuntu系统镜像

Ubuntu有很多个版本，又分为桌面版和服务器版，在这里我们选择下载桌面版Ubuntu 16.04.5 LTS，这里根据CPU选用64bit版的镜像文件ubuntu-16.04.5-desktop-amd64.iso（见图10-2）。

图10-2　Ubuntu 16.04.5官网镜像列表

10.2.2 创建新虚拟主机

在创建Ubuntu虚拟机时建议保留40GB的磁盘存储空间，虚拟机的存储目录需要有足够的磁盘空间。VirtualBox默认的虚拟机存储目录为C盘，如果C盘的磁盘空间不足，可以通过图10-3、图10-4所示的方式更改默认的虚拟机存储目录。

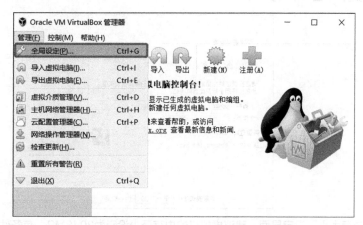

图10-3　VirtualBox"全局设定"菜单

图10-4　在"全局设定"→"常规"中更改默认的虚拟机存储目录

接下来进行虚拟机的安装。

1. 打开VirtualBox主界面，选择"控制"菜单下的"新建"。

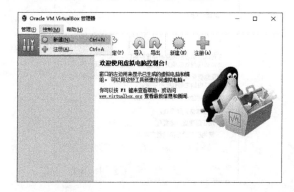

2.弹出"虚拟电脑名称和系统类型"设定界面,为虚拟机设定一个"名称",如"OpenWrt";因为之前已经设定了VirtualBox虚拟机的存储目录,所以此"文件夹"按默认即可,当然也可以变更;"类型"设定为"Linux";"版本"选择"Ubuntu(64-bit)",如果你的计算机是64位的,但此处只显示"Ubuntu(32-bit)",则需要在BIOS中开启虚拟化支持选项。如果没有问题,单击"下一步"。

3.弹出"内存大小"设定界面,建议虚拟机的内存至少设定为2048MB,有条件则可以设定为更大,这里设定为4096MB。单击"下一步"。

4.弹出"虚拟硬盘"设定界面,选择"现在创建虚拟硬盘",单击"创建"。

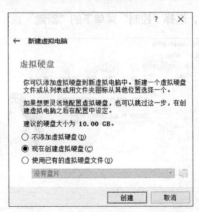

5. 弹出"虚拟硬盘文件类型"设定界面，选择"VDI（VirtualBox 磁盘映像）"，单击"下一步"。

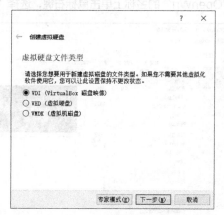

6. 弹出"存储在物理硬盘上"设定界面，选择"动态分配"，单击"下一步"。动态分配的磁盘不会立即占用硬盘容量，它会根据需要动态扩展容量。

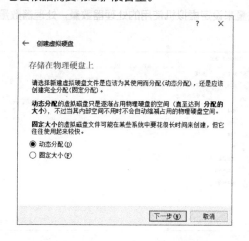

7. 弹出"文件位置和大小"设定界面，文件保存位置按默认设定即可，磁盘容量建议设置为40GB，因为选择了动态分配，所以最大可以动态扩展到40GB。单击"创建"，会创建OpenWrt虚拟机。

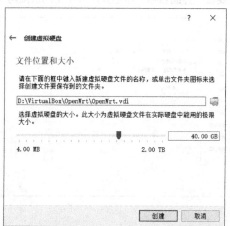

10.2.3 配置新虚拟机

1. 在新创建的虚拟机"OpenWrt"的名称上单击鼠标右键,选择"设置"。

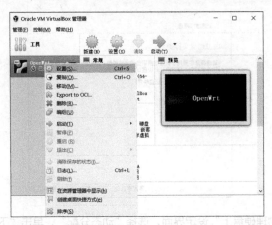

2. 处理器设置:在设置界面左侧选择"系统",在右侧单击"处理器"选项卡。根据自己的计算机处理器(内核)的数量来设定虚拟机使用的处理器数量,处理器数量越多,使用虚拟机进行编译的速度就越快。

3. 网络设置:在设置界面左侧选择"网络",在右侧单击"网卡1"选项卡。"连接方式"选择"桥接网卡",表示虚拟机通过计算机的网卡连接外网,具有独立的IP地址,推荐使用这个模式。在桥接网卡模式下,还要在"界面名称"中选择所要使用的桥接网卡。

4. 存储设置：在设置界面左侧选择"存储"，在"存储介质"中选中光盘图标，在右侧单击光盘图标，选择之前下载的ubuntu-16.04.5-desktop-amd64.iso镜像文件。

5. 单击"OK"，关闭设置窗口。这样当下次启动时，系统就会进入光盘安装状态。

10.2.4 安装Ubuntu系统

VirtualBox虚拟机使用起来就像本地计算机一样，但是其中要注意捕捉模式：单击虚拟机窗口后会进入捕捉模式，此时鼠标和键盘会限定在虚拟机内使用；如果要退出捕捉模式，要按下键盘右边的Ctrl键。接下来为虚拟机安装Ubuntu系统。

1. 在虚拟机列表中选中刚才所创建的虚拟机"OpenWrt"，单击"启动"图标，进行系统安装。

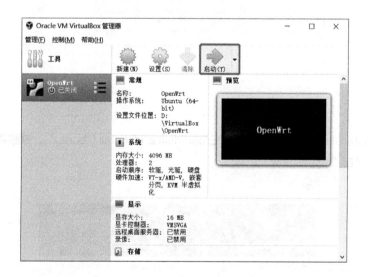

2. 进入 Ubuntu 安装界面，推荐使用英语，选中"English"，然后单击"Install Ubuntu"。

3. 在"Preparing to install Ubuntu"界面直接单击"Continue"。

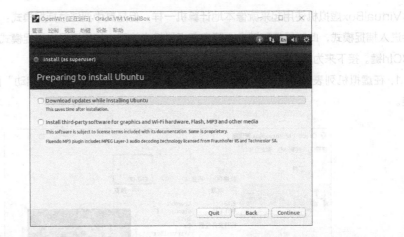

4. 在"Installation type"界面选择"Erase disk and install Ubuntu"，擦除整个虚拟磁盘然后安装系统，单击"Install Now"。

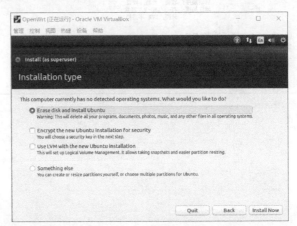

5. 在弹出的"Write the changes to disk？"界面中，单击"Continue"，把分区表写入磁盘。

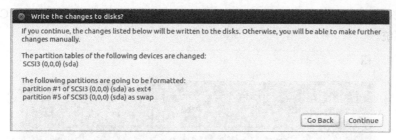

6. 在"Where are you？"界面中设置所在地（以确定时区），单击"Continue"。

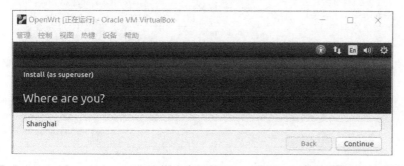

7. 在"Keyboard layout"界面中设置按键布局，默认为"English（US）"，然后单击"Continue"。

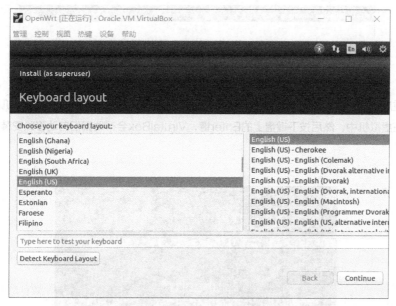

8. 在"Who are you?"界面中设定机器名称、非root用户名和密码。填写完"Your name"后，系统会自动生成"Your computer's name"和"Pick a username"，也可以进行修改。密码要填写两次，两次填写要一致，密码长度至少设置为6位。然后单击"Continue"开始安装系统。

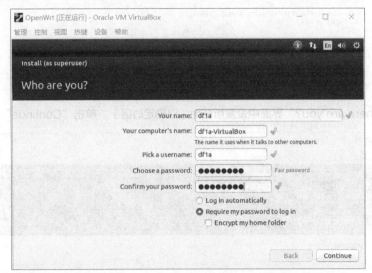

9. 安装系统大约需要等待10~30分钟。安装结束会弹出"Installation Complete"界面，单击"Restart Now"重启系统。

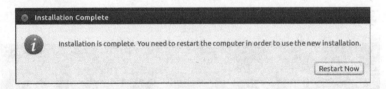

10. 重启过程中，VirtualBox会提示我们要移除安装光盘。用鼠标左键单击虚拟机屏幕，使鼠标指针落在虚拟机中，然后按下键盘上的Enter键，VirutalBox会帮助我们自动移除安装光盘并重启系统。

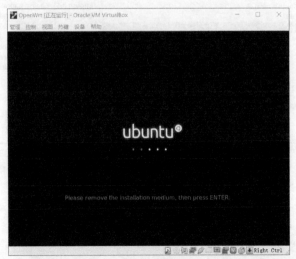

11. 系统重启后，显示登录界面，用户名为df1a，直接输入之前设定的密码，按下Enter键即可登录。

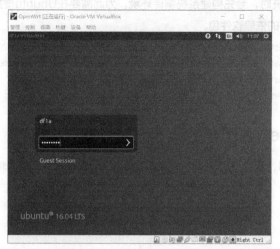

12. 安装增强功能模块。在VirtualBox中"设备"菜单中选择"安装增强功能"，按照提示输入密码，然后单击"授权"，在弹出的安装增强模块提示界面中单击"Run"，等待扩展模块驱动程序安装完毕。

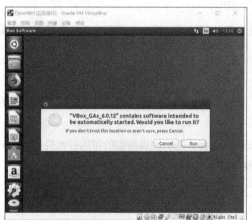 

13. 设置计算机和虚拟机之间可以自由复制、粘贴和拖放文件。在VirtualBox的"设备"菜单中分别选择"共享粘贴板"和"拖放"，均设定为"双向"。

14. 重启系统使设定生效。

10.2.5 配置Ubuntu系统

安装完Ubuntu系统后，我们需要做一些基本的配置并安装一些基础软件包。

（1）登录Ubuntu系统后，按下Ctrl+Alt+T组合键调出终端，后面的命令均在终端中执行。

（2）可选国内源进行更新，由于不同的源在不同区域的使用效果不一致，本文不详细描述，请搜索网上资料来操作。

（3）后续操作需要对Ubuntu系统有一定使用知识积累，关于如何使用Ubuntu系统请另行学习。

10.2.6 安装环境依赖

编译OpenWrt系统时，会先构建系统编译环境，为此需要在Ubuntu虚拟机中，安装交叉编译工具链和系统环境需要的依赖软件包。

```
HOST:~$ sudo apt-get update
HOST:~$ sudo apt-get install build-essential libssl-dev libncurses5-dev unzip gawk
HOST:~$ sudo apt-get install git subversion zlib1g-dev vim
HOST:~$ sudo apt-get install wget mercurial device-tree-compiler bc
```

10.3 编译OpenWrt固件

经过了前面的学习，我们已经积累了很多基础知识。从本节开始，我们将介绍编译OpenWrt系统固件的相关知识。

10.3.1 交叉编译

编译

用源代码生成CPU可以执行的目标程序，这一过程叫作编译，就像不同的语言之间翻译一样。我们通过对OpenWrt系统的编译，可以产生运行在DF1A开发板上的程序。

交叉编译

交叉编译就是在一个平台上生成另一个平台上的可执行代码。平台有两层含义：处理器的体系结构和所运行的操作系统。

宿主机与目标机

● 宿主机（Host）：编辑和编译程序的平台，一般是基于x86的PC，通常也称为主机。此时的宿主机为VirtualBox中的Linux平台。

● 目标机（Target）：开发板系统，通常是非x86平台。主机编译得到的可执行代码在目标机上运行。对于我们来说，目标机就是基于MIPS InterAptiv体系结构的DF1A开发板。

● 调试模型

嵌入式Linux的开发平台示意图如图10-5所示。

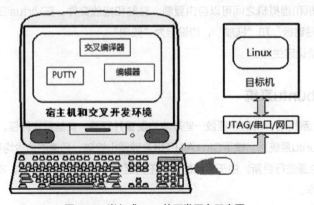

图10-5 嵌入式Linux的开发平台示意图

● 宿主机方面：在宿主机Windows操作系统中安装VirtualBox软件，在VirtualBox中安装Ubuntu操作系统，利用交叉编译器编译适合在目标机中运行的程序。

● 目标机方面：电路板焊接完毕后，存储器中没有可执行程序，此时，宿主机需要通过JTAG、TTL或其他下载调试器下载程序。在宿主机软件辅助下，利用下载调试器将BootLoader（目标机引导代码）下载到目标机，BootLoader的作用是初始化CPU、SDRAM、串口等。当BootLoader下载成功后，目标机在BootLoader的作用下就可以通过网口反复、多次、快速地将Linux内核及文件系统镜像下载到目标机中调试运行。同时BootLoader可以通过串口与宿主机进行通信、输出调试信息等，目标机使用的串口终端软件通常有PUTTY、SecureCRT等。

为何要进行交叉编译？主要原因有以下几点。

（1）目标机上不允许或不能够安装所需要的编译环境。

（2）目标机上的资源贫乏，无法运行所需要的编译程序。

（3）目标机还没有建立，无操作系统及编译器，无法自己开发自己。

10.3.2　固件与OpenWrt发行版

SF16A18的OpenWrt是由该芯片的设计公司维护的一个开源版本。为了方便于教学，本书的编撰团队针对DF1A板经过努力优化出更适用于完成教学内容的版本，也开源提供。为了保证内容的一致性，我们尽量保证所涉及的修改符合开源、OpenWrt、芯片厂原有精神。

DF1A的源代码GitHub托管地址：/hoowa/FreeIRIS-DF1A

DF1A的官网：freeiris项目官网

DF1A的OpenWrt镜像压缩地址：freeiris项目官网/resource/sf16a18-df1a/

OpenWrt发行版的情况如表10-1所示。

表10-1　OpenWrt 发行版的情况

代号	维护组织
Backfire 10.03.1	OpenWrt组织维护，2011年发行
Attitude Adjustment 12.09	OpenWrt组织维护，2013年发行
Barrier Breaker 14.07	OpenWrt组织维护，2014年发行
Chaos Calmer	OpenWrt组织维护，2016年发行
Trunk	OpenWrt组织维护，主线，无明确版本
SF16A18-SDK	SiFlower公司维护（Chaos Calmer改版）
FreeIRIS-DF1A	本书团队维护（Chaos Calmer改版）

10.3.3　准备编译用源代码

请注意，本节所描述的操作，都是在虚拟机下的Ubuntu系统中进行的。

为方便在虚拟机、实体计算机、开发板之间交互数据，我们给虚拟机安装上SSH服务（这样计算机可以通过WinSCP将数据从虚拟机复制到开发板中）。

```
HOST:~$ sudo apt-get install openssh-server
```

在虚拟机中查看网卡IP地址：

```
HOST:~$ ifconfig
```

使用查看到的IP地址，即可使用计算机上的WinSCP登录到虚拟机中获取编译后的固件。

获取源代码的两种方式

（1）通过GitHub获取：

```
HOST:~$ git clone https://GitHub网址/hoowa/sf16a18-sdk-4.2.10.git
```

（2）由于GitHub在国内访问速度过慢，下载几个GB的数据，速度实在难以忍受，这里我们提供了镜像压缩获取的方法：

```
#也可以使用计算机的迅雷下载该地址数据,然后通过WinSCP将其上传至虚拟机中
HOST:~$ wget http://freeiris官网地址/reso urce/sf16a18-df1a/sf16a18-sdk-4.2.10.tgz
#耐心等待下载完成,解压缩该版本
HOST:~$ tar zxf sf16a18-sdk-4.2.10.tgz
#使用git命令将其更新为最新版(如果这个过程中断,可以重复执行git pull直到完成)
HOST:~$ cd sf16a18-sdk-4.2.10/
HOST:sf16a18-sdk-4.2.10$ git pull
```

源代码中有几个主要的目录，其作用如表10-2所示，更详细的说明和定制在后续章节将进行介绍。

表10-2　SF16A18-SDK-4.2.10的主要目录和作用

目录	作用
chaos_calmer_15_05_1	OpenWrt系统环境
image_maker	SF16A18芯片的固件生成脚本
linux-3.18.29-dev	SF16A18芯片使用的内核
README.md	说明文件
uboot	U-Boot程序目录

10.3.4　开始编译

进行编译前的初始化

```
#进入OpenWrt编译目录
HOST:~$ cd sf16a18-sdk-4.2.10/chaos_calmer_15_05_1/
#安装feeds软件包
HOST:chaos_calmer_15_05_1$ ./scripts/feeds install -a
```

自动完成首次编译

```
HOST:chaos_calmer_15_05_1$ ./make.sh df1a fullmask
```

该过程将自动完成针对DF1A板的参数配置及编译的全部流程，消耗时间视计算机性能和虚拟机CPU数量而定，首次编译时间应该在30分钟以上，以后再次编译将进行差异检测，速度非常快。

如中途遇到中断，请重新执行该流程，直到首次编译完成，后续不需要再次执行make.sh步骤。编译后屏幕显示内容如图10-6所示。

图10-6 编译后屏幕显示内容

OpenWrt选项菜单

OpenWrt的选项非常丰富（见图10-7），目前默认已经预设好了，如果只想体验一下编译，可以不做修改。如果某些选项选择错了，可能导致编译失败，解决办法就是耐心地不断尝试！

```
HOST:chaos_calmer_15_05_1$ make menuconfig
```

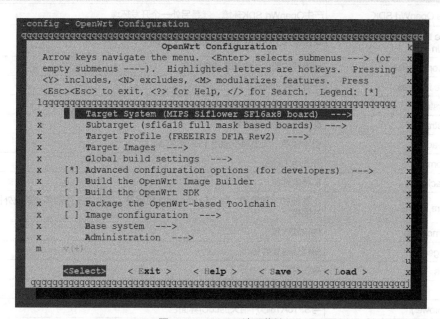

图10-7 OpenWrt选项菜单

选项菜单的使用方法如下。

1. 用键盘上、下方向键可以更换选项。
2. 有"--->"符号的选项代表其有子菜单。
3. 按Enter键可进入选项或修改选项的值。

4．用键盘左、右方向键可以选择菜单屏幕下面的"Select"（选择）、"Exit"（退出）、"Help"（帮助）、"Save"（保存）、"Load"（读取）选项。

5．最终确认并选择"Save"后，再选择"Exit"退出。

6．在配置时，通常有3种选择方式。

（1）Y：将该功能编译进内核，即选择为"*"。

（2）N：不将该功能编译进内核，即选择为"空"。

（3）M：以模块方式编译，需要时可以动态插入内核，即选择为"M"。

（4）按空格键可切换Y、N、M选择。

菜单可用选项如表10-3所示。

表10-3 菜单可用选项

选项	作用
Target System	目标设备芯片类型，SF16A18芯片统一为SF16ax8
Subtarget	目标子类型，fullmask表示全功能
Target Profile	目标硬件，选df1a表示本书所用开发板
Target Images	生成的固件文件采用何种分区格式
Global build settings	编译时的一些全局参数，这些参数与Linux内核或GCC编译器相关
Advanced configuration options	高级选项参数
Build the OpenWrt Image Builder	除了编译固件，再编译一个固件编译环境，可以分发给团队中的其他人使用
Build the OpenWrt SDK	产生OpenWrt SDK环境，就是另外一个开发环境
Package the OpenWrt-based Toolchain	单独编译出OpenWrt的交叉编译工具链
Image configuration	编译好的固件所附带的参数信息
Package features	一些软件包特性
Base system	基本系统命令软件包
Administration	高级管理命令软件包
Boot Loaders	引导程序，该选项不适用于SF16A18芯片
Development	开发用工具包
Extra packages	扩展软件包
Firmware	不要被名字混淆了，这个Firmware的意思是，固件中是否带某些其他外围芯片的固件
Kernel modules	内核软件包
Languages	编程语言软件包
Libraries	库软件包
LuCI	LuCI界面软件包
LuCI(Siflower)	与SF16A18芯片相关的LuCI界面包
Mail	与电子邮件相关的软件包
Multimedia	多媒体软件包
Network	与网络相关的软件包
Sound	与声音相关的软件包
Utilities	工具软件包

并不是所有的选项都要选上,实际上如果所有选项都选上,一定会出问题。一方面编译时间可能要按天计算,另一方面用来存放固件的Flash容量也不足,所以一般要根据系统的专用性、要完成的功能、硬件电路所支持的功能有哪些来选择。OpenWrt的选项有两种选择方式。

● Buildin：绑定,直接将所选择软件包绑定到固件中。

● Module：模块,所选择软件包不绑定到固件中,以模块的形式存在,在需要时候使用opkg命令安装。

更换两种方式很简单,选中要更换策略的软件包,按空格键,如果显示为[*]则表示绑定,如果显示为[M]则表示模块化编译。本书推荐第一次编译时不要去调整软件包,软件包存在很多依赖关系,先成功编译一次,再去选择哪些想要、哪些不想要。

再次编译

在使用make.sh方式自动完成首次编译后,不必再使用该方法编译,该方法将导致重新初始化开始编译,只有修改代码出现问题后才需要重新执行make.sh。

使用make命令即可完成编译。不显示信息,直接编译,命令如下。

```
HOST:chaos_calmer_15_05_1$ make
```

根据虚拟机设置的CPU数量采用多线程（示例采用了2线程）,且显示出详细编译日志,命令如下。

```
HOST:chaos_calmer_15_05_1$ make V=99 -j 2
```

如果编译过程中遇到错误或手动停止编译,不会出现什么问题,只要重复执行编译命令,它会从中断的地方继续编译。

因为某些人为错误导致编译失败,可以使用清理语法:

```
HOST:chaos_calmer_15_05_1$ make clean
```

新固件刷机

编译完成后,固件文件被存放在bin/siflower中：

```
HOST:chaos_calmer_15_05_1$ ls bin/siflower/
```

刷机使用以下固件文件：openwrt-siflower-sf16a18-fullmask-squashfs-sysupgrade.bin

将该文件使用WinSCP传送到开发板/tmp目录下后可以使用sysupgrade完成刷机：

```
DF1A:tmp$ sysupgrade -v -n openwrt-siflower-sf16a18-fullmask-squashfs-sysupgrade.bin
```

10.3.5 固件的OPKG软件仓

现在我们切换到开发板中,当OPKG下载和安装软件包的时候,默认总是访问freeiris官网的下载地址,这一地址我们可以自行完成设定与修改。查看OPKG当前设定可以知晓情况：

```
DF1A:~$ cat /etc/opkg/distfeeds.conf
```

修改OPKG默认软件仓地址

编辑version.mk文件：

```
HOST:chaos_calmer_15_05_1$ vim include/version.mk
```

该文件的第33行内容可以替换成为你的服务器地址：

```
http://freeiris官网/resource/%n/packages
```

删除掉已编译好的OPKG：

```
HOST:chaos_calmer_15_05_1$ rm -f \
build_dir/target-mipsel_mips-interAptiv_uClibc-0.9.33.2/opkg-unsigned
```

重新编译系统（将自动重新编译OPKG）：

```
HOST:chaos_calmer_15_05_1$ make V=99 -j 2
```

搭建自己的OPKG下载服务器

给自己的服务器安装httpd服务（任意一种），并且根据你所设定的下载地址，将bin/siflower目录下的内容复制到该服务器上。

使用开发板测试是否能更新OPKG：

```
HOST:~$ opkg update
```

如果一切显示正确，则表示所有操作全部完成。

10.4 U-Boot固件编译

上一节我们学习了OpenWrt固件编译，这节我们介绍U-Boot的编译步骤。刷写U-Boot的过程十分危险，操作过程中一旦出现错误或断电将导致开发板"成砖"。我们不希望读者遇到这种情况。

一旦出现错误，我们建议参考芯片原厂资料，使用专有工具进行修复。

10.4.1 编译U-Boot固件

编译U-Boot的语法：

```
HOST:sf16a18-sdk-4.2.10$ cd uboot/
HOST:uboot$ ./sf_make.sh prj=df1a
#耐心等待一小段时间
```

查看编译后的文件：

```
HOST:uboot$ ls sfax8/df1a.img
sfax8/df1a.img
```

10.4.2 烧写U-Boot

我们在这里使用串口连接开发板，使用板子上已有的U-Boot刷入新编译的U-Boot。首先需要在计算机中安装Tftpd32软件。Tftpd32的作用是提供TFTP服务，允许开发板上的U-Boot程序通过网线将df1a.img下载到开发板内存中。

在Windows下安装Tftpd32软件

（1）下载Tftpd32软件，进行安装。

（2）开发板断电后，将任意网口与计算机网口直连。在Windows系统中将其他全部网络都禁用，以防止失败。

（3）设置计算机与开发板连接的网口的IP地址为192.168.4.10，具体设置方法在之前章节介绍过，这里不再重复。

（4）开启Tftpd工具，"Server interface"选择"192.168.4.10"（见图10-8），并且在"Settings"中勾选"TFTP Server"（见图10-9）。

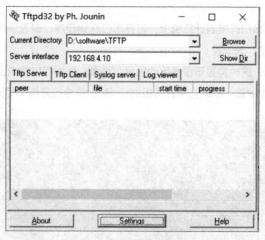

图10-8　Tftpd32的界面

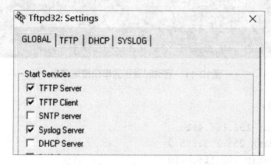

图10-9　Tftpd32的"Settings"

（5）使用WinSCP或其他工具将虚拟机中生成的df1a.img复制到TFTP的Current Directory（同级目录）下（见图10-10）。

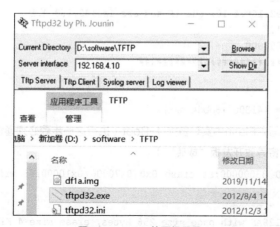

图10-10　TFTP的同级目录

（6）Tftpd32准备完毕。如果计算机有防火墙提醒，请允许Tftpd32访问网络，或关闭Windows系统的防火墙设置。

U-Boot刷机方法

（1）将开发板通过串口连接到计算机上，打开串口终端程序，设置波特率为115 200，数据位为8，停止位为1，无校验，无流控。

（2）将开发板上电，这时将显示很多启动信息，当串口终端程序显示出"Hit any key to stop autoboot:"时，按下Enter键，进入当前开发板的U-Boot命令行环境（见图10-11）。

```
DRAM:  128 MiB
MMC:   emmc@7800000: 0sdio@7c00000: 1
SF: Detected BY25Q128AS with page size 256 Bytes, erase size 4 KiB, t
In:    serial@8300000
Out:   serial@8300000
Err:   serial@8300000
Net:   Registering sfa18 net
Registering sfa18 eth
sf_eth0
Warning: sf_eth0 (eth0) using random MAC address - 4a:0a:ab:7c:96:2f

Hit any key to stop autoboot:  0
sfa18 #
```

图10-11 启动时看到的串口终端信息

（3）设置如下参数：

```
sfa18 # setenv ipaddr 192.168.4.1
sfa18 # setenv netmask 255.255.255.0
sfa18 # setenv serverip 192.168.4.10
sfa18 # saveenv
```

（4）将df1a.img从TFTP下载到开发板内存中：

```
sfa18 # tftp 0x81000000 df1a.img
Using sf_eth0 device
TFTP from server 192.168.4.10; our IP address is 192.168.4.1
Filename 'df1a.img'.
Load address: 0x81000000
Loading: T #############################
         42 KiB/s
done
Bytes transferred = 441300 (6bbd4 hex)
```

（5）将df1a.img写入Flash中实现替换U-Boot（执行这一步骤必须谨慎，在以上步骤没错的情况下才能操作，否则将导致开发板"成砖"）：

```
sfa18 # sf probe 0 33000000;sf erase 0x0 0x70000 0x10000;sf write 0x81000000 0x0 0x70000;
do_spi_flash----cmd = probe
SF: Detected BY25Q128AS with page size 256 Bytes, erase size 4 KiB, total 16 MiB
do_spi_flash----cmd = erase
SF: 458752 bytes @ 0x0 Erased: OK
```

```
do_spi_flash----cmd = write
device 0 offset 0x0, size 0x70000
SF: 458752 bytes @ 0x0 Written: OK
```

（6）如果以上步骤没错（说实话，有错也来不及了），就重启系统吧！

```
sfa18 # reset
```

如果你运气好的话，可以看到系统启动成功！

如果你运气不好的话，就请参考本节最开始所说的方法尝试吧！

11 软件包开发

11.1 软件包构建基础

OpenWrt中包含上千个不同的软件包（Package），我们通过OPKG进行软件包管理，软件包扩展名为.ipk。用户可以自由组合配置选择，也可以移植新的软件包或是定义自己的软件包，来扩展更多的功能。在OpenWrt中增加软件包非常简单和方便，系统为我们规定了软件包的语法格式，还定义了软件包默认的模板，我们只要按照OpenWrt系统规定的方式创建和定义每个软件包，就可以轻松实现软件包开发和移植。

在前面的3.3节中，我们学习了在OpenWrt系统下如何安装、卸载软件包，在下面的章节中，我们进一步探讨这些.ipk软件包在OpenWrt中是如何制作和产生的。学习在OpenWrt下创建软件包之前，我们先了解一下软件包开发的基础知识。

11.1.1 软件包的目录结构

OpenWrt软件包主要存放在工程目录的feeds和package目录下，二者的区别是feeds目录下的软件包可以通过更新来动态获取，package目录下的软件包是OpenWrt某个发行版发行时系统默认自带的。我们编写软件包，在package目录下完成。通常这些目录按照软件包的功能和类别进行划分。当编译系统或软件包时，会根据软件包的配置信息进行软件包的构建，其间会自动解决软件包之间的依赖、从源下载并校验源代码、创建编译目录、交叉编译、打包。

软件包在OpenWrt源码中用一个目录来表示，这个目录中至少包含一个Makefile文件和一些可选目录。一个软件包的目录结构如图11-1所示。

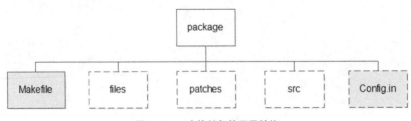

图11-1　一个软件包的目录结构

package目录

package表示软件包,是一个目录,通常以软件包名称命名。

package/Makefile文件

软件包必须包含一个Makefile文件。Makefile文件中记载了软件包变量、软件的名称、校验、获取、构建方法、补丁、打包等信息,make menuconfig中的菜单选项都是通过Makefile传递给OpenWrt编译系统的。Makefile文件是软件包开发的核心,开发软件包的其中一项重要的工作就是写这个Makefile。Makefile是生成整个软件包的"菜谱",有了"菜谱",OpenWrt系统这个"厨师"才能生成对应的"美食"(.ipk文件)。

package/files静态文件目录(可选)

这是存放软件包附带的静态文件的目录。files目录与软件包Makefile文件位于同一父目录下,通常包含软件包的初始化脚本文件、默认配置文件、脚本和其他支持文件(如.ko文件和二进制文件)。files中的文件按照惯例来命名,.conf文件为OpenWrt UCI配置文件,.init文件为初始化脚本。目标系统上files目录内资源的实际放置和命名由源包Makefile文件控制,并且与files目录内的结构和命名无关。files目录中的文件最终会被打包到软件包最终.ipk文件中,会随软件包安装到系统中。软件包编译出的可执行文件和库通常不在此目录中,会在软件包打包时生成。

package/patches补丁目录(可选)

这是软件包的补丁文件存放目录。patches目录与软件包Makefile文件位于同一父目录下。只包含软件包源码的补丁文件,补丁文件必须采用统一的diff格式(使用diff命令生成),并带有扩展名.patch。文件名还必须带有数字前缀,以表示补丁文件的顺序,编译软件包源代码时会把此目录下的补丁文件按照ASCII码排序后,在编译软件包源代码文件之前由OpenWrt系统依次打补丁到源代码文件中。补丁程序文件名应简短明了,并避免使用特殊字符。图11-2所示是Netifd软件包的补丁文件。

```
001-wireless-fix-reload-failure.patch
002-fix-network-reload-interface-issue.patch
003-fix-ap-teardown-boot.patch
004-cancel_retry_limit_restart_hostapd.patch
005-support-group-config-in-bridge.patch
006-fix-interface-up-fail.patch
007-support-multi-macaddrs-in-one-interace.patch
008-fix-autostart-not-update-when-config-change.patch
008-fix-handle-wireless-config-change-crash.patch
009-fix-ubus-connect-fail-when-boot.patch
010-add_netisolate.patch
011-support-bind-dps-netlink-msg-in-netifd.patch
012-add_attr_cond_hidden_for_freq_intergration
013-add_attr_speed_for_network_wan
014-wpas-optimise.patch
015-update-vif_txpower.patch
016-guest-optimise.patch
017-fix-netifd-memory-leak.patch
018-support-rps-xps-config.patch
```

图11-2 Netifd的补丁文件

package/src源代码目录(可选)

这是软件包的源代码目录。src目录与软件包Makefile文件位于同一父目录下。如果想把源代码放到本地，不需要从外部源获取源代码或在编译过程提供其他源代码，可以把源代码放到这个目录。src目录里的文件可以为源代码文件或源代码文件的压缩包，通常为自定义软件包、一些不对外公开的源代码的软件包。源代码目录中通常也包含Makefile文件，源代码目录的Makefile文件是编译源代码用的，不要和package目录下的软件包的Makefile文件混淆。

软件包通常不会包含这个目录，如果存在这个目录，OpenWrt系统构建时会自动把src目录里的内容复制到包的编译暂存目录（构建目录），并保留其内所有文件、目录的结构和命名。

package/Config.in 扩展配置文件（可选）

这是软件包扩展配置文件，与软件包Makefile文件位于同一父目录下。当软件包配置项和配置菜单比较多的时候，通常把配置提取到Config.in文件中，通过source Config.in来引用配置文件。个别配置项比较多的软件包和一些驱动厂商会这么做。

package/network/utils/curl/Config.in的内容如图11-3所示。

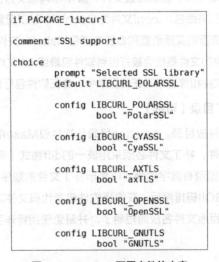

图11-3　Config.in配置文件的内容

11.1.2 软件包的编译命令

在OpenWrt源代码根目录，使用下面的命令对软件包进行单独编译、清理、清理并编译，支持多线程。

编译软件包的命令如下。

```
HOST:chaos_calmer_15_05_1$ make package/<package path>/<package name>/compile [-j <N>] V=99
```

例如：

```
HOST:chaos_calmer_15_05_1$ make package/network/services/dnsmasq/compile V=99
```

清理软件包的命令如下。

```
HOST:chaos_calmer_15_05_1$ make package/<package path>/<package name>/clean [-j <N>] V=99
```

例如：

```
HOST:chaos_calmer_15_05_1$ make package/network/services/dnsmasq/clean V=99
```

清理软件包后进行编译的命令如下:

```
HOST:chaos_calmer_15_05_1$ make package/<package path>/<package name>/
{clean,compile}   [-j <N>]   V=99
```

例如:

```
HOST:chaos_calmer_15_05_1$ make package/network/services/dnsmasq/{clean,compile}
V=99
```

11.1.3 软件包的构建流程

进行软件包编译时,OpenWrt系统会自动根据软件包Makefile文件信息为我们构建软件包,软件包的构建过程如下。

- 获取:获取软件包的源代码。
- 解压:解压源代码。
- 打补丁:修复源代码缺陷和增加新的功能补丁。
- 依赖与配置:根据目标环境准备构建的过程,包括解决软件包依赖并生成软件Makefile文件。
- 构建:进行编译和连接。
- 安装:将二进制文件、库文件和配置文件等复制到安装目录。
- 打包:生成包管理系统软件包,默认打包为.ipkg软件包。

获取

在OpenWrt下使用的软件包,绝大多数为开源软件,开源软件通常有自己的网站、文件服务器或下载地址。下载可以通过一种或更多的协议访问,如HTTP、HTTPS、FTP或其他协议。目前很多开源软件都提供发布版本和Git、Subversion、CVS(Concurrent Versions System,并发版本管理系统)等源代码控制管理(SCM,Source Control Management)系统。

在前面软件包的Makefile文件中,我们描述了软件包源代码通过哪种方式获取、软件包的校验值和版本信息。OpenWrt系统会根据这些描述信息,帮我们到指定URL(Uniform Resource Locator,统一资源定位符)获取相关软件包的压缩包或源代码文件。

以Netifd软件包为例:

- Netifd的源代码采用git方式获取,下载地址为http://git.OpenWrt官网地址/project/netifd.git。git版本日期为2015-12-16,ID为245527193e90906451be35c2b8e972b8712ea6ab。
- 源代码压缩包为netifd-2015-12-16-245527193e90906451be35c2b8e972b8712ea6ab.tar.gz。

Netifd的Makefile描述信息如下:

```
PKG_NAME:=netifd
PKG_VERSION:=2015-12-16
PKG_RELEASE=$(PKG_SOURCE_VERSION)
PKG_SOURCE_PROTO:=git
PKG_SOURCE_URL:=http://git.OpenWrt官网地址/project/netifd.git
PKG_SOURCE_SUBDIR:=$(PKG_NAME)-$(PKG_VERSION)
PKG_SOURCE_VERSION:=245527193e90906451be35c2b8e972b8712ea6ab
PKG_SOURCE:=$(PKG_NAME)-$(PKG_VERSION)-$(PKG_SOURCE_VERSION).tar.gz
```

OpenWrt会自动将源代码下载到dl目录:

```
netifd-2015-12-16-245527193e90906451be35c2b8e972b8712ea6ab.tar.gz
```

对于存放在软件包src目录（与软件包Makefile文件同级目录）的源代码，我们通常会在Makefile文件中实现Build/Prepare函数，手工创建编译目录PKG_BUILD_DIR/PKG_SOURCE_SUBDIR，并将src下的源代码复制到这个编译目录。

解压

源代码压缩包被下载后，OpenWrt系统会自动识别软件包的压缩方式并正确将其解压。具体过程是在系统编译目录（PKG_BUILD_DIR）中创建软件包编译目录（由PKG_SOURCE_SUBDIR定义），并将源代码解压到这个目录。常见的源代码压缩方式有GZIP、BZIP等，打成TAR包。

如Netfid软件包被解压到netifd-2015-12-16目录：

```
build_dir/target-mipsel_mips-interAptiv_uClibc-0.9.33.2/netifd-2015-12-16
```

对于存放在src目录的源代码压缩包，我们可以通过实现Build/Prepare函数手工解压到PKG_BUILD_DIR/PKG_SOURCE_SUBDIR目录。

打补丁

打补丁是通过比较文件之间的差异来实现增加、变更、删除、修改源代码的过程，主要是在已发行的源代码版本中进行漏洞修复、增删功能、提供配置、为交叉编译做出调整等。补丁文件需要有与源代码目录相同的结构和文件，通过diff命令将原始文件和修改过的文件进行比较，创建的补丁文件包含文件名称和文件相对路径、修改位置等差异。补丁文件采用了名为统一格式（unified format）的标准化格式。补丁文件可以包含多个文件的差异信息，所有文件被修改、删除、增加等信息都在补丁文件中进行体现。Netifd补丁文件如图11-4所示。

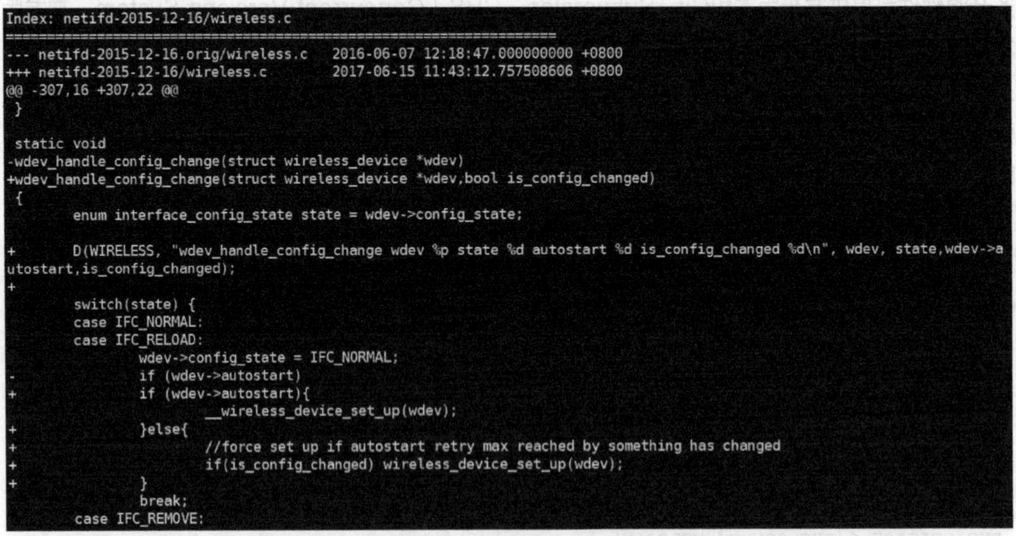

```
Index: netifd-2015-12-16/wireless.c
===============================================================
--- netifd-2015-12-16.orig/wireless.c    2016-06-07 12:18:47.000000000 +0800
+++ netifd-2015-12-16/wireless.c         2017-06-15 11:43:12.757508606 +0800
@@ -307,16 +307,22 @@
 }

 static void
-wdev_handle_config_change(struct wireless_device *wdev)
+wdev_handle_config_change(struct wireless_device *wdev,bool is_config_changed)
 {
        enum interface_config_state state = wdev->config_state;

+       D(WIRELESS, "wdev_handle_config_change wdev %p state %d autostart %d is_config_changed %d\n", wdev, state,wdev->autostart,is_config_changed);
+
        switch(state) {
        case IFC_NORMAL:
        case IFC_RELOAD:
                wdev->config_state = IFC_NORMAL;
-               if (wdev->autostart)
+               if (wdev->autostart){
                        __wireless_device_set_up(wdev);
+               }else{
+                       //force set up if autostart retry max reached by something has changed
+                       if(is_config_changed) wireless_device_set_up(wdev);
+               }
                break;
        case IFC_REMOVE:
```

图11-4　Netifd补丁文件

补丁文件的顺序很重要，补丁文件是有先后顺序的，因为补丁文件有可能是相互依赖的，即只有打过前面的补丁，后面的补丁才能成功打进去。在补丁文件比较多的情况下，如果通过手动方式打补丁，会是非常困难的。OpenWrt采用Quilt工具进行补丁管理，按照补丁文件的ASCII码排序依次打补丁。Quilt的功能非常强大，也支持补丁撤回等操作。

软件包补丁通常放在patches目录下，如果软件包有补丁，则创建这个目录，正确命名补丁文件后，将其放到这个目录中即可。

依赖与配置

对于提供源代码的软件包，我们需要根据目标（Target）系统的CPU体系结构交叉编译和声明依赖检测，当所有显式依赖解决后，会通过源代码的构建工具——通常为GNU构建系统Autotools或CMake来配置依赖检测，如果没问题，会生成源代码的Makefile文件。

构建

OpenWrt系统会进入软件包源代码编译目录，使用make命令（也有使用qmake等其他工具的）构建目标系统二进制文件、可执行程序、库文件及其他文件。

安装

安装分为两部分，一部分是指将软件包源代码编译后生成的可执行文件或库文件安装到软件包编译目录$(PKG_BUILD_DIR)/ipkg-install中；另一部分是指在软件包打包之前，需要按照软件包Makefile文件描述的安装规则将文件复制到打包目录$(PKG_BUILD_DIR)/ipkg-mips_siflower中。

如果软件包选择编译到系统中，则会将二进制文件、库文件、配置文件和其他文件复制到目标rootfs系统，通常安装目录要符合文件系统层级标准（FHS，Filesystem Hierarchy Standard），如配置文件放在/etc目录，库文件放在/lib目录等。安装目录权限使用OpenWrt定义的变量：

```
INSTALL_BIN:=install -m0755
INSTALL_DIR:=install -d -m0755
INSTALL_DATA:=install -m0644
INSTALL_CONF:=install -m0600
```

打包

打包是把软件包预打包目录（已经安装了编译后的二进制文件、配置文件、库文件等文件），按照软件包的格式封装的过程。OpenWrt系统使用OPKG作为包管理软件，生成的包为.ipkg文件。menuconfig选择为Y的软件包，会被安装到根文件系统，也会被打成.ipkg软件包；menuconfig选择为M的软件包，仅会被打成.ipkg软件包。最终所有的软件包会被安装到bin/siflower/packages子目录下。

11.1.4 Makefile语法

软件包的核心Makefile文件及其依赖包含的.mk（Makefile）文件中，涉及一些Makefile的语法，熟悉这些语法可以帮助我们高效地编写和分析软件包Makefile文件。这里我们简单介绍一下，更详细的Makefile语法请参考其他资料。

变量命名规则

- 变量名只使用字符串，可以包含字母、数字、下画线。
- 不能包含":""#""="或空格符。
- 变量名大小写敏感。

常用的变量赋值语法

（1）立即赋值。赋值格式："变量:=值"。定义变量时，赋值立即生效。以这种方式定义的变量会在变量的定义点按照被引用变量的当前值进行展开。

例如Makefile内容如下：

```
A:=netifd
B:=$(A)
A:=procd
all:
        @echo $(A) $(B)
.PHONY:all
```

执行：make。

输出：A的值为procd，B的值为netifd。

（2）递归赋值。赋值格式："变量=值"。这种变量也称延时变量，只有被使用时才展开定义。如果变量的定义引用了其他变量，那么引用会一直展开下去，直至找到被引用变量的最新的定义，并以此作为该变量的值，有时候会陷入无穷递归。

例如Makefile内容如下：

```
A=netifd
B=$(A)
A=procd
all:
        @echo $(A) $(B)
.PHONY:all
```

执行：make。

输出：A的值为procd，B的值为procd。

（3）条件赋值。赋值格式："变量?=值"。当变量值为空时才赋值。如果没有初始化过该变量，则给它赋值；如果该变量已初始化，则赋值无效。

例如Makefile内容如下：

```
A:=netifd
A?=procd
all:
        @echo $(A)
.PHONY:all
```

执行：make。

输出：A的值为netifd。

（4）追加赋值。赋值格式："变量+=值"。为已定义的变量添加新的值。如果变量没有被定义过，"+="和"="是一样的，它定义一个递归展开的变量。如果变量已经被定义了，"+="赋值只是简单地进行字符的添加工作。如果用":="定义变量，那么"+="赋值会将当前值追加到变量。如果用"="定义变量，它并不会在"+="的地方马上展开，而是会把展开工作推后，直至它找到变量最后的定义。

例如Makefile内容如下：

```
A:=netifd
A+=procd
all:
        @echo $(A)
```

```
.PHONY:all
```

执行：make。

输出：A的值为netifd procd。

特殊变量

$+：把所有内容原样输出。

$@--：目标文件。

$^--：所有的依赖文件。过滤重复的文件。

$<--：第一个依赖文件。

$$：类似于转义字符，$在Makefile中具有特殊含义，如果在Makefile中想使用$符号，需要写成$$。

语句的执行

$(call (func))表示要执行func函数。

$(n)表示上下文传递的第n个参数。

例如Makefile内容如下：

```
define func1
      @echo "function name: $(0)"
endef
define func2
      @echo "function name: $(0), param is $(1) $(2)"
endef
all:
      $(call func1)
      $(call func2, hello world)
.PHONY:       all
```

执行：make。

输出：

```
function name: func1
function name: func2, param is hello world
```

条件执行语法

$(if 条件, 成立执行, 失败执行)

例如：

```
$(if $(DUMP),dumpinfo,$(call func1))
```

如果存在变量DUMP，则执行dumpinfo目标，否则执行func1函数。

Makefile条件判断如表11-1所示。

表 11-1 Makefile 条件判断

语法	描述
ifeq(A,B)	判断参数是否相等。相等为true，否则为false
ifneq(A,B)	判断参数是否不相等。不相等为true，否则为false
ifdef A	判断变量是否有值。有值为true，否则为false
ifndef A	判断变量是否没有值。没有值为true，否则为false

循环语句

```
$(foreach 变量,成员列表,执行命令1,执行命令2,…执行命令N)
for in 成员列表;do 执行命令1;执行命令2;…执行命令N done
```

如果一行显示不下，可以在行末尾加上"空格+\"即" \"，换行继续后一行。注意for循环无法支持函数运算，而foreach可以，所以我们看到大部分循环是用foreach。

例如：

```
$(foreach hook,$(Hooks/Prepare/Pre),$(call $(hook))$(sep))
for cmd in $(call QuoteHostCommand,$(3)) $(call QuoteHostCommand,$(4)) \
           $(call QuoteHostCommand,$(5)) $(call QuoteHostCommand,$(6)) \
           $(call QuoteHostCommand,$(7)) $(call QuoteHostCommand,$(8)) \
                $(call QuoteHostCommand,$(9)); do \
...
done
```

包含其他Makefile文件

在Makefile文件中，可以通过include包含其他.mk文件。

```
include $(INCLUDE_DIR)/package.mk
```

常用Makefile函数

- wildcard

```
$(wildcard <pattern1 pattern2 ...>)
```

功能：展开pattern中的通配符，可以有多个，用空格分隔。

返回值：使用空格分隔的匹配对象列表。

举例：

```
$(wildcard include/*)
```

输出：得到include目录下的文件列表。

- filter

```
$(filter <pattern1 pattern2 ...>, <text>)
```

功能：以pattern模式过滤text字符串中的单词，保留符合pattern模式的单词。可以有多个模式，用空格分隔。

返回值：以空格分隔的text字符串中所有符合pattern模式的字符。

举例：

```
$(filter %.c %.cpp, $(wildcard src/*))
```

输出：src目录下所有扩展名是.c和.cpp的文件序列。

- filter-out

```
$(filter-out <pattern1 pattern2 ...>, <text>)
```

功能：与filter相反，以pattern模式过滤text字符串中的单词，保留不符合pattern模式的单词。可以有多个模式，用空格分隔。

返回值：以空格分隔的text字符串中所有不符合pattern模式的字符。

举例：

```
$(filter-out %.c %.cpp, $(wildcard src/*))
```

输出：src目录下所有扩展名不是.c和.cpp的文件序列。

- call

`$(call <expression>,<param1>,<param2>,...)`

功能：call函数可以将多行变量展开，并将传入的参数依次替换到临时变量中。当make执行call函数时，$(0)包含expression，expression参数中的变量，如$(1)、$(2)等，会依次被参数param1、param2等取代。call对参数的数目没有限制，可以没有任何参数。expression的返回值就是call函数的返回值。call函数只是展开返回Makefile中的文本语句，Makefile并未解析，通过调用eval函数执行。

返回值：expression展开后的表达式的值。

例如：

```
define func
        @echo $(0) $(1) $(2)
endef
all:
        $(call func, hello world)
.PHONY: all
```

输出：调用func函数，展开后$(0)=func $(1)=hello $(1)=world。

- eval

`$(eval <text>)`

功能：text的内容将作为Makefile的一部分而被make解析和执行。eval函数可以执行call函数展开返回的Makefile语句，使其生效。

例如：

```
A=foo boo
define func
test:
        @echo $(A) $(1)
endef
$(eval $(call func,hello))
```

输出：调用func函数，执行后输出foo boo hello。

- patsubst

`$(patsubst <pattern>,<replacement>,<text>)`

功能：搜索text中以空格分隔的单词，将符合模式pattern的替换为replacement。参数pattern中可以使用模式通配符"%"来代表一个单词中的若干字符。

返回值：替换后的新的字符串。

例如：

`$(patsubst %.c,%.o,a.c b.c)`

输出：把字串"a.c b.c"中以.c结尾的单词替换成以.o结尾的字符，输出"a.o b.o"。

- subst

`$(subst <from>,<to>,<text>)`

功能：字符串替换函数。把text字符串中的from字符串替换成to。

返回值：返回被替换过的字符串。

例如：

```
$(subst linux,openwrt,linux just linux)
```

输出：把"linux just linux"中的"linux"替换成"openwrt"，输出"openwrt just openwrt"。

- warning

```
$(warning <string>)
```

功能：在函数执行处输出string。常用来调试，可以放在Makefile的任何地方。

例如：

```
$(warning hello i am here)
```

输出：hello i am here。

11.1.5 软件包Makefile文件格式

软件包Makefile文件是构建软件包的核心。一个Makefile文件通常包含版权声明、其他Makefile文件、软件包变量、软件包模板函数、编译入口声明等。

图11-5所示是软件包Makefile文件示例。

图11-5 软件包Makefile文件示例

版权声明

一般在软件包Makefile文件的顶部进行版权（Copyright）声明，也可以没有版权声明。

版权声明格式如下所示：

```
# Copyright (C) 2019 Joe Random <joe@example.org>
```

Include XXX.mk

TOPDIR是OpenWrt源代码或SDK的路径。Makefile中的第一行需要包含rules.mk文件，这个文件包含Makefile的一些全局变量，如ARCH、ARCH_PACKAGES、BORAD、DL_DIR、BUILD_DIR、INCLUDE_DIR、BIN_DIR、HOSTCC、INSTALL_BIN、INSTALL_DIR、INSTALL_DATA、INSTALL_CONF等的定义，如图11-6所示。

```
CFLAGS:=
ARCH:=$(subst i486,i386,$(subst i586,i386,$(subst i686,i386,$(call qstrip,$(CONFIG_ARCH)))))
ARCH_PACKAGES:=$(call qstrip,$(CONFIG_TARGET_ARCH_PACKAGES))
BOARD:=$(call qstrip,$(CONFIG_TARGET_BOARD))
TARGET_OPTIMIZATION:=$(call qstrip,$(CONFIG_TARGET_OPTIMIZATION))
export EXTRA_OPTIMIZATION:=$(call qstrip,$(CONFIG_EXTRA_OPTIMIZATION))
TARGET_SUFFIX:=$(call qstrip,$(CONFIG_TARGET_SUFFIX))
BUILD_SUFFIX:=$(call qstrip,$(CONFIG_BUILD_SUFFIX))
SUBDIR:=$(patsubst $(TOPDIR)/%,%,${CURDIR})
export SHELL:=/usr/bin/env bash

DL_DIR:=$(if $(call qstrip,$(CONFIG_DOWNLOAD_FOLDER)),$(call qstrip,$(CONFIG_DOWNLOAD_FOLDER)),$(TOPDIR)/dl)
BIN_DIR:=$(if $(call qstrip,$(CONFIG_BINARY_FOLDER)),$(call qstrip,$(CONFIG_BINARY_FOLDER)),$(TOPDIR)/bin/$(BOARD))
INCLUDE_DIR:=$(TOPDIR)/include
SCRIPT_DIR:=$(TOPDIR)/scripts
BUILD_DIR_BASE:=$(TOPDIR)/build_dir
BUILD_DIR_HOST:=$(BUILD_DIR_BASE)/host
STAGING_DIR_HOST:=$(TOPDIR)/staging_dir/host

HOSTCC:=gcc
HOSTCXX:=g++
HOST_CPPFLAGS:=-I$(STAGING_DIR_HOST)/include -I$(STAGING_DIR_HOST)/usr/include
HOST_CFLAGS:=-O2 $(HOST_CPPFLAGS)
HOST_LDFLAGS:=-L$(STAGING_DIR_HOST)/lib -L$(STAGING_DIR_HOST)/usr/lib

BASH:=bash
TAR:=tar
FIND:=find
PATCH:=patch
PYTHON:=python

INSTALL_BIN:=install -m0755
INSTALL_DIR:=install -d -m0755
INSTALL_DATA:=install -m0644
INSTALL_CONF:=install -m0600
```

图11-6　rules.mk文件的部分内容

软件包Makefile文件通过include关键字引用扩展名为.mk的其他Makefile文件，来实现对OpenWrt软件包框架的引用。

```
include $(TOPDIR)/rules.mk
include $(INCLUDE_DIR)/package.mk
include $(INCLUDE_DIR)/cmake.mk
include $(INCLUDE_DIR)/kernel.mk
```

cmake.mk用于cmake软件包的编译，kernel.mk用于内核软件包的编译，package.mk是所有软件包都需要包含的文件，最为重要。

package.mk定义了软件包Makefile中用到的默认变量和处理函数，是软件包的框架模板。它主要完成以下任务。

（1）配置默认的PKG*变量，给未定义的变量赋默认值（采用:=变量赋值方式），推导其他变量。

（2）包含其他.mk文件。这些.mk文件正如其名字一样，包含生成最终软件包过程中所需要的各种功能，如下载软件包、解压源代码包、给源代码文件打补丁、生成.ipk文件等。

```
include $(INCLUDE_DIR)/prereq.mk
include $(INCLUDE_DIR)/host.mk
include $(INCLUDE_DIR)/unpack.mk
include $(INCLUDE_DIR)/depends.mk
include $(INCLUDE_DIR)/download.mk
include $(INCLUDE_DIR)/quilt.mk
include $(INCLUDE_DIR)/package-defaults.mk
include $(INCLUDE_DIR)/package-dumpinfo.mk
include $(INCLUDE_DIR)/package-ipkg.mk
include $(INCLUDE_DIR)/package-bin.mk
include $(INCLUDE_DIR)/autotools.mk
include $(INCLUDE_DIR)/prereq.mk
include $(INCLUDE_DIR)/host.mk
include $(INCLUDE_DIR)/unpack.mk
include $(INCLUDE_DIR)/depends.mk
include $(INCLUDE_DIR)/package.mk
include $(INCLUDE_DIR)/cmake.mk
```

（3）定义默认的处理函数。package.mk中定义了很多默认的函数，这些默认的函数在软件包生成过程中有着极其重要的作用。软件包Makefile文件包含package.mk文件后，即使不定义任何模板函数，系统也会调用package.mk中默认的预定义函数。如果软件包中定义了与预定义函数名称相同的函数，则实际执行时会覆盖预定义函数。

软件包变量

软件包变量用于描述软件包信息，常用的变量如表11-2所示。

表11-2 软件包的常用变量

变量名	描述
PKG_NAME	定义软件包的名称（menuconfig和ipkg中显示的软件包的名称）
PKG_VERSION	软件包源代码的版本。下载的软件源代码版本，通常是软件包源代码的主干版本。如openssl软件包编译了已发布的openssl-1.0.2q.tar.gz压缩包，PKG_VERSION则应设置为1.0.2q
PKG_RELEASE	软件包的版本修订版本。软件包在OpenWrt中的版本，通常指软件包Makefile版本
PKG_LICENSE	软件包的授权协议。对于开源软件，一般和原有的授权协议保持一致，如GPL-2.0
PKG_LICENSE_FILES	软件包的授权协议文件
PKG_BUILD_DIR	软件包的源代码编译目录。$(BUILD_DIR)/package
PKG_SOURCE	要下载的软件包的名字，一般由PKG_NAME和PKG_VERSION+软件格式扩展名（.tar.gz、.tar.bz2或其他）组成
PKG_MD5SUM	软件包的MD5值，用于校验软件包。在新版的OpenWrt中，它已经被PKG_HASH替换
PKG_CAT	解压软件包的方法（zcat、bzcat、unzip）
PKG_BUILD_DEPENDS	软件包编译过程中需要的依赖包。在编译软件包之前，会先编译这些依赖的软件包，区别于DEPENDS运行时依赖

续表

变量名	描述
PKG_BUILD_PARALLEL	是否启用并行编译软件包
PKG_CONFIG_DEPENDS	依赖的配置选项将被选中，如PKG_CONFIG_DEPENDS:=CONFIG_IPV6
PKG_INSTALL	如果将这个变量的值设置为1，将会调用这个软件包源代码中的原生make install，并将prefix设置为PKG_INSTALL_DIR的值，如PKG_INSTALL:=1
PKG_INSTALL_DIR	定义了make install把编译好的文件复制到的目录
与获取源代码相关	
PKG_SOURCE_PROTO	获取源代码的协议，包含git、svn、cvs、hg、bzr、sftp等
PKG_SOURCE_URL	源代码仓库地址。要与PKG_SOURCE_PROTO配合使用，如git://、http://、https://、@GNOME、@GNU、@KERNEL、@SF、@SAVANNAH、ftp://、file://等
PKG_SOURCE_VERSION	指定下载软件包源代码的版本，是哈希值或SVN版本。如果proto是git，必须指定这个，表示git中的commit hash（一个tag名或者commit id），用于git的checkout操作，如245527193e90906451be35c2b8e972b8712ea6ab
PKG_SOURCE_DATE	指定软件包源代码的日期格式为YYYY-MM-DD，与PKG_SOURCE_VERSION一起使用
PKG_REV	svn revision用得到，PKG_SOURCE_PROTO是svn时必须指定
PKG_SOURCE_SUBDIR	编译时创建的软件包的目录名称。如果PKG_SOURCE_PROTO是svn或者git，必须要设置，默认值为$(PKG_NAME)-$(PKG_VERSION)
PKG_MAINTAINER	软件包的维护者
PKG_HASH	用于验证下载的软件包压缩包的校验和。它可以是MD5或SHA256校验和
PKG_MD5SUM	软件包的MD5值，用于校验软件包。在新版的OpenWrt中，它已经被PKG_HASH替换
PKG_MIRROR_HASH	源代码仓库中下载的源代码压缩包的SHA256校验值

Package函数模板

1. Package/<package name>

描述软件包的信息，在menuconfig和ipkg中显示出软件包的信息，包含分类、类型、依赖和软件包概要描述。一个Makefile文件中，可以定义多个Package/<packagename>，只要packagename不同即可。这样方便把功能不同，还属于同一套源代码的内容拆分出来，比如一套代码可以编译出多个功能不同的可执行程序，还可以对外提供动态库，那么可以把可执行程序或库文件拆分出不同的软件包。

参数变量：

（1）SECTION：软件包类型（如base、net、utils等）。OpenWrt当前未使用。

（2）CATEGORY：menuconfig菜单中的菜单分类。

（3）SUBMENU：menuconfig→CATEGORY的子菜单。

（4）DEPENDS（可选）：软件包依赖的其他软件包和库，在编译此软件包之前必须先构建或安装的那些软件包。

（5）TITLE：简短的软件包描述。

（6）URL：原始软件包链接。

（7）FILES：软件包内核模块文件。内核软件包使用。

（8）AUTOLOAD：声明自动加载内核模块的顺序和模块的名称，内核软件包使用，例如AUTOLOAD:=$(call AutoLoad,25,nls_cp775)表示自动加载/etc/modules.d/25-nls_cp775文件中的nls_cp775模块。

（9）EXTRA_DEPENDS（可选）：运行时依赖项，仅添加到包control文件中。

（10）PKGARCH（可选）：PKGARCH主要标记软件包应用的平台体系结构，如果软件包仅包含脚本和资源，则用"PKGARCH:= all"进行标记，将可以安装在任何目标体系结构上。

举例：

软件包Netifd。

```
define Package/netifd
  SECTION:=base
  CATEGORY:=Base system
  DEPENDS:=+libuci +libnl-tiny +libubus +ubus +ubusd +jshn +libubox
  TITLE:=OpenWrt Network Interface Configuration Daemon
endef
```

menuconfig中"Base system"（CATEGORY）菜单netifd配置如图11-7所示。

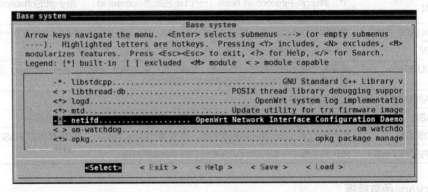

图11-7　netifd配置

将光标移动到netifd上，按下"h"按键显示帮助信息，可以看到Netifd依赖的软件包（DEPENDS），如图11-8所示。

图11-8　netifd菜单帮助信息

2. Package/<package name>/conffiles（可选）

软件包安装的配置文件列表。有多个配置文件时，每个文件一行。

举例：

```
define Package/miniupnpd/conffiles
/etc/config/upnpd
endef
```

3. Package/<package name>/description（可选）

软件包的描述。可以多行。

举例：

```
define Package/libpcap/description
This package contains a system-independent library for user-level network packet
capture.
endef
```

4. Package/<package name>/config（可选）

软件包的配置选项，在menuconfig中能够看到。配置项比较多时，建议创建一个Config.in，把配置项放到里面；配置项较少时，可以直接写。

举例：

配置项较多时，写在Config.in文件中，使用时直接引用Config.in配置文件。

```
define Package/libpcap/config
        source "$(SOURCE)/Config.in"
endef
```

配置项较少时，可以直接写。

```
define Package/procd/config
menu "Configuration"
        depends on PACKAGE_procd
config PROCD_SHOW_BOOT
        bool
        default n
        prompt "Print the shutdown to the console as well as logging it to syslog"
config PROCD_ZRAM_TMPFS
        bool
        default n
        prompt "Mount /tmp using zram."
endmenu
endef
```

5. Package/<package name>/install

软件包文件安装函数，一组用于将文件复制到ipkg的命令。$(1)表示软件包打包目录ipkg-mips_siflower（对于SiFlower芯片体系结构）。$(PKG_INSTALL_DIR)表示软件包编译完最终的要打包的文件目录。

参数变量：

● INSTALL_BIN：安装（复制）可执行程序时使用，通过install命令进行安装，掩码为0755，在rules.mk中定义。

● INSTALL_DIR：在软件包编译目录下ipkg-mips_siflower目录下创建指定目录，通过install命令创建，掩码为0755，在rules.mk中定义。

● INSTALL_DATA：安装（复制）数据时使用，通过install命令进行安装，掩码为0644，在rules.mk中定义。

● INSTALL_CONF：安装（复制）配置文件时使用，通过install命令进行安装，掩码为0600，在rules.mk中定义。

举例：

在根文件系统中创建/sbin，并将netifd编译目录中的netifd可执行文件复制到ipkg-mips_siflower/sbin，将软件包files目录下的文件复制到ipkg-mips_siflower中，将编译目录下scripts目录中的所有内容复制到ipkg-mips_siflower/lib/netifd目录。

```
define Package/netifd/install
    $(INSTALL_DIR) $(1)/sbin
    $(INSTALL_BIN) $(PKG_BUILD_DIR)/netifd $(1)/sbin/
    $(CP) ./files/* $(1)/
    $(CP) $(PKG_BUILD_DIR)/scripts/* $(1)/lib/netifd/
endef
```

6. Package/<package name>/preinst

软件包安装之前要执行的脚本内容。如果要执行Shell，不要忘记包含#!/bin/sh。如果需要中止安装，则让脚本返回false。

举例：

```
define Package/$(PKG_NAME)/preinst
    #!/bin/sh
    # if NOT run buildroot then stop service
    [ -z "$${IPKG_INSTROOT}" ] && /etc/init.d/ddns stop >/dev/null 2>&1
    exit 0  # suppress errors
endef
```

7. Package/<package name>/postinst

软件包安装之后要执行的脚本内容。

举例：

```
define Package/ntfs-3g/postinst
#!/bin/sh
FILE="$${IPKG_INSTROOT}/etc/filesystems"
ID="ntfs-3g"
if ! [ -f '/etc/filesystems' ]; then
    echo "Create '$$FILE'."
    touch "$$FILE"
fi
if ! grep -q -e '^ntfs-3g$$' "$$FILE"; then
    echo "Add '$$ID' to known filesystems."
    echo "$$ID" >> "$$FILE"
fi
endef
```

8. Package/<package name>/prerm

卸载软件包之前要执行的脚本内容。

举例：

```
define Package/$(PKG_NAME)/prerm
    #!/bin/sh
    # if run within buildroot exit
    [ -n "$${IPKG_INSTROOT}" ] && exit 0
    # stop running scripts
    /etc/init.d/ddns stop
```

```
        /etc/init.d/ddns disable
        # clear LuCI indexcache
        rm -f /tmp/luci-indexcache >/dev/null 2>&1
        exit 0  # suppress errors
endef
```

9. Package/<package name>/postrm

卸载软件包之后要执行的脚本内容。

Download函数模板

Download/default（可选）定义下载软件包源文件的信息，用于下载软件包源文件。

参数变量：

（1）FILE：源文件的名称，默认值为$(PKG_SOURCE)。

（2）URL：下载源文件的URL，默认值为$(PKG_SOURCE_URL)。

（3）SUBDIR：创建的源文件的目录名称，默认值为$(PKG_SOURCE_SUBDIR)。

（4）PROTO：下载源文件的协议，默认值为$(PKG_SOURCE_PROTO)。

（5）VERSION：源文件的版本，默认值为$(PKG_SOURCE_VERSION)。

（6）MD5SUM：源文件的MD5校验和，默认值为$(PKG_MD5SUM)。

调用方式：

```
$(eval $(call Download,软件包名称))
```

举例：

```
PKG_DATA_VERSION:=20150115
PKG_DATA_URL:=http://www.××××.de/usb_modeswitch
PKG_DATA_PATH:=usb-modeswitch-data-$(PKG_DATA_VERSION)
PKG_DATA_FILENAME:=$(PKG_DATA_PATH).tar.bz2
define Download/data
  FILE:=$(PKG_DATA_FILENAME)
  URL:=$(PKG_DATA_URL)
  MD5SUM:=662bcd56a97e560ea974bc710822de51
endef
$(eval $(call Download,data))
```

Build函数模板

1. Build/Prepare（可选）

这是编译软件包源代码前执行的函数，包含软件包解压、打补丁、将src下的源代码复制到软件包源代码编译目录等操作。一般不用定义这个函数。

```
define Build/Prepare
        mkdir -p $(PKG_BUILD_DIR)
        $(CP) ./src/* $(PKG_BUILD_DIR)/
endef
```

2. Build/Configure（可选）

如果使用configure进行源代码配置，可以调用"$(call Build/Configure/Default，XXX)"传递configure配置脚本参数。否则，不用定义这个函数。

举例：
```
define Build/Configure
        $(call Build/Configure/Default, \
                --with-kernel-support \
                --without-x \
        )
endef
```

3. Build/Compile（可选）

它定义了如何编译源代码。一般不需要定义这个函数，因为这时将使用默认值，即调用make。如果要传递特殊参数，调用"$(call Build /Compile /Default, XXX)"传递参数。

举例：
```
define Build/Compile
        $(call Build/Compile/Default, \
                DESTDIR="$(PKG_INSTALL_DIR)" \
                CC="$(TARGET_CC)" \
                all \
        )
endef
```

4. Build/Install（可选）

它定义了如何安装已编译的源代码，默认值为"make install"。如果要传递特殊参数或目标，调用"$(call Build/Install/Default，XXX)"。

举例：
```
define Build/Install
        $(MAKE) -C $(PKG_BUILD_DIR) \
                ARCH="$(LINUX_KARCH)" \
                CROSS_COMPILE="$(TARGET_CROSS)" \
                DESTDIR="$(PKG_INSTALL_DIR)" \
                DEPMOD="/bin/true" \
                install
endef
```

5. Build/InstallDev（可选）

它定义了依赖于这个软件包的其他软件包、需要这个软件包安装的库和头文件。在目标设备上没有用处。

举例：
```
define Build/InstallDev
        $(INSTALL_DIR) $(1)/usr/include/iwinfo
        $(CP) $(PKG_BUILD_DIR)/include/iwinfo.h $(1)/usr/include/
        $(CP) $(PKG_BUILD_DIR)/include/iwinfo/* $(1)/usr/include/iwinfo/
        $(INSTALL_DIR) $(1)/usr/lib
        $(INSTALL_BIN) $(PKG_BUILD_DIR)/libiwinfo.so $(1)/usr/lib/libiwinfo.so
        $(INSTALL_DIR) $(1)/usr/lib/lua
        $(INSTALL_BIN) $(PKG_BUILD_DIR)/iwinfo.so $(1)/usr/lib/lua/iwinfo.so
endef
```

6. Build/Clean（可选）

清理软件包。

举例：

```
define Build/Clean
        rm -rf $(BUILD_DIR)/$(PKG_NAME)/
endef
```

依赖类型

软件包之间相互依赖，需要经常定义软件包之间的相互依赖类型。为了清晰，假设我们要做的软件包叫packageA，表11-3所示是其依赖类型。常用依赖配置符号如表11-4所示。

表11-3　常用依赖类型

依赖类型	含义
+<foo>	packageA依赖软件包foo，当packageA被选中时，也会自动选中软件包foo
<foo>	packageA是否可见取决于软件包foo是否被选中，如果foo没有被选中，则packageA不可见
@FOO	packageA是否显示完全依赖于CONFIG_FOO是否配置，这经常用于特定的Linux内核版本或目标。如@TARGET_foo表示仅在foo可用时，软件包才可用。可以用复杂的表达式（&&、||、！等）来表示依赖关系，如@(!TARGET_foo&&!TARGET_bar)表示仅当foo和bar不可用时，软件包才可用
+FOO:<bar>	如果设置了CONFIG_FOO，则packageA依赖于bar，如果packageA被选择，则bar被自动选中。这通常用于依赖外部库的情况
@FOO:<bar>	如果设置了CONFIG_FOO，则packageA是否可见取决于bar，选择了bar，则packageA可见

表11-4　常用依赖配置符号

依赖符号	含义
TARGET_<foo>	目标foo被选择。对于DF1A开发板，foo为siflower
LINUX_3_X	使用Linux 3.X内核版本
USE_UCLIBC,USE_GLIBC,USE_EGLIBC	依赖uclibc、glibc、eglic库

11.1.6　软件包配置菜单

某些功能复杂的软件包，需要在OpenWrt配置选项菜单时设置软件包的一些选项参数，从而编译出某些特定功能版本，例如BusyBox。软件包配置菜单的编写语法同Linux内核的Kconfig文件语法。要实现这个功能，需要在软件包的Makefile文件中重写Package/<package name>/config函数。下面我们详细介绍一下如何编写软件包的配置菜单。

配置选项

关键字config定义了配置选项。所属配置选项下面的代码（TAB缩进）作为配置选项的属性。属性包含类型、默认值、数据范围、提示、依赖关系、帮助信息等。

● 属性类型：配置选项类型包括bool（布尔）、string（字符串）、hex（十六进制）、int（整型）、tristate（内置、模块、移除3种状态）。bool类型的选项只包含选中或不选中状态，显示为[ ]。tristate类型的选项显示为< >，<*>表示编译到固件中，<M>表示编译为模块，< >表示不编译此模块。类型定义后面会紧跟着输入提示和条件表达式。

例如：只有FUNC_B_SUPPORT被选中后，才显示FUNC_A_SUPPORT配置选项。

```
define Package/demo/config
```

```
        config FUNC_A_SUPPORT
                bool "support funcA" if FUNC_B_SUPPORT

        config FUNC_B_SUPPORT
                bool
                prompt "support funcB"
endef
```

效果如图11-9所示。

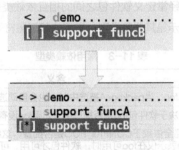

图11-9 菜单配置属性示例

● 输入提示：关键字为prompt，格式如下。

```
prompt <expression > [if <expression >]
```

if后面跟着依赖表达式。prompt与类型后面直接写提示字符是等价的。

例如下面两段配置等价。

```
config FUNC_A_SUPPORT
                bool "support funcA"
```

```
config FUNC_A_SUPPORT
                bool
prompt "support funcA"
```

● 默认值：关键字为default，格式如下。

```
default <expression > [if <expression >]
```

if后面跟着依赖表达式。如果用户不设置配置选项，配置选项的值为默认值。

例如：

```
config PROCD_SHOW_BOOT
     bool
     default y
     prompt "Print the shutdown to the console as well as logging it to syslog"
```

● 依赖关系：关键字为depend on或requires，格式如下。

```
default on <expression >
requires <expression >
```

如果有多重依赖，可以用&&进行连接。

例如：

```
config FUNC_A_SUPPORT
                bool "support funcA" if FUNC_B_SUPPORT
```

```
        config FUNC_C_SUPPORT
                bool "support funcC"
                depends on FUNC_B_SUPPORT && FUNC_A_SUPPORT
        config FUNC_B_SUPPORT
                bool
                default n
                prompt "support funcB"
```

效果如图11-10所示。

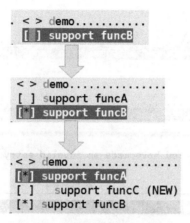

图11-10 菜单依赖关系配置示例

● 选择关系：关键字为select，也称为反向依赖，当配置选项被选中时，symbol也会被自动选中，格式如下。

```
select <symbol> <expression >
```

例如：当FUNC_B_SUPPORT被选中时，FUNC_D_SUPPORT也会被自动选中。

```
        config FUNC_D_SUPPORT
                bool "support funcD"
        config FUNC_B_SUPPORT
                bool
                default n
                prompt "support funcB"
                select FUNC_D_SUPPORT
```

效果如图11-11所示。

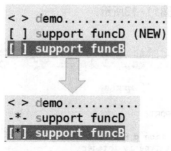

图11-11 菜单选择关系配置示例

- 数据范围：关键字为range，限制用户输入数据的范围，常用于限定int和hex类型输入值的范围，可接受值的范围为大于等于symbol1到小于等于symbol2的范围，格式如下。

```
range <symbol 1> <symbol 2> [if <expression >]
```

例如：配置可接受范围为[0-10]。

```
config FUNC_D_SUPPORT
        int "configure input integer value range [0-10]:"
        range 0 10
```

效果如图11-12所示。

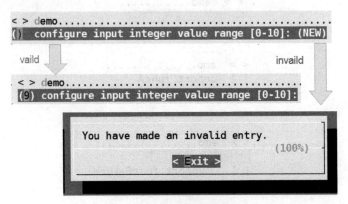

图11-12 数据范围配置示例

- 帮助信息：关键字为help或---help---，可以定义多段帮助信息，在menuconfig中按下h键可以显示出帮助信息，格式如下。

```
help (或---help---)
...
...
```

例如：

```
config FUNC_D_SUPPORT
        int "configure input integer value range [0-10]:"
        range 0 10
        help
                configure the function d PWM control
                range,accept input type is integer
                ,accpet value range is [0-10].
```

按下h键后显示的帮助信息如图11-13所示。

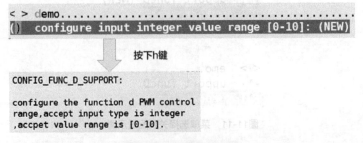

图11-13 帮助信息示例

- 表达式：即expression，表达式的内容可以为symbol（符号，即配置项定义的常数，如FUNC_A_SUPPORT）、symbol相等、symbol不相等、expression的赋值（如n或y，及数值等）以及expression间的与、或、非运算。

假设A和B是config定义的配置选项，表达式的内容可以为以下样子。

（1）A：A被选中时，显示配置项。
（2）A=B：即A和B都被选中或者都没被选中时，才显示配置项。
（3）A!=B：A和B不同时被选中或者不同时没被选中时，才显示配置项。
（4）!A：A没被选中时，选中配置项。
（5）A=y或A!=n：A被选中时，选中配置项。
（6）A && B：A和B同时被选中时，选中配置项。
（7）A||B=!n：A被选中并且B也被选中时，才选中配置选项。

菜单

菜单使用关键字menu和endmenu表示。处于menu和endmenu之间的配置选项会成为菜单的子菜单，并且所有子菜单都会继承父菜单的menu的依赖关系。格式如下：

```
menu "父菜单"
...
config 配置选项
...
endmenu
```

例如："DF1A Device Support"对DF1A的依赖，也会被DF1A_I2C_SUPPORT和DF1A_SPI_SUPPORT继承。

```
menu "DF1A Device Support"
depends on DF1A
     config DF1A_I2C_SUPPORT
          bool
          default n
          prompt "support i2c device"
     config DF1A_SPI_SUPPORT
          bool
          default n
          prompt "support SPI device"
endmenu
```

效果如图11-14所示。

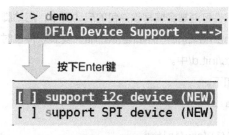

图11-14 菜单示例

多项选择

多项选择使用关键字choice和endchoice表示，支持选项列表，格式如下：

```
choice
    ...
    ...
endchoice
```

例如：支持2种固件选择。

```
choice
        prompt "DF1A firmware version"
        default DF1A_DEFAULT if !TARGET_siflower
        default DF1A_1_2_1_34 if TARGET_adm5120
        help
          This option allows you to select the version of the DF1A firmware.
    config DF1A_DEFAULT
        bool "Default"
        help
          Default firmware for DF1A devices.
          If unsure, select this.
    config ACX_1_2_1_34
        bool "1.2.1_34"
        help
          1.2.1_34 firmware for DF1A devices. Works with Siflower Panel device.
          If unsure, select the "default" firmware.
endchoice
```

效果如图11-15所示。

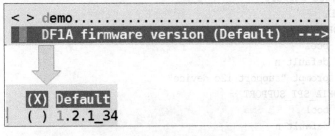

图11-15　多项选择示例

11.1.7 软件包启动脚本

在OpenWrt中，编写软件包通常会编写软件包的启动脚本，启动脚本放在软件包files目录中，约定命名规则为<packagename>.init，例如netifd.init。安装软件包时，把<packagename>.init复制到目标文件系统中的/etc/init.d/中。

foo软件包的启动脚本如下。

```
define Package/foo/install
...
        $(INSTALL_DIR) $(1)/etc/init.d
```

```
        $(INSTALL_BIN) ./files/foo.init $(1)/etc/init.d/foo
endef
```

启动脚本的格式

启动脚本有两种格式：一种是基于BusyBox init的旧格式SysV风格的init脚本，另一种是基于procd新格式的init脚本。OpenWrt对于两种风格的脚本都支持。

推荐采用procd格式。procd是OpenWrt下的进程守护程序，通过procd管理的启动服务进程，可以在配置和环境没有改变的情况下避免重启服务。我们可用ubus call service lis命令查看通过procd启动的所有服务程序。

启动脚本声明

创建启动脚本，首先要包含初始化脚本模板文件/etc/rc.common，并声明使用procd方式。

```
!/bin/sh /etc/rc.common
USE_PROCD=1
```

1. /etc/rc.common模板

/etc/rc.common模板中定义了启动脚本默认支持的命令及函数模板。/etc/rc.common是通过调用其他脚本来实现功能的。

```
. $IPKG_INSTROOT/lib/functions.sh
. $IPKG_INSTROOT/lib/functions/service.sh
```

默认的命令有start、stop、restart、reload、enable、disable等。

这些命令由/etc/rc.common中的函数定义，如果用户启动脚本实现了对应的函数，则调用用户定义的函数，否则调用模板默认函数。rc.common实现的模板函数如表11-5所示。

表11-5 rc.common 实现的模板函数

命令	实现函数	含义
start	start()	启动服务
stop	stop()	停止服务
reload	reload()	重新加载配置文件
restart	restart()	重启服务
	boot()	系统启动时会执行boot()，内部会调用start()函数
shutdown	shutdown()	系统关机时会执行shutdown()，内部会调用stop()函数
disable	disable()	根据初始化脚本STOP和START编号，删除/etc/rc.d/K<STOP><name>和/etc/rc.d/S<START><name>
enable	enable()	根据初始化脚本STOP和START编号，创建/etc/rc.d/K<STOP><name>和/etc/rc.d/S<START><name>
start	start_service()	以procd方式启动服务
stop	stop_service()	以procd方式停止服务
	service_triggers()	以procd方式监控引起服务变化的触发条件，如配置文件发生变化、网络状态发生变化等
reload	reload_service()	如果服务没有任何配置文件或参数，但是在reload被调用时依然需要强制服务进程重启，需要复写本函数

2. USE_PROCD=1

表明使用procd格式实现初始化脚本，要实现start_service()、stop_service()、service_triggers()等函数。

启动服务（start_service）

启动一个服务，需要实现start_service函数()，调用procd API函数。procd支持的API函数在/lib/functions/procd.sh中定义。常用的API有以下几个。

1. 创建实例

```
procd_open_instance [name]
```

说明：创建一个procd实例。name表示创建实例的名称，也可以省略。可以在start_service()中创建一个或多个实例。我们用ubus call service list命令可以查看服务及实例信息。

示例：/etc/init.d/samba init脚本。

```
start_service() {
    init_config
    procd_open_instance
    procd_set_param command /usr/sbin/smbd -F
    procd_set_param respawn
    procd_close_instance
    procd_open_instance
    procd_set_param command /usr/sbin/nmbd -F
    procd_set_param respawn
    procd_close_instance
}
```

2. 关闭实例

```
procd_close_instance
```

说明：创建完一个procd实例要关闭实例。

3. 设置参数

```
procd_set_param type [value...]
```

说明：设置传递给启动的服务参数，支持的参数如表11-6所示。

表11-6 procd_set_param 支持的参数

参数	数据类型	描述	示例
respawn [threshold timeout retry]	数字列表	设置启动服务崩溃后自动重启。respawn后面跟着3个连续的数字。threshold（阈值）表示重新启动尝试限制的时间范围（单位为秒）。timeout表示服务在尝试重启之前等待的时间（单位为秒）。retry表示服务崩溃之前将进行多少次重新启动尝试；除非执行显式重新启动，如restart，否则不会对此崩溃的服务进行重新启动尝试。将重试值设置为0将导致procd尝试无限期重新启动服务。respawn后面的参数可以省略，默认值是"3600 5 5"	procd_set_param respawn 300 5 5 procd_set_param respawn
env	键值对列表	设置派生进程的环境变量	procd_set_param env VARIABLE=4

data	键值对列表	将key=value表示法中的任意用户数据设置为ubus服务状态。即为ubus中这个服务添加额外数据，以便其他服务使用	procd_set_param data hello=world
			通过"ubus call service list"查看服务节点，会多出data节点:
			"data": {
			    "hello": "world"
			}

续表

参数	数据类型	描述	示例
file	file	将文件名列表（通常为UCI配置文件）传递给procd以监视更改。启动服务时，每个传递的文件的内容都会计算校验和，并作为procd的内存中服务状态的一部分存储。处理服务重新加载请求时，如果其中一个关联文件的校验和已更改，或文件列表本身已更改，则运行的服务状态将失效，procd将重新启动关联的进程或向其传递UNIX信号	procd_set_param file /etc/config/system
netdev	列表	将Linux网络设备名称的列表传递给要监视其更改的procd。在启动服务时，每个网络设备名称的接口索引将被解析并存储为procd的内存中服务状态的一部分。处理服务重新加载请求时，如果任何关联网络设备的接口索引发生更改，或者列表本身发生更改，则运行的服务状态将失效，procd将重新启动关联的进程或向其传递UNIX信号	
command	列表	设置用于执行进程的argv，即命令行参数	procd_set_param command "$PROG" -f -c /etc/crontabs -l ${loglevel:-5}
limits	键值对列表	通过key=value传递给派生进程设置ulimit值。procd可识别以下限制名称：as(RLIMIT_as)、core(RLIMIT_are)、cpu(RLIMIT_arecpu)、data(RLIMIT_aredata)、fsize(RLIMIT_arefsize)、memlock(RLIMIT_arememlock)、nofile(RLIMIT_arenofile)、nproc(RLIMIT_arenproc)、rss(RLIMIT_arerss)、stack(RLIMIT_arestack)、nice(RLIMIT_arenice)、rtprio(RLIMIT_arertprio)、msgqueue(RLIMIT_aremsgqueue)、信号等待(RLIMIT_sigpinding)	procd_set_param limits core="unlimited"
watch	列表	将ubus命名空间的列表传递给watch-procd将向每个名称空间订阅并等待传入的ubus事件，这些事件随后被转发到注册的JSON脚本触发器进行计算	procd_set_param watch network.interface
nice	整型	设置生成的进程的调度优先级。有效值范围为-20（高）~19（低）	
user	字符串	指定要生成进程的用户名。procd将在/etc/passwd中查找给定的名称，并相应地设置派生进程的有效uid和主指南。如果省略，进程将生成为根（uid 0, gid 0）	
stdout	布尔	如果设置为1，则procd将进程的stdout派生到系统日志。以command的第一个参数的name作为id，以LOG_INFO为优先级，使用LOG_DAEMON发送到syslog	procd_set_param stdout 1
stderr	布尔	如果设置为1，则procd将进程的stderr派生到系统日志。以command的第一个参数的name作为id，以LOG_INFO为优先级，使用LOG_DAEMON发送到syslog	procd_set_param stderr 1

停止服务（stop_service）

通常可以不写stop_service()函数，也可以根据情况复写stop_service()函数。

例如：/etc/init.d/dnsmasq

```
stop_service() {
    [ -f /tmp/resolv.conf ] && {
            rm -f /tmp/resolv.conf
            ln -s /tmp/resolv.conf.auto /tmp/resolv.conf
    }
    rm -f /var/run/dnsmasq.*.dhcp
}
```

重载服务（service_triggers）

当服务依赖外界环境变化时，可以在init脚本中添加一个service_triggers()函数，把依赖变化的内容写在里面，procd会监控这些变化，执行触发动作。

例如：/etc/init.d/system添加了2个触发，其中当/etc/config/system配置文件发生变化时执行reload。

```
service_triggers()
{
        procd_add_reload_trigger "system"
        procd_add_validation validate_system_section
}
```

重载服务（reload_service）

可以设定"procd_set_param file /etc/config/yourconfig"来监控服务的配置文件是否发生变化，procd会检测配置文件的MD5值是否发生变化，当文件变化时才会触发reload命令，会执行一次stop/start调用。对于没有配置文件的服务，procd不会自动触发reload命令。如果想强制执行服务进程的重启，可以实现reload_service()函数，内部调用stop()/start()函数。

```
reload_service()
{
        echo "Explicitly restarting service, are you sure you need this?"
        stop
        start
}
```

启动脚本调试

要跟踪脚本的所有调用，在执行的脚本前面加上INIT_TRACE=1。

例如：

```
DF1A:~$ INIT_TRACE=1 /etc/init.d/rpcd start
+ . /etc/init.d/rpcd
+ START=12
+ PROCD_DEBUG=1
+ USE_PROCD=1
+ NAME=rpcd
+ PROG=/sbin/rpcd
+ [ -n 1 ]
+ EXTRA_COMMANDS= running trace
+ . /lib/functions/procd.sh
+ . /usr/share/libubox/jshn.sh
...
```

procd调试：在脚本中添加PROCD_DEBUG=1，可以在start/stop时打印procd调试信息。

```
DF1A:~$ /etc/init.d/foo start
{ "name": "foo", "script": "\/etc\/init.d\/foo", "instances": { "lala": { "command": [ "\/sbin\/foo" ] } }, "triggers": [ [ "config.change", [ "if", [ "eq", "package", "system" ], [ "run_script", "\/etc\/init.d\/foo", "reload" ] ] ] ] }
DF1A:~$ /etc/init.d/foo stop
{ "name": "foo" }
```

实例

下面以创建/etc/init.d/foo启动脚本为例，来说明创建启动脚本的过程。

1. 编写程序

编写可执行程序，确定程序支持的参数。为了调试和演示方便，假设我们的foo程序是一段Shell程序，内容如下。

```sh
#!/bin/sh
[ -n "$1" ] && name=$1 || {
        name="df1a"
}
while [ 1 ]; do
        echo "Hello, $name ,welcome to OpenWrt World! VAR=$VAR"
        sleep 1
done
exit 0
```

把foo.sh程序放到/sbin/目录下并重命名为foo，设定权限为0755：

```
DF1A:~$ cp foo.sh /sbin/foo
DF1A:~$ chmod 755 /sbin/foo
```

我们先测试一下这段程序，不加任何参数，每隔1s输出欢迎默认用户df1a的信息：

```
DF1A:~$ /sbin/foo
Hello, df1a ,welcome to OpenWrt World! VAR=
Hello, df1a ,welcome to OpenWrt World! VAR=
...
```

添加参数lele，再次执行程序，会根据传递的参数，打印欢迎信息：

```
DF1A:~$ /sbin/foo lele
Hello, lele ,welcome to OpenWrt World! VAR=
Hello, lele ,welcome to OpenWrt World! VAR=
...
```

2. 编写初始化脚本

```sh
#!/bin/sh /etc/rc.common
START=10
STOP=15
USE_PROCD=1
NAME=foo
PROG=/sbin/foo
start_service() {
        echo "start $NAME service"
        procd_open_instance
        procd_set_param command "$PROG" zhengqw
        procd_set_param respawn
        procd_set_param env VAR=2
        procd_set_param data NAME=ZQW
        procd_set_param file /etc/config/system
        procd_set_param limits core "unlimited"
        procd_set_param stderr 1
        procd_set_param stdout 1
        procd_close_instance
}
```

```
service_triggers()
{
    procd_add_reload_trigger "system"
}
stop_service()
{
    echo "stop $NAME service"
}
reload_service()
{
    echo "reload $NAME service"
    stop
    start
}
```

3. 开机启动服务

设置开机启动服务，如图11-16所示。

```
DF1A:$ /etc/init.d/foo enable
DF1A:$ ls /etc/rc.d/
K10ddns          S12nlpkd         S60samba
K15foo           S12rpcd          S60xl2tpd
K15miniupnpd     S13ts-init       S61syncservice
K50dropbear      S14acl           S62subservice
K85odhcpd        S19network       S80relayd
K89log           S20firewall      S90advanced_wifi
K90network       S35odhcpd        S94miniupnpd
K98boot          S40fstab         S95ddns
K99umount        S50cron          S95done
S00sysfixtime    S50dropbear      S96led
S10boot          S50telnet        S96set_default_cpufreq
S10foo           S50uhttpd        S98sysntpd
S10system        S50vsftpd        S99device_listen
S11sysctl        S51init_pctl
S12log           S60dnsmasq
```

图11-16　设置开机启动

4. 启动服务

```
DF1A:~$ /etc/init.d/foo start
```

我们通过"logread -f"监控输出，可以看到环境变量VAR和用户名参数已经传递给服务进程，如图11-17所示。

```
start foo service
DF1A:~$ logread -f
Thu Nov 21 09:46:49 2019 daemon.info foo[1137]: Hello, zhengqw ,welcome to OpenWrt World! VAR=2
Thu Nov 21 09:46:50 2019 daemon.info foo[1137]: Hello, zhengqw ,welcome to OpenWrt World! VAR=2
Thu Nov 21 09:46:51 2019 daemon.info foo[1137]: Hello, zhengqw ,welcome to OpenWrt World! VAR=2
Thu Nov 21 09:46:52 2019 daemon.info foo[1137]: Hello, zhengqw ,welcome to OpenWrt World! VAR=2
Thu Nov 21 09:46:53 2019 daemon.info foo[1137]: Hello, zhengqw ,welcome to OpenWrt World! VAR=2
```

图11-17　通过"logread -f"监控输出

5. 停止服务

我们通过kill命令杀死foo进程，procd也会"重新复活"（respawn）foo进程，除非执行停止服务：

```
DF1A:~$ /etc/init.d/foo stop
```

11.2 创建常规软件包

有了前面的软件包构建基础，下面我们通过实际的案例来学习如何创建软件包。软件包分为常规软件包和内核软件包。常规软件包主要面向应用层程序；内核软件包为内核功能模块，面向内核层程序。创建内核软件包将在下节中介绍，本节我们来学习如何创建一个常规软件包。

11.2.1 软件包的信息

假设要创建的常规软件包名称为nlpk，其基本信息如表11-7所示。

表 11-7 nlpk 软件包的信息

基本信息	说明
软件包名称	nlpk
软件包版本	1.0
软件包功能	实现配置参数传递、信号注册、UCI文件读取、ubus注册和功能调用、系统帮助
源码获取方式	源代码存放在本地，即软件包src目录，采用.tar.gz格式压缩
源码构建方式	CMake
包含两个子软件包	nlpk和libnlpk。nlpk为常规软件包，libnlpk为提供动态库支持的软件包
补丁	提供补丁文件
配置文件和启动脚本	UCI配置文件为/etc/config/nlpk，启动脚本为/etc/init.d/nlpkd
可执行文件和库	提供/usr/bin/nlpkd和/lib/libnlpk.so
软件包依赖	+libuci +libubus +ubus +ubusd +jshn +libubox +libnlpk

创建软件包

创建软件包目录：

```
HOST:chaos_calmer_15_05_1$ mkdir -p package/df1a/nlpk
```

创建Makefile文件，src、patches、files目录：

```
HOST:chaos_calmer_15_05_1$ cd package/df1a/nlpk
HOST:nlpk$ touch Makefile
HOST:nlpk$ mkdir src patches files
```

准备源代码的全部过程都要在packages/df1a/nlpk中来完成，请提前进入该目录，如图11-18所示。

图11-18 nlpk软件包的目录和文件

11.2.2 准备软件包源代码

nlpk源代码为演示工程,该工程为假设演示,将这些内容放置于src目录下。这个源代码包含了两个项目,分别为nlpk与libnlpk,其关系如图11-19所示。src目录下的源代码文件如图11-20所示。

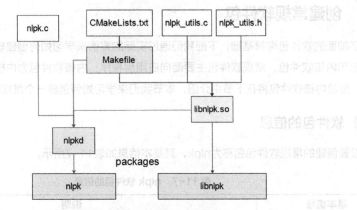

图11-19 nlpk和libnlpk软件包的关系

图11-20 src目录下的源代码文件

1. 创建src/nlpk.c文件:

```c
#include <stdio.h>
#include <stdlib.h>
#include <string.h>
#include <getopt.h>
#include <signal.h>
#include <stdarg.h>
#include <syslog.h>
#include "nlpk_utils.h"
#include <libubox/uloop.h>
#include <libubox/utils.h>
#include <libubus.h>
#include <uci.h>
enum {
    NLPK_SAY_HELLO_USER_NAME,
    __NLPK_SAY_HELLO_MAX
};
extern void say_hello(char * name);
static struct ubus_context *ctx;
static const struct blobmsg_policy rpc_api_say_hello_policy[__NLPK_SAY_HELLO_MAX]
= {
```

```c
        [NLPK_SAY_HELLO_USER_NAME]        = { .name = "username", .
type = BLOBMSG_TYPE_STRING }
};
static struct blob_buf buf;
static void handle_signal(int signo)
{
        uloop_end();
}
static void setup_signals(void)
{
        struct sigaction s;
        memset(&s, 0, sizeof(s));
        s.sa_handler = handle_signal;
        s.sa_flags = 0;
        sigaction(SIGINT, &s, NULL);
        sigaction(SIGTERM, &s, NULL);
        sigaction(SIGUSR1, &s, NULL);
        sigaction(SIGUSR2, &s, NULL);
        s.sa_handler = SIG_IGN;
        sigaction(SIGPIPE, &s, NULL);
}
static int rpc_api_say_hello(struct ubus_context *ctx, struct ubus_object *obj,
                    struct ubus_request_data *req, const char *method,
                    struct blob_attr *msg)
{
        struct blob_attr *tb[__NLPK_SAY_HELLO_MAX];
        blobmsg_parse(rpc_api_say_hello_policy, __NLPK_SAY_HELLO_MAX, tb,
                    blob_data(msg), blob_len(msg));
        if (!tb[NLPK_SAY_HELLO_USER_NAME])
                return UBUS_STATUS_INVALID_ARGUMENT;
        blob_buf_init(&buf, 0);
        blobmsg_add_string(&buf,blobmsg_get_string(tb[NLPK_SAY_HELLO_USER_NAME])
,"welcome to here!");
        ubus_send_reply(ctx, req, buf.head);
        return UBUS_STATUS_OK;
}
static int nlpk_api_init(struct ubus_context *ctx)
{
        static const struct ubus_method api_methods[] = {
                UBUS_METHOD("say_hello", rpc_api_say_hello, rpc_api_say_hello_
policy),
        };
        static struct ubus_object_type api_type =
                UBUS_OBJECT_TYPE("nlpk-api", api_methods);
        static struct ubus_object obj = {
                .name = "nlpk-rpc-api",
                .type = &api_type,
                .methods = api_methods,
                .n_methods = ARRAY_SIZE(api_methods),
        };
```

```c
        return ubus_add_object(ctx, &obj);
}
int main(int argc, char **argv)
{
        const char *ubus_socket = NULL;
        int ch;
        int rv = 0;
        while ((ch = getopt(argc, argv, "s:")) != -1) {
                switch (ch) {
                case 's':
                        ubus_socket = optarg;
                        break;
                default:
                        break;
                }
        }

#ifdef DEBUG
        printf("Debug On\n");
#else
        printf("Debug Off\n");
#endif
#ifdef DF1A_FW_VER
        printf("fw ver define\n");
#else
        printf("fw ver un define\n");
#endif
#if DF1A_FW_VER == 3
        printf("Match firmware version 3.0\n");
#else
        printf("Match firmware version default 1.0\n");
#endif
        setup_signals();
        /*uci*/
        struct uci_context *uci = NULL;
        struct uci_package *p = NULL;
        uci = uci_alloc_context();
        if (!uci) {
                rv = UBUS_STATUS_UNKNOWN_ERROR;
                goto out;
        }
        struct uci_ptr ptr = { .package = "nlpk" };
        struct uci_element *e;
        struct uci_section *s;
        uci_load(uci, ptr.package, &p);
        if (!p){
                rv = UBUS_STATUS_UNKNOWN_ERROR;
                goto out;
        }
        uci_foreach_element(&p->sections, e)
```

```
            {
                    s = uci_to_section(e);
                    if (strcmp(s->type, "sayhello"))
                            continue;
                    ptr.section = s->e.name;
                    ptr.s = NULL;
                    /* test for matching username */
                    ptr.option = "username";
                    ptr.o = NULL;
                    if (uci_lookup_ptr(uci, &ptr, NULL, true))
                            continue;
                    if (ptr.o->type != UCI_TYPE_STRING)
                            continue;
                    if(ptr.o->v.string)
                            break;
            }

       say_hello(ptr.o->v.string);
       uloop_init();
       ctx = ubus_connect(ubus_socket);
       if (!ctx) {
                fprintf(stderr, "Failed to connect to ubus\n");
                return -1;
       }
       printf("Connected to ubus, id=%08x\n", ctx->local_id);
       ubus_add_uloop(ctx);
       nlpk_api_init(ctx);
       uloop_run();
       ubus_free(ctx);
       uloop_done();
out:
       if (uci)
                uci_free_context(uci);
       return rv;
}
```

2. 创建src/nlpk_utils.c文件：

```
#include <stdio.h>
void say_hello(char * name);
void say_hello(char * name)
{
       printf("Hello,%s\n",name);
}
```

3. 创建src/nlpk_utils.h文件：

```
#ifdef __NLPK_UTILS_H__
extern void say_hello(char *name);
#endif
```

4. 创建src/CMakeLists.txt文件：

```
cmake_minimum_required(VERSION 2.6)
```

```
PROJECT(nlpk C)
ADD_DEFINITIONS(-Os -Wall -Werror -Wmissing-declarations)
SET(CMAKE_SHARED_LIBRARY_LINK_C_FLAGS "")
OPTION(DEBUG "option for debug" OFF)
OPTION(DF1A_FW_VER "match version" 1)
IF(DEBUG)
  ADD_DEFINITIONS(-DDEBUG -g3)
ENDIF(DEBUG)
IF(DF1A_FW_VER)
  ADD_DEFINITIONS(-DDF1A_FW_VER=${DF1A_FW_VER})
ENDIF(DF1A_FW_VER)
#MESSAGE(${DEBUG})
#MESSAGE(${DF1A_FW_VER})
# install libnlpk header files
INSTALL(FILES nlpk_utils.h DESTINATION include/libnlpk)
SET(LIB_SOURCES nlpk_utils.c)
ADD_LIBRARY(nlpk SHARED ${LIB_SOURCES})
INSTALL(TARGETS nlpk
        LIBRARY DESTINATION lib
)
SET(SRCS nlpk.c)
SET(LIBS ubox ubus uci json-c blobmsg_json nlpk)
ADD_EXECUTABLE(nlpkd ${SRCS})
TARGET_LINK_LIBRARIES(nlpkd ${LIBS})
```

准备软件源代码

将src目录下的文件打包为nlpk-2019-1.tar.gz文件（稍后该名字要用于软件包Makefile的PKG_SOURCE），如图11-21所示。

```
PKG_SOURCE:=$(PKG_NAME)-$(PKG_VERSION)-$(PKG_RELEASE).tar.gz
```

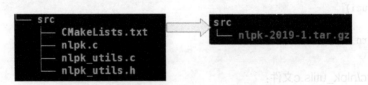

图11-21 打包src下所有源代码文件为.tar.gz压缩包

压缩软件包（这一过程将删除掉已经准备好的nplk文件，我们可以提前备份nplk文件，用于重复实验）：

```
#根据当前所在目录，进入nlpk的src目录
HOST:src$ ls
CMakeLists.txt  nlpk.c  nlpk_utils.c  nlpk_utils.h
HOST:src$ tar -czvf nlpk-2019-1.tar.gz * --remove-files
CMakeLists.txt
nlpk.c
nlpk_utils.c
```

计算nlpk-2019-1.tar.gz的md5sum值:

```
HOST:src$ md5sum nlpk-2019-1.tar.gz
b993d734ec7672f96cd0d7166a69217d  nlpk-2019-1.tar.gz
```

MD5值每次都不同,记录下最后一次生成的该值用于软件包Makefile的PKG_MD5SUM:

```
PKG_MD5SUM:=b993d734ec7672f96cd0d7166a69217d
```

准备nlpk的patch文件

为了演示patch功能,我们创建一个补丁,将其放到patches目录中,让它支持usage帮助功能。

创建patches/001-support-help.patch文件:

```
--- a/nlpk.c    2019-11-14 11:51:06.866391040 +0800
+++ b/nlpk.c    2019-11-14 11:52:15.104820859 +0800
@@ -89,17 +89,29 @@
        return ubus_add_object(ctx, &obj);
 }
+static int usage(void)
+{
+       printf("nlpkd is an normal package example\n");
+       printf("Usage nlpkd [options]\n");
+       printf("\t-s <path> Path to ubus socket file\n");
+       printf("\t-h help\n");
+
+       return 1;
+}
+
 int main(int argc, char **argv)
 {
        const char *ubus_socket = NULL;
        int ch;
        int rv = 0;
-       while ((ch = getopt(argc, argv, "s:")) != -1) {
+       while ((ch = getopt(argc, argv, "s:h")) != -1) {
                switch (ch) {
                case 's':
                        ubus_socket = optarg;
                        break;
+               case 'h':
+                       return usage();
                default:
                        break;
                }
```

11.2.3 编写软件包

上面已经准备完成nplk的全部源代码,现在我们开始编写用于OpenWrt系统的软件包。

创建Makefile文件（请注意PKG_SOURCE、PKG_MD5SUM以上面的值为准）：

```
#
# Copyright (C) 2019 Freeirs
#
# This is free software, licensed under the GNU General Public License v2.
# See /LICENSE for more information.
#
include $(TOPDIR)/rules.mk
PKG_NAME:=nlpk
PKG_VERSION:=2019
PKG_RELEASE:=1
PKG_BUILD_DIR:=$(BUILD_DIR)/$(PKG_NAME)-$(PKG_VERSION)-$(PKG_RELEASE)
PKG_SOURCE:=$(PKG_NAME)-$(PKG_VERSION)-$(PKG_RELEASE).tar.gz
PKG_MD5SUM:=b993d734ec7672f96cd0d7166a69217d
CMAKE_INSTALL:=1
PKG_LICENSE:=GPL-2.0
PKG_LICENSE_FILES:=
PKG_MAINTAINER:=Zheng QiWen <zhengqwmail@gmail.com>
include $(INCLUDE_DIR)/package.mk
include $(INCLUDE_DIR)/cmake.mk
define Package/$(PKG_NAME)
SECTION:=example
  CATEGORY:=DF1A Packages
  DEPENDS:= +libuci +libubus +ubus +ubusd +jshn +libubox +libnlpk
  TITLE:=df1a normal package example
endef
define Package/lib$(PKG_NAME)
SECTION:=example
  CATEGORY:=DF1A Packages
  DEPENDS:=+libubox +libuci
  TITLE:=df1a normal package library example
endef
define Package/$(PKG_NAME)/config
menu "Configuration"
        depends on PACKAGE_$(PKG_NAME)
config NLPG_DEBUG
        bool
        default n
        prompt "Debug On"
help
        Open nlpkg Debug On/Off
endmenu
choice
        prompt "Adapter firmware version"
        default NLPK_MATCH_DF1A_FW_VER_DEFAULT if !TARGET_siflower
        default NLPK_MATCH_DF1A_FW_VER_3 if TARGET_arm64
        help
         This option allows you to select the adapter  version of the DF1A firmware.
        config NLPK_MATCH_DF1A_FW_VER_DEFAULT
            bool "Default"
```

```
                    help
                        Default firmware for DF1A devices.
                        If unsure, select this.
            config NLPK_MATCH_DF1A_FW_VER_3
                    bool "3.0"
                    help
                        3.0 firmware for DF1Ai ARM64 devices.
                        If unsure, select the "default" firmware.
endchoice
endef
ifeq ($(CONFIG_NLPK_DEBUG),y)
  CMAKE_OPTIONS += -DDEBUG=ON
endif
ifeq ($(CONFIG_NLPK_MATCH_DF1A_FW_VER_DEFAULT),y)
  CMAKE_OPTIONS += -DDF1A_FW_VER=1
endif
ifeq ($(CONFIG_NLPK_MATCH_DF1A_FW_VER_3),y)
  CMAKE_OPTIONS += -DDF1A_FW_VER=3
endif
define Build/Prepare
        rm -rf $(PKG_BUILD_DIR)
        mkdir -p $(PKG_BUILD_DIR)
        $(TAR) -C $(PKG_BUILD_DIR) -xzf ./src/$(PKG_SOURCE)
        $(Build/Patch)
endef
define Package/$(PKG_NAME)/install
        $(INSTALL_DIR) $(1)/usr/bin $(1)/usr/lib
        $(INSTALL_BIN) $(PKG_BUILD_DIR)/nlpkd $(1)/usr/bin
        $(INSTALL_DIR) $(1)/etc/init.d
        $(INSTALL_BIN) ./files/nlpk.init $(1)/etc/init.d/nlpkd
        $(INSTALL_DIR) $(1)/etc/config
        $(INSTALL_CONF) ./files/nlpk.config $(1)/etc/config/nlpk
endef
define Package/lib$(PKG_NAME)/install
        $(INSTALL_DIR) $(1)/usr/lib
        $(INSTALL_DATA) $(PKG_INSTALL_DIR)/usr/lib/libnlpk.so $(1)/usr/lib
endef
$(eval $(call BuildPackage,nlpk))
$(eval $(call BuildPackage,libnlpk))
```

Makefile脚本实现的内容如下。

（1）定义了软件包的基本信息。

（2）使用Package/nlpk和Package/libnlpk函数定义nlpk和libnlpk两个软件包的基本信息，这些信息将在menuconfig菜单中显示。

（3）使用Package/nlpk/config函数定义nlpk软件包的配置菜单，这些信息将在menuconfig菜单中显示，nlpk软件包依赖libnlpk软件包，当选中nlpk软件包后，会自动选中libnlpk软件包。

（4）根据nlpk软件包设定的配置，生成CMAKE_OPTIONS选项。

（5）使用Build/Prepare函数定义软件包在编译前要执行的动作。由于nlpk源码压缩包放在src目录，所以在编译前，首先要创建编译目录nlpk-2019-1，再解压源代码文件到编译目录，然后打上

patches目录下的补丁文件。

（6）使用Package/nlpk/install和Package/libnlpk/install函数实现软件包文件（配置文件、可执行文件、启动脚本、动态库）的安装。

（7）定义软件包编译入口$(eval $(call BuildPackage,nlpk))和$(eval $(call BuildPackage,libnlpk))。

启动脚本与UCI配置

files目录下的脚本和配置文件如图11-22所示。

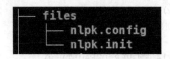

图11-22　files目录下的脚本和配置文件

创建files/nlpk.init文件：

```
#!/bin/sh /etc/rc.common

START=12
STOP=99
USE_PROCD=1
NAME=nlpkd
PROG=/usr/bin/nlpkd
start_service() {
    procd_open_instance
    procd_set_param command "$PROG"
    procd_close_instance
}
stop() {
    service_stop ${PROG}
}
reload() {
    service_reload ${PROG}
}
```

创建src/nlpk.config文件：

```
config sayhello
    option username 'admin'
```

nlpk软件包目录所有目录和文件如图11-23所示。

图11-23　nlpk软件包目录所有目录和文件

11.2.4 编译测试nlpk

使用make menuconfig命令配置软件包:

```
#请先切换到OpenWrt的根目录,然后操作如下命令
HOST:chaos_calmer_15_05_1$ make menuconfig
```

nlpk和libnlpk软件包配置如图11-24所示。

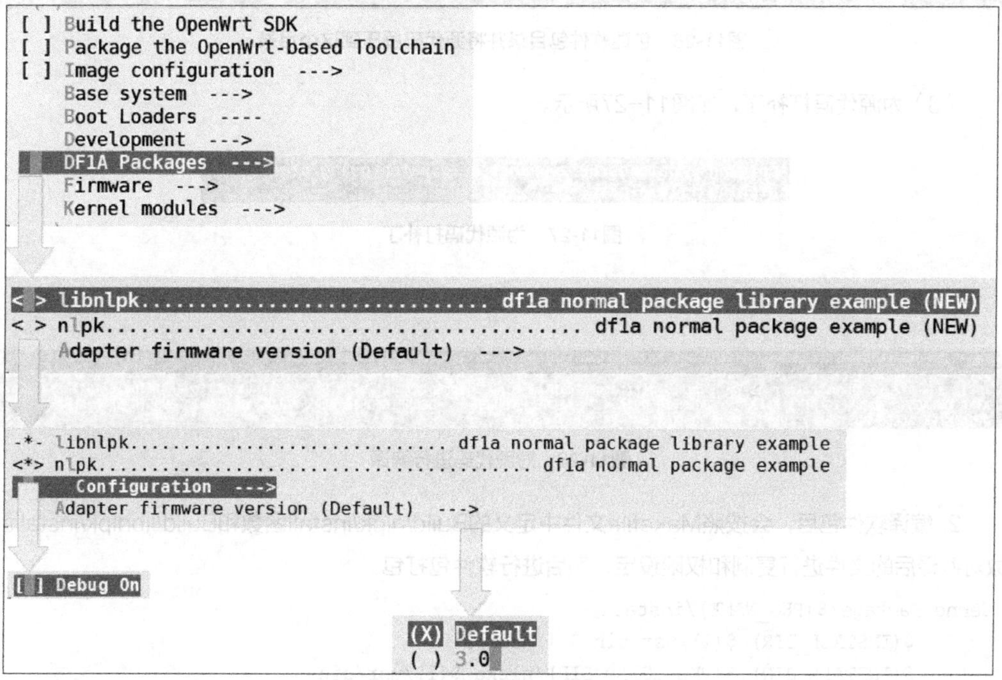

```
[ ] Build the OpenWrt SDK
[ ] Package the OpenWrt-based Toolchain
[ ] Image configuration  --->
    Base system  --->
    Boot Loaders  ---->
    Development  --->
    DF1A Packages  --->
    Firmware  --->
    Kernel modules  --->


< > libnlpk........................... df1a normal package library example (NEW)
< > nlpk.............................. df1a normal package example (NEW)
    Adapter firmware version (Default)  --->


-*- libnlpk........................... df1a normal package library example
<*> nlpk.............................. df1a normal package example
    Configuration  --->
    Adapter firmware version (Default)  --->


[ ] Debug On

                          (X) Default
                          ( ) 3.0
```

图11-24 nlpk和libnlpk软件包配置

编译软件包

当对OpenWrt进行固件编译时会自动完成,这里使用单独编译用于测试。

```
HOST:chaos_calmer_15_05_1$ make package/df1a/nlpk/{clean,compile} V=99
```

编译流程说明

1. 编译软件包前,会按照软件包Makefile文件中定义的Build/Prepare函数执行下面的内容。

```
define Build/Prepare
    rm -rf $(PKG_BUILD_DIR)
    mkdir -p $(PKG_BUILD_DIR)
    $(TAR) -C $(PKG_BUILD_DIR) -xzf ./src/$(PKG_SOURCE)
    $(Build/Patch)
endef
```

编译前执行的过程输出如下。

(1)删除软件包编译目录,如图11-25所示。

```
rm -f /home/zhengqw/dragonfly/sf16a18-sdk-4.2.10/chaos_calmer_15_05_1/bin/siflower/packages/base/nlpk_*
rm -f /home/zhengqw/dragonfly/sf16a18-sdk-4.2.10/chaos_calmer_15_05_1/bin/siflower/packages/base/libnlpk_*
```

图11-25　删除build_dir下的软件包编译目录

（2）创建软件包编译目录并将软件包源代码解压到这个目录，如图11-26所示。

```
mkdir -p /home/zhengqw/dragonfly/sf16a18-sdk-4.2.10/chaos_calmer_15_05_1/build_dir/target-mipsel_mips-interAptiv_uClibc-0.9.33.2/nlpk-2019-1
tar -C /home/zhengqw/dragonfly/sf16a18-sdk-4.2.10/chaos_calmer_15_05_1/build_dir/target-mipsel_mips-interAptiv_uClibc-0.9.33.2/nlpk-2019-1 -xzf ./src/nlpk-2019-1.tar.gz
```

图11-26　创建软件包目录并将源代码解压到这个目录

（3）为源代码打补丁，如图11-27所示。

```
Applying ./patches/001-support-help.patch using plaintext:
patching file nlpk.c
```

图11-27　为源代码打补丁

（4）对源代码进行编译，如图11-28所示。

```
make -C "/home/zhengqw/dragonfly/sf16a18-sdk-4.2.10/chaos_calmer_15_05_1/build_dir/target-mipsel_mips-interAptiv_uClibc-0.9.33.2/linux-siflower_sf16a18-fullmask/linux-3.18.29" ARCH="mips" CROSS_COMPILE="mipsel-openwrt-linux-uclibc-" SUBDIRS="/home/zhengqw/dragonfly/sf16a18-sdk-4.2.10/chaos_calmer_15_05_1/build_dir/target-mipsel_mips-interAptiv_uClibc-0.9.33.2/linux-siflower_sf16a18-fullmask/klpk" EXTRA_CFLAGS=" -DCONFIG_DF1A_KLPK=1 -DCONFIG_DF1A_KLPK_ALLOC_PHY_MEM=1 -DCONFIG_DF1A_KLPK_ALLOC_VM_MEM=1" CONFIG_DF1A_KLPK=m CONFIG_DF1A_KLPK_ALLOC_PHY_MEM=y CONFIG_DF1A_KLPK_ALLOC_VM_MEM=y modules
```

图11-28　对源代码进行编译

2. 编译软件包后，会按照Makefile文件中定义的Build/nlpk/install函数和Build/libnlpk/install函数对编译后的文件进行复制和权限设定，然后进行软件包打包。

```
define Package/$(PKG_NAME)/install
    $(INSTALL_DIR) $(1)/usr/bin $(1)/usr/lib
    $(INSTALL_BIN) $(PKG_BUILD_DIR)/nlpkd $(1)/usr/bin
    $(INSTALL_DIR) $(1)/etc/init.d
    $(INSTALL_BIN) ./files/nlpk.init $(1)/etc/init.d/nlpkd
    $(INSTALL_DIR) $(1)/etc/config
    $(INSTALL_CONF) ./files/nlpk.config $(1)/etc/config/nlpk
endef
define Package/lib$(PKG_NAME)/install
    $(INSTALL_DIR) $(1)/usr/lib
    $(INSTALL_DATA) $(PKG_INSTALL_DIR)/usr/lib/libnlpk.so $(1)/usr/lib
endef
```

（1）打包libnlpk文件，如图11-29所示。

```
install -d -m0755 /home/zhengqw/dragonfly/sf16a18-sdk-4.2.10/chaos_calmer_15_05_1/build_dir/target-mipsel_mips-interAptiv_uClibc-0.9.33.2/nlpk-2019-1/ipkg-mips_siflower/libnlpk/usr/lib
install -m0644 /home/zhengqw/dragonfly/sf16a18-sdk-4.2.10/chaos_calmer_15_05_1/build_dir/target-mipsel_mips-interAptiv_uClibc-0.9.33.2/nlpk-2019-1/ipkg-install/usr/lib/libnlpk.so /home/zhengqw/dragonfly/sf16a18-sdk-4.2.10/chaos_calmer_15_05_1/build_dir/target-mipsel_mips-interAptiv_uClibc-0.9.33.2/nlpk-2019-1/ipkg-mips_siflower/libnlpk/usr/lib
```

图11-29　打包libnlpk文件

（2）打包nlpk文件，如图11-30所示。

```
install -d -m0755 /home/zhengqw/dragonfly/sf16a18-sdk-4.2.10/chaos_calmer_
v_uClibc-0.9.33.2/root-siflower/tmp-nlpk/usr/bin /home/zhengqw/dragonfly/s
ir/target-mipsel_mips-interAptiv_uClibc-0.9.33.2/root-siflower/tmp-nlpk/us
install -m0755 /home/zhengqw/dragonfly/sf16a18-sdk-4.2.10/chaos_calmer_15_
ibc-0.9.33.2/nlpk-2019-1/nlpkd /home/zhengqw/dragonfly/sf16a18-sdk-4.2.10/
ips-interAptiv_uClibc-0.9.33.2/root-siflower/tmp-nlpk/usr/bin
install -d -m0755 /home/zhengqw/dragonfly/sf16a18-sdk-4.2.10/chaos_calmer_
v_uClibc-0.9.33.2/root-siflower/tmp-nlpk/etc/init.d
install -m0755 ./files/nlpk.init /home/zhengqw/dragonfly/sf16a18-sdk-4.2.1
_mips-interAptiv_uClibc-0.9.33.2/root-siflower/tmp-nlpk/etc/init.d/nlpkd
install -m0755 -d /home/zhengqw/dragonfly/sf16a18-sdk-4.2.10/chaos_calmer_
v_uClibc-0.9.33.2/root-siflower/tmp-nlpk/etc/config
install -m0600 ./files/nlpk.config /home/zhengqw/dragonfly/sf16a18-sdk-4.2
el_mips-interAptiv_uClibc-0.9.33.2/root-siflower/tmp-nlpk/etc/config/nlpk
```

图11-30　打包nlpk文件

3. 编译后的软件包存放在OpenWrt根目录的bin/siflower/packages/base目录下：

```
HOST:chaos_calmer_15_05_1$ find ./bin -name *nlpk*
./bin/siflower/packages/base/nlpk_2019-1_mips_siflower.ipk
./bin/siflower/packages/base/libnlpk_2019-1_mips_siflower.ipk
```

软件包安装测试

通过scp命令把nlpk和libnlpk两个软件包上传到开发板的/tmp目录。假设开发板的IP地址为172.16.10.126。

1. 使用scp命令将软件包上传到开发板。

```
HOST:chaos_calmer_15_05_1$ cd bin/siflower/packages/base
chaos_calmer_15_05_1/bin/siflower/packages/base$ scp *nlpk* admin@172.16.10.126:/tmp
admin@172.16.10.126's password:
libnlpk_2019-1_mips_siflower.ipk
100%   1974      1.9KB/s   00:00
nlpk_2019-1_mips_siflower.ipk
100%   3974      3.9KB/s   00:00
```

2. 安装软件包（因其他依赖已经安装，不用独立安装）。软件包nlpk依赖于软件包libnlpk，所以先要安装libnlpk，再安装nlpk。

可以通过opkg命令对软件包进行安装：

```
DF1A:tmp$ opkg install libnlpk_2019-1_mips_siflower.ipk
Installing libnlpk (2019-1) to root...
Configuring libnlpk.
DF1A:tmp$ opkg install nlpk_2019-1_mips_siflower.ipk
Installing nlpk (2019-1) to root...
Configuring nlpk.
```

3. 测试。

安装完nlpk软件包后，procd会自动启动nlpkd程序并在后台运行。我们先停止nlpkd程序：

```
DF1A:$ /etc/init.d/nlpkd stop
```

（1）查看补丁是否生效，是否支持了usage函数。

```
DF1A:$ nlpkd -h
nlpkd is an normal package example
Usage nlpkd [options]
      -s <path> Path to ubus socket file
      -h help
```

（2）查看参数是否传递，查看UCI配置文件和ubus支持，执行完毕后按Ctrl+C组合键终止运行。

```
DF1A:$ nlpkd
Debug Off
fw ver define
Match firmware version default 1.0
Hello,admin
Connected to ubus, id=dbc83161
```

关于常规软件包开发，我们就介绍到这里，我们通过一个综合的实例进行了演示，相信大家已经对常规软件包的开发有了一定的认识，大家可以"依葫芦画瓢"，尝试开发一个自己的软件包，在package和feeds目录中有很多软件包可以参考，也可以移植开源软件包或对已有的开源软件包进行升级。

11.3　创建内核软件包

内核软件包是用于扩展内核功能的模块，常以内核模块（modules）形式存在，通常在内核启动后按照模块加载顺序进行自动加载，或者采用insmod或modprobe命令进行手动加载。

内核软件包通常存放在package/kernel目录。DF1A开发板环境提供的内核软件包如图11-31所示。

```
acx-mac80211       button-hotplug        lantiq          mwlwifi           sf_eth_led     spi-gpio-custom
ar7-atm            ep80579-drivers       linux           om-watchdog       sf_gmac        trelay
avila-wdt          gpio-button-hotplug   mac80211        reset-button      sf_smac        w1-gpio-custom
brcm2708-gpu-fw    hostap-driver         mmc_over_gpio   rotary-gpio-custom sf_switch     wrt55agv2-spidevs
broadcom-wl        i2c-gpio-custom       mt76            rtc-rv5c386a      sf-ts
```

图11-31　DF1A开发板环境提供的内核软件包

内核软件包的创建

内核软件包的创建和常规软件包的创建有一些区别。内核软件包特有的一些内容如下。

内核软件包Makefile文件除了需要包含rules.mk和package.mk这样的.mk文件外，还需要包含kernel.mk文件，kernel.mk文件是内核软件包的模板。

```
include $(INCLUDE_DIR)/kernel.mk
```

1. KernelPackage函数

常规软件包Makefile文件中的define Package/XXX 系列函数，在内核软件包Makefile文件中被替换为define KernelPackage/XXX系列函数。常用的KernelPackage函数如表11-8所示。

表 11-8 常用的 KernelPackage 函数说明

KernelPackage 函数	说明
KernelPackage/<package name>	内核软件包信息
KernelPackage/<package name>/conffiles	安装内核软件包的配置文件列表
KernelPackage/<package name>/description	内核软件包描述信息
KernelPackage/<package name>/config	内核软件包配置选项
KernelPackage/<package name>/install	内核软件包安装文件函数

KernelPackage/<package name>用于描述内核软件包信息，新增FILES和AUTOLOAD配置参数。

（1）FILES：软件包内核模块文件，包含编译出的模块.ko文件。

```
FILES:= \
    $(PKG_BUILD_DIR)/compat/compat.ko \
    $(PKG_BUILD_DIR)/net/wireless/cfg80211.ko
```

（2）AUTOLOAD：声明自动加载内核模块的顺序和模块的名称。AUTOLOAD会调用kernel.mk的AutoProbe函数。AutoProbe函数语法格式如下。

```
define KernelPackage/内核软件包名称
...
$(call AutoProbe,模块加载序号,模块1,模块2…模块N)
endef
```

```
define KernelPackage/内核软件包名称
...
$(call AutoProbe,模块加载序号,模块1 模块2…模块N,1)
endef
```

（3）AutoProbe会根据模块加载序号和模块名称，在/etc/modules.d/目录下生成"模块加载序号-内核软件包名称"，内容如下。

```
模块1
模块2
...
模块N
```

例如：有内核软件包foo，以下列格式使用AUTOLOAD命令：

```
define KernelPackage/foo
...
AUTOLOAD:=$(call AutoLoad,30,a_mod b_mod c_mod)
endef
```

则会生成/etc/modules.d/30-foo文件，内容如下。

```
a_mod
b_mod
c_mod
```

（4）如果AUTOLOAD函数的最后一个参数值为1，表示在创建/etc/modules.d/"模块加载序号-内核软件包名称"文件后，还要在/etc/modules-boot.d下创建"模块加载序号-内核软件包名称"软连接文件，用于系统启动早期的模块加载。

```
define KernelPackage/scsi-core
…
  FILES:= \
        $(LINUX_DIR)/drivers/scsi/scsi_mod.ko \
        $(LINUX_DIR)/drivers/scsi/sd_mod.ko
  AUTOLOAD:=$(call AutoProbe,40,scsi_mod sd_mod,1)
endef
```

会生成/etc/modules.d/40-scsi-core文件，文件内容如下。

```
scsi_mod
sd_mod
```

也会生成/etc/modules-boot.d/40-scsi-core软连接文件。

2. 内核软件包入口

内核软件包入口函数为 $(eval $(call KernelPackage，<package name>))。

3. 内核软件包安装目录

内核软件包提供的模块.ko文件默认被安装到/lib/modules/3.18.29/目录下，如图11-32所示。

```
DF1A:$ ls /lib/modules/3.18.29/
act_mirred.ko          ip_set_bitmap_ip.ko        mac80211.ko               sd_mod.ko
act_skbedit.ko         ip_set_bitmap_ipmac.ko     mbcache.ko                sf-ts.ko
button-hotplug.ko      ip_set_bitmap_port.ko      nf_conntrack_ftp.ko       sf16a18_rf.ko
cfg80211.ko            ip_set_hash_ip.ko          nf_conntrack_ipv6.ko      sf16a18_smac.ko
cls_flow.ko            ip_set_hash_ipmark.ko      nf_conntrack_rtcache.ko   sfax8-reset-button-top.ko
cls_fw.ko              ip_set_hash_ipport.ko      nf_defrag_ipv6.ko         sfax8eth.ko
cls_route.ko           ip_set_hash_ipportip.ko    nf_log_common.ko          sgmac.ko
cls_tcindex.ko         ip_set_hash_ipportnet.ko   nf_log_ipv4.ko            sha1_generic.ko
cls_u32.ko             ip_set_hash_mac.ko         nf_log_ipv6.ko            slhc.ko
compat.ko              ip_set_hash_net.ko         nf_nat_ftp.ko             startcore.ko
crc-ccitt.ko           ip_set_hash_netiface.ko    nf_reject_ipv4.ko         tun.ko
crc16.ko               ip_set_hash_netnet.ko      nf_reject_ipv6.ko         udp_tunnel.ko
crypto_hash.ko         ip_set_hash_netport.ko     nls_cp437.ko              vfat.ko
ecb.ko                 ip_set_hash_netportnet.ko  nls_cp850.ko              xt_CT.ko
em_u32.ko              ip_set_list_set.ko         nls_cp936.ko              xt_LOG.ko
ext4.ko                ip_tunnel.ko               nls_iso8859-1.ko          xt_REDIRECT.ko
fat.ko                 ipt_REJECT.ko              nls_iso8859-15.ko         xt_TCPMSS.ko
fuse.ko                iptable_filter.ko          nls_utf8.ko               xt_comment.ko
gre.ko                 iptable_mangle.ko          ntfs.ko                   xt_conntrack.ko
ip6_tables.ko          iptable_raw.ko             ppp_async.ko              xt_id.ko
ip6_udp_tunnel.ko      ipv6.ko                    ppp_generic.ko            xt_limit.ko
ip6t_REJECT.ko         jbd2.ko                    ppp_mppe.ko               xt_mac.ko
ip6table_filter.ko     klpk-bar.ko                pppoe.ko                  xt_multiport.ko
```

图11-32 /lib/modules/3.18.29目录下的模块文件

内核软件包的加载顺序(AutoLoad)文件放在/etc/modules.d目录下，每个文件内包含加载模块的名称列表，在实际加载时会按照列表顺序依次加载，如图11-33所示。

```
admin@SiWiFic04c:/# ls /etc/modules.d/
02-crypto-hash   25-nls-cp437       30-fs-ext4       32-l2tp          55-sf_switch     lib-crc-ccitt    pppoe
09-crypto-aead   25-nls-cp850       30-fs-ntfs       32-udptunnel4    60-sf_gmac       mppe             pppol2tp
09-crypto-arc4   25-nls-cp936       30-fs-vfat       32-udptunnel6    70-sched-core    nf-conntrack     pptp
09-crypto-ecb    25-nls-iso8859-1   30-tun           39-gre           80-fuse          nf-conntrack6
09-crypto-sha1   25-nls-iso8859-15  31-iptunnel      40-scsi-core     ipt-conntrack    nf-ipt6
20-ipv6          25-nls-utf8        31-reset-button  42-ip6tables     ipt-core         nf-nathelper
20-lib-crc16     30-button-hotplug  31-sf-ts         49-ipt-ipset     ipt-nat          ppp
```

图11-33 /etc/modules.d/目录下的加载顺序文件

系统启动前期自动加载的模块描述文件放在/etc/modules-boot.d/目录下，这些文件为/etc/modules.d目录下同名文件的软连接，如图11-34所示。

```
02-crypto-hash      25-nls-cp850        30-fs-vfat        32-udptunnel6     80-fuse           nf-ipt6
09-crypto-aead      25-nls-cp936        30-klpk           39-gre            ipt-conntrack     nf-nathelper
09-crypto-arc4      25-nls-iso8859-1    30-tun            40-scsi-core      ipt-core          ppp
09-crypto-ecb       25-nls-iso8859-15   31-iptunnel       42-ip6tables      ipt-nat           pppoe
09-crypto-sha1      25-nls-utf8         31-reset-button   49-ipt-ipset      lib-crc-ccitt     pppol2tp
20-ipv6             30-button-hotplug   31-sf-ts          55-sf_switch      mppe              pptp
20-lib-crc16        30-fs-ext4          32-l2tp           60-sf_gmac        nf-conntrack
25-nls-cp437        30-fs-ntfs          32-udptunnel4     70-sched-core     nf-conntrack6
```

图11-34　/etc/modules-boot.d目录中的软连接文件

11.3.1 软件包信息

和学习常规软件包一样，我们以一个实例来学习如何创建内核软件包。假设要创建的内核软件包名称叫作klpk(kernel package)。klpk软件包的基本信息如表11-9所示。

表 11-9　klpk 软件包的基本信息

基本信息	说明
软件包名称	klpk
软件包版本	1.0
软件包功能	包含两个内核模块，有两个内核配置选项，分别用于开启物理内存和虚拟内存分配功能。如果没有开启选项，模块仅打印启动和退出日志，如果开启了对应选项，则打印对应分配内存的地址信息
源代码获取方式	源代码存放在本地，即软件包src目录
源代码构建方式	Make
包含软件包	klpk
补丁	提供补丁文件
配置文件和启动脚本	无
模块文件	klpk-foo.ko和klpk-bar.ko
软件包依赖	+kernel

创建内核软件包

创建内核软件包目录。

```
HOST:chaos_calmer_15_05_1$ mkdir -p package/kernel/klpk
```

创建Makefile文件，src、patches目录。

```
HOST:chaos_calmer_15_05_1$ cd package/kernel/klpk
HOST:klpk$ touch Makefile
HOST:klpk$ mkdir src patches
```

准备源代码的全部过程都要在packages/kernel/klpk中完成，操作前需提前进入该目录，如图11-35所示。

图11-35　klpk软件包的目录和文件

11.3.2 准备软件包源代码

klpk的源代码为演示用的文件,我们将这些内容放置于src目录中,如图11-36所示。

图11-36 src目录下的源代码文件

src目录中各源代码文件的作用如下。

- Kconfig:定义软件包Kconfig(用于内核配置菜单,在make menuconfig中不显示)。
- klpk-bar.c:支持虚拟内存分配的模块。
- klpk-foo.c:支持物理内存分配的模块。
- Makefile:模块源代码编译的Makefile文件,对要编译哪些文件进行定义。

准备软件包源代码的操作步骤如下。

1. 创建src/Kconfig文件。

```
config DF1A_KLPK
        tristate "DF1A Kernel Package Demo Support"
```

2. 创建src/klpk-bar.c文件。

```c
#include <linux/err.h>
#include <linux/init.h>
#include <linux/module.h>
#include <linux/vmalloc.h>
static int __init klpk_bar_init(void);
static void __exit klpk_bar_exit(void);
#define DRV_NAME "klpk-bar"
#define DRV_VERSION "0.0.1"
#define MEM_VMALLOC_SIZE 8092
char * mem_vp;
int __init klpk_bar_init(void)
{
      printk("klpk bar module init!\n");
#ifdef CONFIG_DF1A_KLPK_ALLOC_VM_MEM
      printk("support CONFIG_DF1A_KLPK_ALLOC_VM_MEM\n");
      mem_vp = (char *)vmalloc(MEM_VMALLOC_SIZE);
      if(mem_vp == NULL)
              printk("vmalloc failed\n");
      else
              printk("vmalloc successfully \n\taddr =0x%1x\n",(unsigned long)mem_vp);
#endif
      return 0;
}
```

```
void __exit klpk_bar_exit(void)
{
#ifdef CONFIG_DF1A_KLPK_ALLOC_VM_MEM
        if(!mem_vp){
                vfree(mem_vp);
                printk("vfree ok!\n");
        }
#endif
        printk("klpk bar module exit!\n");
}
module_init(klpk_bar_init);
module_exit(klpk_bar_exit);
MODULE_DESCRIPTION(DRV_NAME);
MODULE_VERSION(DRV_VERSION);
MODULE_AUTHOR("Zheng QiWen <zhengqwmail@gmail.com>");
MODULE_LICENSE("GPL");
```

3. 创建src/klpk-foo.c文件。

```
#include <linux/err.h>
#include <linux/init.h>
#include <linux/module.h>
#include <linux/slab.h>
static int __init klpk_foo_init(void);
static void __exit klpk_foo_exit(void);
#define DRV_NAME "klpk-foo"
#define DRV_VERSION "0.0.1"
#define MEM_KMALLOC_SIZE 8092
char * mem_kp;
int __init klpk_foo_init(void)
{
        printk("klpk foo module init!\n");
#ifdef CONFIG_DF1A_KLPK_ALLOC_PHY_MEM
        printk("support CONFIG_DF1A_KLPK_ALLOC_PHY_MEM\n");
        mem_kp = (char *)kmalloc(MEM_KMALLOC_SIZE, GFP_KERNEL);
        if(mem_kp == NULL)
                printk("kmalloc failed\n");
        else
                printk("kmalloc successfully \n\taddr =0x%1x\n",(unsigned long)mem_kp);
#endif
        return 0;
}
void __exit klpk_foo_exit(void)
{
#ifdef CONFIG_DF1A_KLPK_ALLOC_PHY_MEM
        if(!mem_kp){
                kfree(mem_kp);
                printk("kfree ok!\n");
        }
#endif
        printk("klpk foo module exit!\n");
```

```
}
module_init(klpk_foo_init);
module_exit(klpk_foo_exit);
MODULE_DESCRIPTION(DRV_NAME);
MODULE_VERSION(DRV_VERSION);
MODULE_AUTHOR("Zheng QiWen <zhengqwmail@gmail.com>");
MODULE_LICENSE("GPL");
```

4. 创建src/Makefile文件。

```
obj-$(CONFIG_DF1A_KLPK) +=klpk-foo.o klpk-bar.o
```

准备klpk的patch文件

接下来，为了演示patch功能，创建一个补丁文件放到patches目录中。创建patches/001-test-patch.patch文件。

```
--- a/klpk-bar.c    2019-11-19 12:58:23.076905569 +0800
+++ b/klpk-bar.c    2019-11-19 13:00:54.770459324 +0800
@@ -15,6 +15,7 @@
 int __init klpk_bar_init(void)
 {
     printk("klpk bar module init!\n");
+    printk("klpk bar patch!\n");
 #ifdef CONFIG_DF1A_KLPK_ALLOC_VM_MEM
     printk("support CONFIG_DF1A_KLPK_ALLOC_VM_MEM\n");
     mem_vp = (char *)vmalloc(MEM_VMALLOC_SIZE);
diff -uNr a/klpk-foo.c b/klpk-foo.c
--- a/klpk-foo.c    2019-11-19 12:58:23.076905569 +0800
+++ b/klpk-foo.c    2019-11-19 12:59:48.815532128 +0800
@@ -15,6 +15,7 @@
 int __init klpk_foo_init(void)
 {
     printk("klpk foo module init!\n");
+    printk("klpk foo patch!\n");
 #ifdef CONFIG_DF1A_KLPK_ALLOC_PHY_MEM
     printk("support CONFIG_DF1A_KLPK_ALLOC_PHY_MEM\n");
     mem_kp = (char *)kmalloc(MEM_KMALLOC_SIZE, GFP_KERNEL);
```

11.3.3 编写软件包

创建Makefile文件。

```
include $(TOPDIR)/rules.mk
PKG_NAME:=klpk
PKG_BUILD_PARALLEL:=1
PKG_MAINTAINER:=Zheng QiWen <zhengqwmail@gmail.com>
include $(INCLUDE_DIR)/kernel.mk
include $(INCLUDE_DIR)/package.mk
define KernelPackage/$(PKG_NAME)
  SUBMENU:=DF1A Modules
  TITLE:=Kernel Pacakge Demo Foo/Bar Module
  DEPENDS:=+kernel
```

```
    FILES:=$(PKG_BUILD_DIR)/klpk-foo.ko \
          $(PKG_BUILD_DIR)/klpk-bar.ko
    AUTOLOAD:=$(call AutoLoad,30,$(PKG_NAME)-foo $(PKG_NAME)-bar,1)
endef
define KernelPackage/$(PKG_NAME)/description
  Kernel module for kernel package demo,
  Include alloc physical memory and alloc virtual memory test.
endef
define KernelPackage/$(PKG_NAME)/config
        if PACKAGE_kmod-klpk
                config PACKAGE_DF1A_KLPK_ALLOC_PHY_MEM
                        bool "Alloc Physical Memory Test"
                        default "n"
                config PACKAGE_DF1A_KLPK_ALLOC_VM_MEM
                        bool "Alloc Virutal Memory Test"
                        default "n"
        endif
endef
NOSTDINC_FLAGS = \
        -I$(PKG_BUILD_DIR)
EXTRA_KCONFIG:=CONFIG_DF1A_KLPK=m
ifdef CONFIG_PACKAGE_DF1A_KLPK_ALLOC_PHY_MEM
EXTRA_KCONFIG += CONFIG_DF1A_KLPK_ALLOC_PHY_MEM=y
endif
ifdef CONFIG_PACKAGE_DF1A_KLPK_ALLOC_VM_MEM
EXTRA_KCONFIG += CONFIG_DF1A_KLPK_ALLOC_VM_MEM=y
endif
EXTRA_CFLAGS:= \
        $(patsubst CONFIG_%, -DCONFIG_%=1, $(patsubst %=m,%,$(filter %=m,$(EXTRA_KCONFIG)))) \
        $(patsubst CONFIG_%, -DCONFIG_%=1, $(patsubst %=y,%,$(filter %=y,$(EXTRA_KCONFIG))))
MAKE_OPTS:= \
        ARCH="$(LINUX_KARCH)" \
        CROSS_COMPILE="$(TARGET_CROSS)" \
        SUBDIRS="$(PKG_BUILD_DIR)" \
        EXTRA_CFLAGS="$(EXTRA_CFLAGS)" \
        $(EXTRA_KCONFIG)
define Build/Prepare
        rm -rf $(PKG_BUILD_DIR)
        mkdir -p $(PKG_BUILD_DIR)
        $(CP) ./src/* $(PKG_BUILD_DIR)/
        $(Build/Patch)
endef
define Build/Compile
        $(MAKE) -C "$(LINUX_DIR)" \
                $(MAKE_OPTS) \
                modules
endef
$(eval $(call KernelPackage,$(PKG_NAME)))
```

Makefile脚本实现的内容如下。

（1）定义软件包的基本信息。

（2）KernelPackage/klpk函数：定义内核软件包的基本信息，这些信息将在menuconfig菜单中显示。

（3）KernelPackage/klpk/description函数：定义内核软件包的详细描述信息。在menuconfig的klpk菜单中，按h键会显示详细的软件包描述信息。

（4）Package/klpk/config 函数：定义klpk软件包的配置菜单。此函数支持两个选项，一个是PACKAGE_DF1A_KLPK_ALLOC_PHY_MEM，另一个是PACKAGE_DF1A_KLPK_ALLOC_VM_MEM，分别表示开启物理内存分配和开启虚拟内存分配。

（5）配置头文件编译路径。设置NOSTDINC_FLAGS，即非标准路径，在本实例中设置为软件包编译目录。由于本实例没有使用头文件，所以这个选项并没有实际用途，仅作为演示使用。

（6）定义扩展KCONFIG内容。使用EXTRA_KCONFIG定义，执行make命令时，会把这个选项传递给内核。CONFIG_DF1A_KLPK=m 告诉内核要把这个软件包编译成模块，再传递给src/Makefile。

```
obj-$(CONFIG_DF1A_KLPK) +=klpk-foo.o klpk-bar.o
```

接下来，根据以上对menuconfig菜单的介绍，介绍怎样添加两个配置选项CONFIG_DF1A_KLPK_ALLOC_PHY_MEM和CONFIG_DF1A_KLPK_ALLOC_VM_MEM。

```
ifdef CONFIG_PACKAGE_DF1A_KLPK_ALLOC_PHY_MEM
EXTRA_KCONFIG += CONFIG_DF1A_KLPK_ALLOC_PHY_MEM=y
endif
ifdef CONFIG_PACKAGE_DF1A_KLPK_ALLOC_VM_MEM
EXTRA_KCONFIG += CONFIG_DF1A_KLPK_ALLOC_VM_MEM=y
endif
```

（7）根据CONFIG_XXX选项内容和设定值，生成-DCONFIG_XXX=1，赋值给EXTRA_CFLAGS，最终通过make命令传递给内核。

（8）设置编译选项MAKE_OPTS。

（9）使用Build/Prepare函数，定义软件包在编译前要执行的动作。执行的动作包括：创建编译目录$(PKG_BUILD_DIR)，将软件包src/目录下所有的文件复制到编译目录，并打上patches目录下的所有补丁文件。

（10）使用Build/Compile函数，执行软件包的编译。

（11）定义软件包编译入口$(eval $(call KernelPackage,klpk))。

klpk软件包中包含的所有文件如图11-37所示。

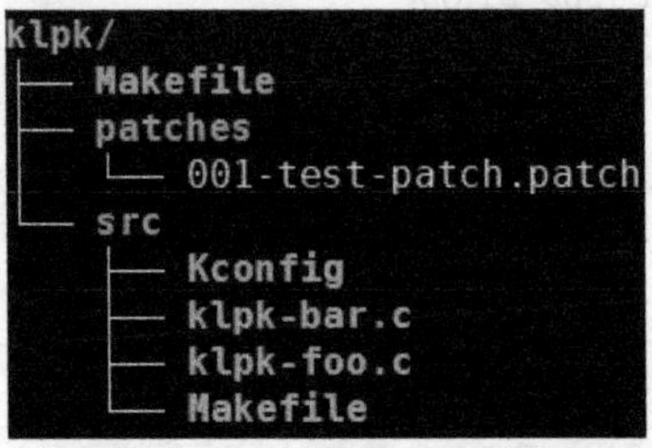

```
klpk/
├── Makefile
├── patches
│   └── 001-test-patch.patch
└── src
    ├── Kconfig
    ├── klpk-bar.c
    ├── klpk-foo.c
    └── Makefile
```

图11-37　klpk软件包中包含的所有文件

11.3.4 编译测试klpk

使用make menuconfig命令配置软件包的操作如下，klpk软件包的配置如图11-38所示。

```
#请先切换到OpenWrt的根目录后操作如下命令
HOST:chaos_calmer_15_05_1$ make menuconfig
```

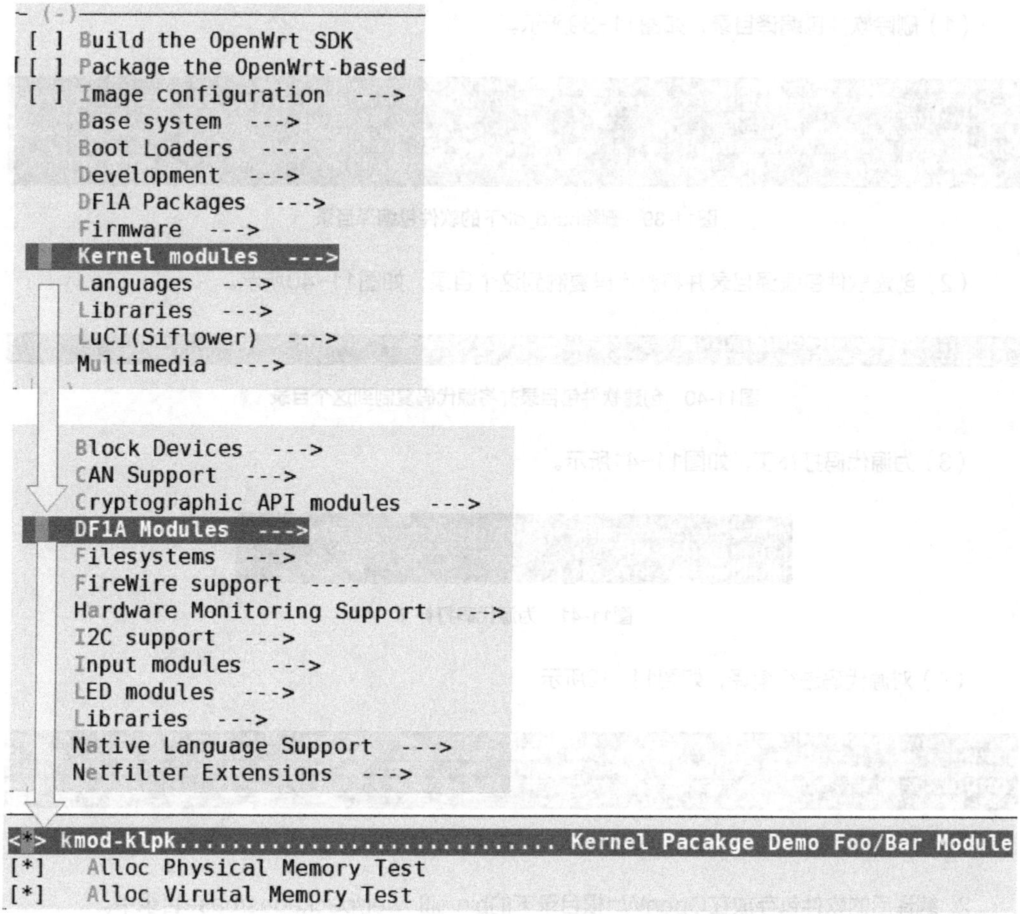

```
    ( - )
[ ] Build the OpenWrt SDK
[ ] Package the OpenWrt-based
[ ] Image configuration      --->
    Base system       --->
    Boot Loaders      ----
    Development       --->
    DF1A Packages     --->
    Firmware          --->
    Kernel modules    --->
    Languages         --->
    Libraries         --->
    LuCI(Siflower)    --->
    Multimedia        --->

    Block Devices          --->
    CAN Support            --->
    Cryptographic API modules  --->
    DF1A Modules           --->
    Filesystems            --->
    FireWire support       ----
    Hardware Monitoring Support  --->
    I2C support            --->
    Input modules          --->
    LED modules            --->
    Libraries              --->
    Native Language Support   --->
    Netfilter Extensions   --->

<*> kmod-klpk.......................... Kernel Pacakge Demo Foo/Bar Module
[*]     Alloc Physical Memory Test
[*]     Alloc Virutal Memory Test
```

图11-38　klpk软件包的配置

编译软件包

此步骤在对OpenWrt进行固件编译时自动完成，在以下的操作中使用单独编译的方式用于测试：

```
HOST:chaos_calmer_15_05_1$ make package/kernel/klpk/{clean,compile} V=99
```

编译流程说明

1. 编译软件包前，会按照软件包Makefile文件中定义的Build/Prepare函数执行下列内容。

```
define Build/Prepare
```

```
        rm -rf $(PKG_BUILD_DIR)
        mkdir -p $(PKG_BUILD_DIR)
        $(CP) ./src/* $(PKG_BUILD_DIR)/
        $(Build/Patch)
endef
```

编译前执行的过程输出按下列步骤实现。

(1) 删除软件包编译目录,如图11-39所示。

```
rm -f /home/zhengqw/dragonfly/sf16a18-sdk-4.2.10/chaos_calmer_15_05_1/bin/siflower/packages/base/kmod-klpk_*
rm -f /home/zhengqw/dragonfly/sf16a18-sdk-4.2.10/chaos_calmer_15_05_1/staging_dir/target-mipsel_mips-interAptiv_uClibc-0.9.33.2/stamp/.klpk_installed
rm -f /home/zhengqw/dragonfly/sf16a18-sdk-4.2.10/chaos_calmer_15_05_1/staging_dir/target-mipsel_mips-interAptiv_uClibc-0.9.33.2/packages/klpk.list /home/zhengqw/dragonfly/sf16a18-sdk-4.2.10/chaos_calmer_15_05_1/staging_dir/host/packages/klpk.list
rm -rf /home/zhengqw/dragonfly/sf16a18-sdk-4.2.10/chaos_calmer_15_05_1/build_dir/target-mipsel_mips-interAptiv_uClibc-0.9.33.2/linux-siflower_sf16a18-fullmask/klpk
```

图11-39 删除build_dir下的软件包编译目录

(2) 创建软件包编译目录并将源代码复制到这个目录,如图11-40所示。

```
cp -fpR ./src/* /home/zhengqw/dragonfly/sf16a18-sdk-4.2.10/chaos_calmer_15_05_1/build_dir/target-mipsel_mips-interAptiv_uClibc-0.9.33.2/linux-siflower_sf16a18-fullmask/klpk/
```

图11-40 创建软件包目录并将源代码复制到这个目录

(3) 为源代码打补丁,如图11-41所示。

```
Applying ./patches/001-test-patch.patch using plaintext:
patching file klpk-bar.c
patching file klpk-foo.c
```

图11-41 为源代码打补丁

(4) 对源代码进行编译,如图11-42所示。

```
make -C "/home/zhengqw/dragonfly/sf16a18-sdk-4.2.10/chaos_calmer_15_05_1/build_dir/target-mipsel_mips-interAptiv_uClibc-0.9.33.2/linux-siflower_sf16a18-fullmask/linux-3.18.29" ARCH="mips" CROSS_COMPILE="mipsel-openwrt-linux-uclibc-" SUBDIRS="/home/zhengqw/dragonfly/sf16a18-sdk-4.2.10/chaos_calmer_15_05_1/build_dir/target-mipsel_mips-interAptiv_uClibc-0.9.33.2/linux-siflower_sf16a18-fullmask/klpk" EXTRA_CFLAGS=" -DCONFIG_DF1A_KLPK=1 -DCONFIG_DF1A_KLPK_ALLOC_PHY_MEM=1 -DCONFIG_DF1A_KLPK_ALLOC_VM_MEM=1" CONFIG_DF1A_KLPK=m CONFIG_DF1A_KLPK_ALLOC_PHY_MEM=y CONFIG_DF1A_KLPK_ALLOC_VM_MEM=y modules
```

图11-42 对源代码进行编译

2. 编译后的软件包存放在OpenWrt根目录下的bin/siflower/packages/base目录中。

```
HOST:chaos_calmer_15_05_1$ find ./bin -name *klpk*
./bin/siflower/packages/base/kmod-klpk_3.18.29-1_mips_siflower.ipk
```

软件包安装测试

通过scp命令把klpk软件包上传到开发板的/tmp目录中。在下列的操作实例中,假设开发板的IP地址为172.16.10.126。

(1) 使用scp命令将软件包上传到开发板中。

```
HOST:chaos_calmer_15_05_1$ cd bin/siflower/packages/base/
HOST:base$ scp kmod-klpk* admin@172.16.10.126:/tmp
admin@172.16.10.126's password:
kmod-klpk_3.18.29-1_mips_siflower.ipk
100% 2298     2.2KB/s   00:00
```

(2) 安装软件包。在开发板中,通过opkg命令对软件包进行安装。

```
DF1A:tmp$ opkg install kmod-klpk_3.18.29-1_mips_siflower.ipk
```

（3）进行测试。安装后的模块会自动加载，可采用lsmod命令查看模块是否加载：

```
DF1A:$ lsmod |grep klpk*
klpk_bar                 656  0
klpk_foo                 656  0
```

查看模块安装文件。/lib/modules/3.18.29/是模块的默认安装路径。/etc/modules.d/30-klpk表示模块的加载顺序。/etc/modules-boot.d/30-klpk表示模块会在系统启动时实现加载：

```
DF1A:$ opkg files kmod-klpk
Package kmod-klpk (3.18.29-1) is installed on root and has the following files:
/etc/modules.d/30-klpk
/lib/modules/3.18.29/klpk-bar.ko
/lib/modules/3.18.29/klpk-foo.ko
/etc/modules-boot.d/30-klpk
```

klpk软件包内的模块加载顺序如下。

```
DF1A:$ cat /etc/modules.d/30-klpk
klpk-foo
klpk-bar
```

查看模块加载信息如下。

```
DF1A:$ dmesg
...
[535604.105028] klpk foo module init!
[535604.108555] klpk foo patch!
[535604.111445] support CONFIG_DF1A_KLPK_ALLOC_PHY_MEM
[535604.116333] kmalloc successfully
[535604.116333] addr =0x86e6c000
[535604.130927] klpk bar module init!
[535604.134378] klpk bar patch!
[535604.137414] support CONFIG_DF1A_KLPK_ALLOC_VM_MEM
[535604.142255] vmalloc successfully
[535604.142255] addr =0xc4c1a000
```

12 硬件定制

12.1 源代码结构

DF1A开发板的SDK工程源代码以芯片原厂SDK为基础，满足本书教程需要。接下来对其源代码目录结构进行介绍。

整个源代码工程分为4个部分：chaos_calmer_15_05_01、image_maker、linux-3.18.29-dev、uboot。FreeIRIS-DF1A的源代码结构如图12-1所示。

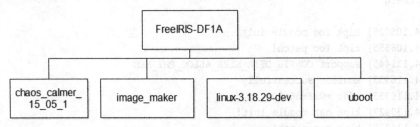

图12-1　FreeIRIS-DF1A源代码结构

DF1A开发板源代码下载地址：freeiris官网地址/resource/sf16a18-df1a/

chaos_calmer_15_05_01

chaos_calmer_15_05_01是OpenWrt官方在2015年5月发布的Release版，就是业内通常提到的CC版，矽昌团队在该版本的基础上添加了SiFlower芯片支持包和一些SiFlower特有的软件包。图12-2所示为OpenWrt系统的根目录结构。

第三篇 深入浅出OpenWrt系统

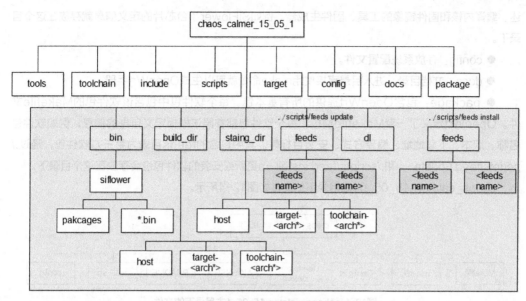

图12-2 OpenWrt系统的根目录结构

chaos_calmer_15_05_01主目录结构如下。

● tools：获取代码和编译时使用的主机端工具。编译时，主机需要使用一些工具软件，例如autoconfig、automake等，tools目录包含了获取和编译这些工具的命令，构建系统的第一步就是获取和编译这些软件包。

● toolchain：包括内核头文件、C库、交叉编译器、调试器，如binutils、gcc、glibc等。

编译时，tools与toolchain会生成3个目录，如图12-3所示。

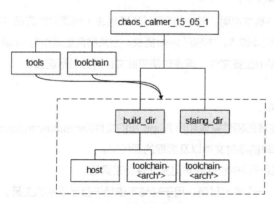

图12-3 工具生成流程

● include：包含主要的Makefiles文件和编译规则。其内容由其他Makefile文件包含。例如软件包Makefile文件需要包含package.mk文件。

● scripts：包括配置脚本、补丁脚本、软件源脚本，使用Shell、Python、Perl等多种脚本语言编写。在编译过程中用到的脚本也统一放在这个目录下。另外，软件包feeds的源安装脚本也放在这个目录下。

● target：定义供应商平台文件和镜像工具。包含不同平台固件镜像产生过程和内核编译的描

· 295 ·

述、编译内核和固件镜像的工具、固件生成器。例如SiFlower平台芯片的定义信息就存放在这个目录下。

● config：存放系统配置文件。
● docs：文档目录。进入目录直接使用make命令就可以生成OpenWrt手册。
● package：包含OpenWrt提供的所有基本包。每个软件包中包含该软件包的Makefile文件。OpenWrt定义了一套Makefile模板，每个软件包都参照该模板定义自身的信息，例如软件包名称、版本、下载地址、编译方式、安装目标等。其中包含的feeds目录为第三方软件包，通过./scripts/feed update -a和./scripts/feed install -a更新或安装的软件包也会存放在这个目录下。

chaos_calmer_15_05_01主目录下的文件如图12-4所示。

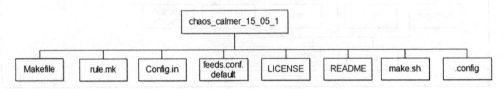

图12-4　chaos_calmer_15_05_1主目录下的文件

这些文件的内容如下。
● Makefile：在顶层目录执行make命令的入口文件。
● rules.mk：定义Makefile中使用的一些通用变量和函数。每个软件包都需要包含此文件。
● Config.in：菜单配置。和make menuconfig相关联的文件。
● feeds.conf.default：下载第三方一些软件包时所使用的地址。进行./scripts/feeds update操作时使用。
● LICENSE：软件许可。
● README：软件基本说明。其中的README描述了编译软件的基本过程和依赖文件。
● make.sh：工程编译脚本，由SiFlower提供。相关的典型操作有./make.sh df1a fullmask。
● .config：OpenWrt配置文件。编译时使用此文件进行目标配置。

编译后生成的目录如下。
● bin：存放最终生成的固件镜像和所有.ipk 包的文件(bin/siflower/packages)。
● build_dir：编译时的临时文件以及提取的源代码。
● staing_dir：编译环境中包括的常见头文件和工具链。
● build_dir/host：一个临时目录，用于存储不依赖于目标平台的工具。
● build_dir/target-<arch> *：包源代码目录。
● build_dir/toolchain-<arch>*：用于存储依赖于指定平台的编译链。
● build_dir/linux-<platform>：包含内核解压后的源代码。
● staing_dir/toolchain-<arch>*：编译链的最终安装位置。
● feeds：所有可选软件包由OpenWrt或第三方提供。这些包并不是在主分支中维护的，而是由第三方维护，用于扩展基本系统的功能。当执行./script/feeds update -a命令时，会从feed.conf.default配置文件指定的路径进行下载（实际上是索引）。当从menuconfig中选择对应的package后,进行make操作时,会根据package的Makefile文件中描述的地址,下载对应的packages。packages的打包文件被下载到dl目录。

● package/feeds:<buildroot>/feeds/下packages的软连接。在执行.scripts/feeds install -a命令后生成。

构建系统时,会按照图12-5中所示的顺序进行构建。

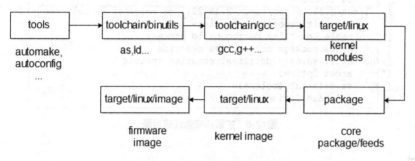

图12-5 CC SDK系统的构建流程

image_maker

image_maker是固件合并工具的目录。操作者可以使用工具把U-Boot和OpenWrt固件合并为一个Flash镜像。Freeiris提供的SDK代码中包含了U-Boot和OpenWrt(CC),能够生成一个U-Boot镜像和一个OpenWrt镜像,而Flash中最终只使用一个镜像,所以需要用image_maker工具将各个小镜像合并成一个最终的Flash镜像。此功能主要在工厂生产等场合应用。

工具主要有以下文件。

● sf-makeimage.sh:执行脚本文件。
● factory_default.bin:全由0xff组成的4KB大小的镜像,默认的factory分区镜像。
● expa_fac.bin:外置PA的factory分区射频校准经验值。
● inpa_fac.bin:内置PA的factory分区射频校准经验值。
● README:参考文档。

linux-3.18.29-dev

这是Linux内核3.18.29源代码目录,默认集成了SiFlower相关的设备驱动。我们对源代码的目录结构不逐一介绍。

在OpenWrt中,可以通过Make menuconfig配置OpenWrt的Linux扩展内核源代码目录(CONFIG_EXTERNAL_KERNEL_TREE=../linux-3.18.29-dev/linux-3.18.29)。扩展内核源代码目录如图12-6所示。

```
- | - Advanced configuration options (for developers)
[ ]    Show broken platforms / packages
( )    Binary folder
( )    Download folder
( )    Local mirror for source packages
[*]    Automatic rebuild of packages
( )    Build suffix to append to the target BUILD_DIR variable
( )    Override the default TARGET_ROOTFS_DIR variable
[ ]    Use ccache
(../linux-3.18.29-dev/linux-3.18.29) Use external kernel tree
( )    Enter git repository to clone
[ ]    Enable log files during build process
[ ]    Enable package source tree override
(-fno-caller-saves) Additional compiler options
[*]    Target Options  --->
[ ]    Use external toolchain  ----
[ ]    Toolchain Options  ----
```

图12-6 扩展内核源代码目录

uboot

uboot目录下是SiFlower定制的通用引导加载程序。除标准U-Boot功能以外，还集成了UIP协议栈、bare_spl程序等功能。其目录结构如下。

- api：外部扩展程序的API。
- arch：与体系结构相关的代码。
- bare_spl：SiFlower提供的裸机spl程序，用于引导U-Boot。
- board：根据不同开发板定制的代码。此目录与开发板相关，其下每一个子目录代表一个芯片厂家，芯片厂家目录下存放该厂家不同开发板的代码。
- common：存放与处理器体系结构无关的通用代码，如bootm、console等。
- configs：存放所有与目标板相关的配置文件。
- disk：存放磁盘驱动程序的分区处理代码。
- doc：存放U-Boot的说明文档。
- drivers：存放所有外围芯片的驱动程序，如网卡、USB、串口、LCD、Nand Flash等的驱动程序。
- dts：设备树支持。
- fs：存放与文件系统相关的代码，每一个子目录代表一种文件系统。
- httpd：移植了UIP协议栈，支持HTTP服务，用于U-Boot Web升级。
- include：头文件，包括各种CPU的寄存器、FDT、Flash、文件系统、网络等定义。
- example：示例文件。
- lib：通用库文件，包含libfdt设备树相关代码，加密算法、压缩算法、字符处理相关代码等。
- license：协议。
- net：存放与网络协议相关的代码，例如BOOTP协议、TFTP协议、RARP协议和NFS等。
- post：存放上电自检程序。
- scripts：辅助脚本。
- sign：对目标进行RSA私钥签名。
- test：与单元测试相关。

- toolchain：交叉编译工具链，包含ARM和MIPS的工具链。
- tools：辅助工具集合，用于生成U-Boot镜像，如mkimage等。
- sf_make.sh：适配不同版本型号的编译辅助脚本。

12.2 定制案例

在实际开发中，用户通常需要根据硬件规格对开发板进行系统定制。假如用户自己修改了硬件设计或自己根据DF1A开发板重新设计了一款硬件，就需要对新硬件进行系统定制。我们将在本章节中通过一个实际案例学习如何基于SF16A18芯片的OpenWrt系统定制出自己的独立产品。

12.2.1 新硬件规格

对新硬件进行系统定制的第一步是要明确新硬件的规格。为了简化学习，我们将以DF1A开发板为原型，只改变其中的某些硬件规格，其他接口保持与DF1A开发板相同，在此过程中展示如何重新定制一款硬件产品。

新定义的硬件规格如表12-1所示：

表 12-1 新定义的硬件规格

项目	说明	参数
硬件名称	用于编译固件时使用	MY888
RAM	DDR3内存颗粒	128MB(1024Mbit)
以太网接口	CPU内置交换机 1×WAN 1×LAN	WAN = ETH_PHY0 LAN = ETH_PHY1 其他PORT不定义
LED	各类指示灯	ETH_LED0 = GPIO 55 ETH_LED1 = GPIO 56 SYS_LED = GPIO 25

12.2.2 让U-Boot支持MY888

以下的操作以DF1A SDK源代码为基础定义新硬件系统。

第一步操作是定制U-Boot，因为新硬件MY888与DF1A基础硬件信息相比没有变化，所以在U-Boot中修改的地方非常少，只需要在工程脚本中增加MY888的型号描述即可。

切换到U-Boot源代码目录(sf16a18-sdk-4.2.10/uboot)，后面的操作都是在这个目录中进行，具体步骤如下。

1. 增加MY888版本型号。

编辑vi sf_make.sh工程编译文件，修改内容如下。

（1）在第26行"df1a"后添加字符"|my888"。

```
show_help() {
 25         echo "Usage: $0"
 26         echo "       prj=p10[b/m/flash]|p20[b]|wrt|evb|86v|ac|x10|p10h|evb_v5|df1a|my888|air001|cp e|ott_router"
```

（2）在第180行"df1a"后添加字符"|my888"。

```
180       evb_v5 | df1a | my888)
```

```
181                    DEFCONFIG="sfa18_"$ver"_p20b"
182                    [ -z $ddr3 ] && ddr3=m15t1g1664a
183                    ;;
```

2. 编译新型号,操作如下。

```
HOST:uboot$ ./sf_make.sh prj=my888
```

编译成功后,会生成sfax8/my888.img文件。

```
HOST:uboot$ ls sfax8/*
sfax8/my888.img    sfax8/uboot_full.img    sfax8/u-boot.img    sfax8/u-boot-spl.img
```

3. 烧写测试

烧写U-Boot固件的方法请参考10.4.2节。烧写成功,系统重启后,即可在U-Boot启动信息中看到U-Boot固件编译日期已经更新,如图12-7所示。

```
U-Boot 2016.07-rc2 (Dec 13 2019 - 10:33:02 +0800)

Board: MIPS sfa18 FULLMASK P20B
       Watchdog enabled
DRAM:  128 MiB
MMC:   emmc@7800000: 0sdio@7c00000: 1
SF: Detected BY25Q128AS with page size 256 Bytes, erase size 4 KiB,
```

图12-7 U-Boot启动信息

12.2.3 设备树

设备树(Device Tree,简称DT)是描述硬件的数据结构。它采用类似于JSON格式的语法风格,是一个简单的树型结构,由节点(Node)和属性(Property)组成。节点本身可以包含子节点和各种属性。属性包含名称(Name)和属性值(Value),属性值可以为空。设备树来源于开放固件(Open Firmware),使操作系统不用包含这些信息的硬编码。

设备树用于描述设备的信息,包括CPU的数量和类别、内存及地址和大小、总线和桥、外设、中断控制器和中断使用情况、GPIO控制器和GPIO使用情况、时钟控制器和时钟使用情况等。

通常使用DTS(Device Tree Source)文件(以.dts结尾)以文本方式对系统设备树进行描述,通用的部分由DTSI(Device Tree Source Include)文件(以.dtsi结尾)描述。使用DTC(Device Tree Compiler)工具将.dts文件编译成以.dtb结尾的二进制文件DTB(Device Tree Blob),.dtb文件可以由Linux内核解析。在MIPS体系架构中,DTS描述信息放在arch/mips/boot/dts/目录。对于已经使用DT的设备,可以通过访问/sys/firmware/devicetree/目录查看设备树的信息。

芯片厂一般会提供参考板的DTS文件,对该文件进行修改,即可实现目标硬件的DTS文件编写。

设备树的语法

设备树有着简单的树形结构,包含节点和属性。属性是键值对,节点包含属性和子节点。

节点语法格式如下。

```
<名称>[@<设备地址>]
```

● <名称>：是一个不超过31位的ASCII字符串，用于表示设备的类型，如switch、ethernet等。

● <设备地址>：用来访问该设备的主地址，该地址在节点的reg属性中列出。属于同一个父节点的子节点命名必须是唯一的。可以支持地址不同但名称相同的节点命名方式。设备地址是可选的，可以没有设备地址。节点语法的实例如下。

```
grfgpio: syscon@19e3f000 {
        compatible = "siflower,sfax8-syscon";
        reg = <0x19e3f000 0x1000>;
};
```

● 树中表示的每一个设备节点都需要一个compatible属性。

属性语法

属性用键值对表示，它的值可以为空或者包含任意字节流。设备树中包含几种基本的数据表示形式。

● 文本字符串：没有特定结束符，用双引号括起来。

```
a-string-property = "A string";
```

● 32位无符号整数：用尖括号括起来，常用来表示设备地址。

```
cell-property = <0xbeef 123 0xabcd1234>;
```

● 二进制数据：用中括号括起来。

```
binary-property = [0x01 0x23 0x45 0x67];
```

● 混合形式的数据：用逗号连接在一起。

```
mixed-property = "a string", [0x01 0x23 0x45 0x67], <0x12345678>;
```

● 字符串列表：用逗号连接在一起。

```
string-list = "red fish", "blue fish";
```

以下是一个DTS文件的示例。

```
/dts-v1/;
/ {
    node1 {
        a-string-property = "A string";
        a-string-list-property = "first string", "second string";
        // hex is implied in byte arrays. no '0x' prefix is required
        a-byte-data-property = [01 23 34 56];
        child-node1 {
            first-child-property;
            second-child-property = <1>;
            a-string-property = "Hello, world";
        };
        child-node2 {
        };
    };
    node2 {
        an-empty-property;
        a-cell-property = <1 2 3 4>; /* each number (cell) is a uint32 */
        child-node1 {
```

```
                };
        };
};
```

上述示例中包含以下内容。
- 一个根节点"/"。
- 两个子节点"node1"和"node2"。
- "node1"子节点下还包含有两个子节点"child-node1"和"child-node2"。
- 属性包含在各节点中。

常见节点和属性

- compatible属性

它指定系统的名称，是一个字符串列表，在代码中会对字符串进行匹配，包含一个"<制造商>, <型号>"形式的字符串。

```
/dts-v1/;
/ {
        compatible = "siflower,sf16a18";
...
};
```

- CPUs节点

它用于描述每一个CPU，在节点名"cpus"下，根据CPU数量为每一个CPU增加节点。

```
/dts-v1/;
/ {
        cpus{
                        cpu0: cpu@0 {
                                        compatible = "mips,interAptiv";
                                        ...
                        };
                        cpu1: cpu@1 {
                                        compatible = "mips,interAptiv";
                                        ...
                        };
                        cpu2: cpu@2 {
                                        compatible = "mips,interAptiv";
                                        ...
                        };
                        cpu3: cpu@3 {
                                        compatible = "mips,interAptiv";
                                        ...
                        };
        };
        ...
};
```

- 设备节点

可以增加设备节点，用于描述设备。它通过compatible匹配设备。

```
/dts-v1/;
/ {
        uart0: serial@8300000 {
                compatible = "siflower,sfax8-uart";
                ...
        };
        uart1: serial@8301000 {
                compatible = "siflower,sfax8-uart";
                ...
        };
        pcm0: pcm@8400000 {
                compatible = "siflower,sfax8-pcm";
                ...
        };
        watchdog: watchdog@8700000 {
                compatible = "siflower,sfax8-wdt";
                ...
        };
        pinctrl: pinctrl {
                compatible = "siflower,sfax8-pinctrl";
                ...
        };
        gic: gic@1bdc0000 {
                compatible = "siflower,sfax8-gic";
                ...
        };
        palmbus@10000000 {
                compatible = "palmbus";
                ...
                leds: gpio-leds {
                        compatible = "gpio-leds";
                        ...
                };
                gpio_keys: gpio-keys {
                        compatible = "gpio-keys";
                        ...
                };
                ethernet: ethernet@0000000 {
                        compatible = "siflower,sfax8-eth";
                        ...
                };
                switch: switch@0000000 {
                        compatible = "siflower,sfax8-switch";
                        ...
                };
                ...
        };
};
```

● address属性

address属性用于限定子节点地址信息，用#address-cells和#size-cells表示。我们要在父节点中对其进行定义，用于约定子节点的地址信息。#address-cells表示子节点reg属性需要用多少个cell（位置）来表示。#size-cells表示子节点reg属性需要用多少个cell来表示地址长度。如果#size-cells为0，则reg中不会出现地址长度信息。

例如图12-8中父节点external-bus约定了子节点的地址信息中需要用2个cell表示地址，需要用1个cell表示地址长度。

```
external-bus {
        #address-cells = <2>;
        #size-cells = <1>;

        ethernet@0,0 {
            compatible = "smc,smc91c111";
            reg = <0 0 0x1000>;
        };

        i2c@1,0 {
            compatible = "acme,a1234-i2c-bus";
            reg = <1 0 0x1000>;
            rtc@58 {
                compatible = "maxim,ds1338";
            };
        };

        flash@2,0 {
            compatible = "samsung,k8f1315ebm", "cfi-flash";
            reg = <2 0 0x4000000>;
        };
};
```

图12-8　address属性

● reg属性

reg属性表示设备使用的地址信息，格式为reg=<address1 length1 [address2 length2]…>。其中每一组address length表示设备使用的一个地址范围。address和length由父节点#address-cells和#size-cells来约束。reg属性的示例如下。

```
/dts-v1/;
/ {
        palmbus@10000000 {
                    compatible = "palmbus";
                    #address-cells = <1>;
                    #size-cells = <1>;
                    ...
            timer0: timer@8600000 {
                    compatible = "siflower,sfax8-timer";
                    reg = <0x8600000 0x14>, <0x8600014 0x14>, <0x8601000 0x14>, <0x8601014 0x14>;
                    ...
            };
```

```
        ...
    };
};
```

● interrupts属性

interrupts属性用于描述中断信息,具体通过4个属性来表示。

■ interrupt-controller:一个空属性,用于定义一个接收中断信号的设备。通常是中断控制器,如GIC。

■ #interrupt-cells:中断控制器节点的属性,用于声明该中断控制器的终端指示符中cell的个数,与#address-cells和#size-cells属性类似。

■ interrupt-parent:设备节点的属性,包含一个指向该设备连接的中断控制器的phandle(指向或引用&)。没有interrupt-parent的节点,则会从其父节点继承该属性。

■ interrupts:一个设备节点属性,包含一个终端指示符列表,与该设备上的每个中断输出信号相对应。如果只有1个cell,则表示中断控制器的索引号,即中断号。如果有2个cell,则第一个cell表示中断号,第二个cell表示中断触发方式。

```
#linux-3.18.29/Documentation/devicetree/bindings/interrupt-controller
bits[3:0] trigger type and level flags
1 = low-to-high edge triggered
2 = high-to-low edge triggered
4 = active high level-sensitive
8 = active low level-sensitive
```

例如:以下示例中的gic节点为中断控制器,指定cell为2。palmbus总线节点属性interrupt-parent引用了gic节点,所以palmbus下面的所有子节点都会继承该属性。switch节点定义了具体中断,其中使用了25号中断,后面的0表示中断触发方式为none。

```
/dts-v1/;
/ {
    gic: gic@1bdc0000 {
        #interrupt-cells = <2>;
        interrupt-controller;
        ...
    };
    palmbus@10000000 {
        compatible = "palmbus";
        reg = <0x10000000 0x10000000>;
        ranges = <0x0 0x10000000 0xFFFFFFFF>;
        #address-cells = <1>;
        #size-cells = <1>;
        interrupt-parent = <&gic>;
        switch: switch@0000000 {
            compatible = "siflower,sfax8-switch";
            reg = <0x0000000 0x6CFFFF>;
            interrupts = <25 0>;
            max-speed = <100>;
            sfax8,port-map = "lllllw";
            status = "disabled";
        };
```

```
    ...
};
```

● ranges属性

在设备树中描述的地址是从CPU视角的地址空间来看的,当描述的设备不是本地设备时,就需要把设备地址空间与CPU地址空间进行映射,这时就需要用到ranges属性。ranges属性为一个地址转换表,每一行都包含"子节点地址""父节点地址""子节点地址空间大小",分别由子节点#address-cells值、父节点#address-cells值、子节点#size-cells值来决定。

以下示例的external-bus中定义了ranges属性用于扩展设备,其中:

芯片选择0的偏移量0映射到地址范围0x10100000~0x1010ffff(网卡);

芯片选择1的偏移量0映射到地址范围0x10160000~0x1016ffff(I^2C控制器);

芯片选择2的偏移量0映射到地址范围0x30000000~0x30ffffff(NOR Flash)。

```
/dts-v1/;
/ {
    compatible = "acme,coyotes-revenge";
    #address-cells = <1>;
    #size-cells = <1>;
    ...
    external-bus {
        #address-cells = <2>;
        #size-cells = <1>;
        ranges = <0 0  0x10100000  0x10000     // Chipselect 1, Ethernet
                  1 0  0x10160000  0x10000     // Chipselect 2, i2c controller
                  2 0  0x30000000  0x1000000>; // Chipselect 3, NOR Flash
        ethernet@0,0 {
            compatible = "smc,smc91c111";
            reg = <0 0 0x1000>;
        };
        i2c@1,0 {
            compatible = "acme,a1234-i2c-bus";
            #address-cells = <1>;
            #size-cells = <0>;
            reg = <1 0 0x1000>;
            rtc@58 {
                compatible = "maxim,ds1338";
                reg = <58>;
            };
        };
        flash@2,0 {
            compatible = "samsung,k8f1315ebm", "cfi-flash";
            reg = <2 0 0x4000000>;
        };
    };
};
```

ranges属性与address属性的对应关系如图12-9所示。

```
/dts-v1/;
/ {
    compatible = "acme,coyotes-revenge";
    #address-cells = <1>;
    #size-cells = <1>;
    ...
    external-bus {
        #address-cells = <2>
        #size-cells = <1>;
        ranges = <0 0  0x10100000  0x10000    // Chipselect 1, Ethernet
                  1 0  0x10160000  0x10000    // Chipselect 2, i2c controller
                  2 0  0x30000000  0x1000000>; // Chipselect 3, NOR Flash
        ...
    };
};
```

图12-9 ranges属性与address属性的对应关系

12.2.4 新硬件的OpenWrt系统定制

本节中，我们将进行新硬件的OpenWrt系统的定制。

SF16A18芯片配置目录如图12-10所示。

```
base-files                              sf16a18_p10_fullmask_8m.config
base-files.mk                           sf16a18_p10_fullmask_def.config
image                                   sf16a18_p10_fullmask_flash.config
Makefile                                sf16a18_p10_fullmask_gmac.config
patches                                 sf16a18_p10_fullmask_x10.config
sf16a18_86v_c2_fullmask_def.config      sf16a18_p10h_fullmask_gmac.config
sf16a18_86v_fullmask_def.config         sf16a18_p10_mpw0_autotest.config
sf16a18_ac_fullmask_def.config          sf16a18_p10_mpw0_def.config
sf16a18_AiRouter_fullmask_def.config    sf16a18_p10_mpw0_flash.config
sf16a18_cpe_fullmask_def.config         sf16a18_p20_fullmask_def.config
sf16a18_df1a_fullmask_def.config        sf16a18_p20_mpw0_autotest.config
sf16a18_evb_v5_fullmask_def.config      sf16a18_p20_mpw0_def.config
sf16a18-fullmask                        sf16a18_rep_fullmask_def.config
sf16a18-mpw0                            sf16a18_rep_nopa_fullmask_def.config
sf16a18-mpw1
```

图12-10 chaos_calmer_15_05_1/target/Linux/siflower/目录

- sf16a18-fullmask/：SF16A18芯片的配置目录。
  - config-3.18-*/：表示不同硬件对应的Kernel config。
  - base-files-SF16A18-*/：表示针对该硬件的专用脚本或数据文件，即最终存放在rootfs目录下的文件。
  - profile/：该目录下面存放每个硬件对应的.mk文件，其中可以设定硬件对应的核心软件包选型，包含选择软件包和config项，其中包括了该硬件的名称和描述内容。
- base-files/：存放SF16A18芯片各硬件通用的脚本或数据文件。

- patches/：存放针对SF16A18芯片内核的补丁文件。
- images/：存放固件文件生成规格的文件。
- sf16a18_*_fullmask_*.config文件：OpenWrt编译时预设的配置文件。

```
#当使用该方法初始化时,会自动将sf16a18_df1a_fullmask_def.config复制到OpenWrt主目录下并修改
为.config文件
./make.sh df1a fullmask
```

SF16A18芯片Kernel目录(linux-3.18.29-dev/linux-3.18.29/)

在Kernel中修改有关版本型号配置时,主要修改DTS文件。该文件位于linux-3.18.29/arch/mips/boot/dts目录中。

MY888的OpenWrt定制

1. 创建profile/sf16a18-my888.mk文件,修改所有的相应名称为指定型号,名称需要和文件名统一,注意大小写。

```
#复制DF1A开发板的profile文件,在此基础上修改
HOST:chaos_calmer_15_05_1$ cd target/linux/siflower/sf16a18-fullmask/profiles/
#从df1a模板中复制
HOST:profiles$ cp sf16a18-df1a.mk sf16a18-my888.mk
#修改sf16a18-my888.mk
HOST:profiles$ vim sf16a18-my888.mk
#修改内容如下
define Profile/SF16A18-MY888
        NAME:= SF16A18 MY888
        PACKAGES:=\
                iperf \
                samba36-server luci-app-samba vsftpd vsftpd-tls \
                libwebsockets ssst subcloud libcurl ndscan tc curl netdetect netdiscover\
                iwinfo luasql-sqlite3 luci-ssl luci-lib-json \
                block-mount fstools badblocks ntfs-3g \
                kmod-fs-vfat kmod-fs-ntfs kmod-fs-ext4 kmod-nls-base kmod-nls-utf8 kmod-nls-cp936 kmod-scsi-core \
                kmod-nls-cp437 kmod-nls-cp850 kmod-nls-iso8859-1 kmod-nls-iso8859-15 kmod-nls-cp950 \
                openssl-util p2p tcpdump rwnxtools wandetect luci-app-upnp miniupnpd kmod-sf-ts kmod-ipt-ipset ipset
endef
define Profile/SF16A18-MY888/Description
        Support for my888 boards
endef
define Profile/SF16A18-MY888/Config
    select BUSYBOX_DEFAULT_FEATURE_TOP_SMP_CPU
    select BUSYBOX_DEFAULT_FEATURE_TOP_DECIMALS
    select BUSYBOX_DEFAULT_FEATURE_TOP_SMP_PROCESS
    select BUSYBOX_DEFAULT_FEATURE_TOPMEM
    select BUSYBOX_DEFAULT_FEATURE_USE_TERMIOS
    select BUSYBOX_DEFAULT_CKSUM
    select TARGET_ROOTFS_SQUASHFS
```

```
select LUCI_LANG_zh-cn
endef
$(eval $(call Profile,SF16A18-MY888))
```

2. 创建base-files-SF16A18-MY888目录。

```
HOST:chaos_calmer_15_05_1$ cd target/linux/siflower/sf16a18-fullmask/
#复制DF1A开发板的base-files,用于存放额外的文件,最终会放到rootfs中
HOST:sf16a18-fullmask$ cp -af base-files-SF16A18-DF1A/ base-files-SF16A18-MY888
```

3. 创建Kernel Config文件。

```
HOST:chaos_calmer_15_05_1$ cd target/linux/siflower/sf16a18-fullmask/
HOST:sf16a18-fullmask$ cp config-3.18_df1a config-3.18_my888
```

4. 创建MY888的预设配置。

```
HOST:chaos_calmer_15_05_1$ cd target/linux/siflower/
#复制DF1A开发板的预设配置。后期如果修改预设配置,可用.config文件覆盖该文件
HOST:siflower$ cp sf16a18_df1a_fullmask_def.config sf16a18_my888_fullmask_def.config
#编辑该文件
HOST:siflower$ vim sf16a18_my888_fullmask_def.config
#修改第50行
CONFIG_TARGET_siflower_sf16a18_fullmask_SF16A18-DF1A=y
为
CONFIG_TARGET_siflower_sf16a18_fullmask_SF16A18-MY888=y
#CONFIG_TARGET_siflower_sf16a18_fullmask_SF16A18-DF1A
```

5. 创建内核通用配置文件。

```
HOST:chaos_calmer_15_05_1$ cd target/linux/generic/
#复制DF1A开发板的内核通用配置文件
HOST:generic$ cp config-3.18_df1a config-3.18_my888
```

6. 建立profile和Kernel config的对应关系。

```
#进入include目录
HOST:chaos_calmer_15_05_1$ cd include/
#编辑target.mk文件
HOST:chaos_calmer_15_05_1$ vim target.mk
#在
ifneq ($(findstring DF1A,$(PROFILE)),)
  __config_name_list = $(1)/config-$(KERNEL_PATCHVER)_df1a $(1)/config-default
endif
#下面添加
ifneq ($(findstring MY888,$(PROFILE)),)
  __config_name_list = $(1)/config-$(KERNEL_PATCHVER)_my888 $(1)/config-default
endif
```

7. 在SF16A18的编译前脚本中增加对MY888的支持。

```
#进入OpenWrt主目录,修改make.sh文件,增加对于新版本型号的编译指令
HOST:chaos_calmer_15_05_1$ vim make.sh
#在第132行后增加
  df1a)
            target_board=target/linux/siflower/sf16a18_df1a_${chip}_def.config
            ;;
  my888)
```

```
                target_board=target/linux/siflower/sf16a18_my888_${chip}_def.config
                ;;
```

8. 生成新的预设配置。

```
#获得当前的预设配置
HOST:chaos_calmer_15_05_1$ cp target/linux/siflower/sf16a18_my888_fullmask_def.config
.config
HOST:chaos_calmer_15_05_1$ make menuconfig
#选择配置,保存退出
        Target System (MIPS Siflower SF16ax8 board)     --->
        Subtarget (sf16a18 full mask based boards)     --->
        Target Profile (SF16A18 MY888)     --->
#将保存好的.config文件覆盖回预设配置
HOST:chaos_calmer_15_05_1$ cp .config target/linux/siflower/sf16a18_my888_fullmask_
def.config
```

Linux内核适配

Linux内核适配的操作主要是根据新硬件信息修改DTS硬件的描述文件,步骤如下。

1. 生成和修改DTS文件。

linux-3.18.29/arch/mips/boot/dts/目录存放MIPS芯片相关的DTS描述文件。sf16a18_full_mask.dtsi是SF16A18芯片的通用设备树描述文件。生成和修改DTS文件的过程如下。

```
HOST:sf16a18-sdk-4.2.10$ cd linux-3.18.29-dev/linux-3.18.29/arch/mips/boot/dts/
HOST:dts$ cp sf16a18_fullmask_df1a.dts sf16a18_fullmask_my888.dts
HOST:dts$ vim sf16a18_fullmask_my888.dts
#内容修改如下:
/dts-v1/;
#include "sf16a18_full_mask.dtsi"
#include "sf16a18-thermal.dtsi"
/ {
        #address-cells = <1>;
        #size-cells = <1>;
        compatible = "siflower,sf16a18-soc";
        eth0-led {
                compatible = "eth-led";
                pinctrl-names = "default",  "gpio";
                pinctrl-0 = <&eth0_led>;
                pinctrl-1 = <&eth0_led_gpio>;
                eth0 {
                        label = "eth_led0";
                        gpios = <&gpio 55 1>;
                        default-state = "on";
                };
        };
        eth1-led {
                compatible = "eth-led";
                pinctrl-names = "default",  "gpio";
                pinctrl-0 = <&eth1_led>;
                pinctrl-1 = <&eth1_led_gpio>;
                eth1 {
```

```
                        label = "eth_led1";
                        gpios = <&gpio 56 1>;
                        default-state = "on";
                };
        };
        sf_mm_led: sf-mm-led {
                compatible = "sf-mm-led";
                blue-led-gpio = <&gpio 55 0>;
                green-led-gpio = <&gpio 56 0>;
                red-led-gpio = <&gpio 59 0>;
                default-mode = <0x39>;
                default-interval = <500>;
        };
        w18_wifi_rf: w18_wifi-rf@7A00000{
                compatible = "siflower,pistachio-uccp";
                phy = <&phy1>;
                #address-cells = <1>;
                #size-cells = <0>;
                phy1: w18_wifi-phy@1 {
                        reg = <1>;
                        max-speed = <100>;
                };
        };
};
&usb_phy{
        status = "okay";
};
&ethernet {
        shutdown-portlist = "wl***";
        led-on-off-time = /bits/ 16 <0xff 0x2ff>;
        status = "okay";
        smp-affinity = <2>;
};
&switch{
        sfax8,port-map = "wl***";
        status = "okay";
        smp-affinity = <2>;
};
&gmac {
        status = "okay";
        smp-affinity = <0>;
};
&gdu {
        status = "okay";
        num-windows = <2>;
        rgb_order = <0 1 2>;
        power_gpio = <26 37>;
        bpp-mode = "RGB888";
        display-timings {
                native-mode = <&timing0>;
```

```
                timing0: gm05004001q {
                        clock-frequency = <34539600>;
                        hactive = <800>;
                        vactive = <480>;
                        hback-porch = <46>;
                        hfront-porch = <210>;
                        vback-porch = <23>;
                        vfront-porch = <22>;
                        hsync-len = <20>;
                        vsync-len = <10>;
                };
        };
};
&sham {
        status = "okay";
};
&cipher {
        status = "okay";
};
&rng {
        status = "okay";
};
&gdma {
        status = "okay";
};
&usb {
        status = "okay";
};
&emmc {
        status = "disabled";
};
&sdio {
        status = "okay";
};
&i2s_master {
        status = "disabled";
};
&i2s_slave {
        status = "okay";
};
/* This is used to set voltage during dvfs.
 * You should first know which pmu is in use, and
 * remove the default core-voltage set in pmu dts.
 */
/*
&cpu0 {
        cpu0-supply = <&rn5t567_core>;
        dcdc0-maxv = <1000000>;
};
*/
&i2c0 {
```

```
        status = "okay";
        #address-cells = <1>;
        #size-cells = <0>;
};
&i2c1 {
        status = "disabled";
};
&i2c2 {
        status = "disabled";
};
&spi0 {
        status = "okay";
        //use-dma;
        dmas = <&gdma 10
                &gdma 11>;
        dma-names = "tx", "rx";
        #address-cells = <1>;
        #size-cells = <0>;
        w25q128@0 {
                compatible = "w25q128";
                reg = <0>;        /* chip select */
                spi-max-frequency = <33000000>;
                bank-width = <2>;
                device-width = <2>;
                #address-cells = <1>;
                #size-cells = <1>;
                partition@0 {
                        label = "spl-loader";
                        reg = <0x0 0x20000>; /* 128k */
                        read-only;
                };
                partition@20000 {
                        label = "u-boot";
                        reg = <0x20000 0x60000>; /* 384k */
                };
                partition@80000 {
                        label = "u-boot-env";
                        reg = <0x80000 0x10000>; /* 64k */
                };
                factory:partition@90000 {
                        label = "factory";
                        reg = <0x90000 0x10000>; /* 64k */
                };
                partition@a0000 {
                        label = "firmware";
                        reg = <0xa0000 0xf60000>; /* 640k-16M */
                };
        };
        spidev: spi@8200000 {
                compatible = "rohm,dh2228fv";
```

```
                reg = <1>;
                clock = <50000000>;
                spi-cpha;
                spi-cpol;
                spi-max-frequency=<12000000>;
        };
};
&uart0 {
        status = "okay";
};
&uart1 {
        status = "disabled";
};
&uart2 {
        pinctrl-0 = <&uart2_tx &uart2_rx>;
        status = "okay";
};
&uart3 {
        pinctrl-0 = <&uart3_tx &uart3_rx>;
        status = "okay";
};
&pcm0 {
        status = "okay";
};
&pwm0 {
        status = "okay";
};
&pwm1 {
        status = "okay";
};
&timer0 {
        status = "okay";
};
&watchdog {
        status = "okay";
};
&spdif
{
        status = "disabled";
};
&wifi_rf {
        status = "okay";
        gpio-expa = <&gpio 51 0>;
        /*for new evb board,gpio use different map with that in p10h board*/
        expa_map_type = <2>;
};
&wifi_lb {
        status = "okay";
        #address-cells = <1>;
        #size-cells = <0>;
        gpio-leds = <&gpio 61 0>;
```

```
        smp-affinity = <2>;
};
&wifi_hb {
        status = "okay";
        #address-cells = <1>;
        #size-cells = <0>;
        gpio-leds = <&gpio 36 0>;
        smp-affinity = <3>;
};
&leds {
        status = "okay";
        sys_led {
                label = "sys_led";
                gpios = <&gpio 25 1>;
        };
};
```

2. 添加MY888的DTS文件编译宏。

```
HOST:sf16a18-sdk-4.2.10$ cd linux-3.18.29-dev/linux-3.18.29/arch/mips/boot/dts/
#编辑Makefile文件
HOST:dts$ vi Makefile
#内容如下:
在第33行
dtb-$(CONFIG_DT_SF16A18_FULLMASK_DF1A)   += sf16a18_fullmask_df1a.dtb
下面添加
dtb-$(CONFIG_DT_SF16A18_FULLMASK_MY888)  += sf16a18_fullmask_my888.dtb
```

3. 增加Kconfig的菜单。

```
HOST:sf16a18-sdk-4.2.10$ cd linux-3.18.29-dev/linux-3.18.29/arch/mips/siflower/
#编辑Kconfig文件
HOST:siflower$ vi Kconfig
#在config DT_SF16A18_FULLMASK_DF1A前一行添加MY888信息
config DT_SF16A18_FULLMASK_MY888
        bool "Built-in device tree for sf16a18 fullmask my888 boards"
        default n
        select BUILTIN_DTB
        help
          Add an FDT blob for XLP EVP boards into the kernel.
          This DTB will be used if the firmware does not pass in a DTB
          pointer to the kernel.  The corresponding DTS file is at
          arch/mips/boot/dts/sf16a18_fullmask_my888.dts
```

4. 选择DTS配置。

```
#进入OpenWrt主目录
HOST:chaos_calmer_15_05_1$ make kernel_menuconfig
Machine selection--->
[*] Built-in device tree for sf16a18 fullmask my888 boards
```

5. 再次保存预配置。

```
#进入OpenWrt主目录后覆盖如下文件
HOST:chaos_calmer_15_05_1$ cp .config target/linux/siflower/sf16a18_my888_fullmask_def.config
```

编译固件镜像

首次编译时采用下列方式。

```
#该过程将自动使用预配置生成.config文件并且自动完成整个编译流程,最终生成固件文件
HOST:chaos_calmer_15_05_1$ ./make.sh my888 fullmask
```

后期编译时采用下列方式。

```
#如果修改过OpenWrt软件包选择,将更新.config文件,这时再使用make.sh将导致所做的选择丢失
HOST:chaos_calmer_15_05_1$ make V=99 -j 4
#也可以通过前面介绍的写回方法用新的.config文件覆盖预设配置
```

如果需要刷机,请参考10.3.4节的介绍进行固件刷机。

查看MY888的信息。

```
admin@SiWiFia12d:/# cat /etc/openwrt_release
DISTRIB_ID='SiWiFi'
DISTRIB_RELEASE='Chaos Calmer'
DISTRIB_REVISION='unknown'
DISTRIB_CODENAME='sf16a18-my888'
DISTRIB_TARGET='siflower/sf16a18-fullmask'
DISTRIB_DESCRIPTION='SiWiFi SF16A18-MY888 openwrt_master_my888_fullmask_rel_'
DISTRIB_TAINTS='no-all busybox'


admin@SiWiFia12d:/# cat /sys/kernel/debug/gpio
GPIOs 0-70, platform/pinctrl, gpio:
 gpio-5   (?                   ) out hi
 gpio-6   (?                   ) out hi
 gpio-25  (sys_led             ) out hi
 gpio-55  (eth_led0            ) in  lo
 gpio-56  (eth_led1            ) in  lo
```

13 总线原理分析

13.1 系统启动原理

OpenWrt系统的启动过程和大多数Linux系统的启动过程类似，都经过U-Boot引导、加载内核、挂载文件系统、启动init进程、加载模块和启动应用进程等步骤。

基于芯片和OpenWrt本身特点，OpenWrt系统的启动过程有一些特有的处理步骤。大部分OpenWrt系统的启动过程基本类似，但是不同的版本会有一些差别。例如我们使用的OpenWrt发行版为chaos_calmer 15.05.1，init由procd包提供，而在较老的版本中init由busybox提供。对于DF1A开发板，在开发板上电时，其U-Boot启动前还需要使用SPL先进行内存等初始化，然后才切换到U-Boot引导。整体上，不同芯片的OpenWrt系统启动过程差别不大，可以在学习本章节后触类旁通，结合本章内容分析自己的开发板的启动过程。

首先介绍DF1A开发板的主要启动流程。

（1）系统上电，SPL启动，进行内存等硬件核心初始化，然后引导U-Boot。

（2）U-Boot启动，完成CPU、内存、网络驱动、堆栈等初始化，为引导内核做准备，然后引导内核，并传递内核参数。

（3）内核自解压和启动。内核执行CPU、内存、子系统初始化和驱动程序加载，扫描mtd的rootfs分区，挂载SquashFS分区，启动/sbin/init进程。

（4）/sbin/init进程执行一系列处理，完成系统启动的后续过程。包括加载文件系统内的内核驱动模块；调用/etc/preinit，执行/sbin/mount_root，挂载JFFS2分区（/overlay），并与SquashFS分区（/rom）进行合并，创建新的虚拟根文件系统（/）；调用/sbin/procd，完成rcS系列脚本启动。

接下来，我们对启动过程中涉及的主要流程进行详细分析。

13.1.1 SPL引导

通过3.1节的介绍可以知道：spl-loader（下文中简称为SPL）是系统上电后执行的第一个引导程序，它是介于芯片内部ROM程序与U-Boot之间的一个BootLoader，其主要功能为初始化DDR、系统管理器、时钟，以及加载U-Boot。SPL程序本身需要加载到系统内部RAM中运行，是一个轻量级的U-Boot。

U-Boot开源程序本身是包含SPL选项的。Uboot-spl是用U-Boot同一套代码，通过

CONFIG_SPL_BUILD宏分割编译出的结果。由于SF16A18芯片内部RAM只有64KB,不足以同时支持Uboot-spl的RAM+RAM的需求,因此需要采用以裸机实现的bare_spl。bare_spl与原Uboot-spl相比,更加简单可控,也同样可以实现引导U-Boot的功能,并且能减少存储空间的使用,为Flash优化、分区的重新定义提供支持。

SPL执行流程如图13-1所示。

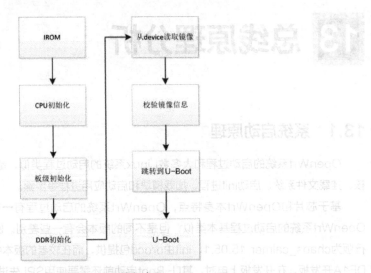

图13-1 SPL执行流程

SPL与U-Boot相同,引导的镜像需要包含一个uImage的header,其中会包含所引导镜像的类型。因此,SPL不仅可以引导U-Boot进而启动内核,还可以引导开发板工厂测试程序。我们通过串口工具连接开发板的串口调试接口,重启开发板,就可以看到以下输出信息。

```
Booting...
eth check connect ret=1 cost time=1440 ms
IROM DONE!
SiFlower SFAX8 Bootloader (Jul  9 2019 - 14:46:03)
ddr3 nt5cc128m16ip init start
MEM_PHY_CLK_DIV = 0x3
DDR training success
now ddr frequency is 400MHz!!!
ddr test
DR1BW a0000000 OK
DR1BR OK
DR2BW a0000000 OK
DR2BR OK
DR4BW a0000000 OK
DR4BR OK
DR8BW a0000000 OK
DR8BR OK
Boot from spi-flash
U-image: U-Boot 2016.07-rc2 for sfa18_p2, size is 310044
loaded - jumping to U-Boot 0xa0000000...
```

SPL初始化结束后,会加载和引导U-Boot启动。

13.1.2 U-Boot启动

U-Boot的启动分为两个阶段：stage1和stage2。可以认为stage1是U-Boot的ROM阶段，stage2是RAM阶段。stage1向stage2转换的主要标志就是代码段的relocation，即将代码段从ROM复制到RAM。

U-Boot引导过程如图13-2所示。

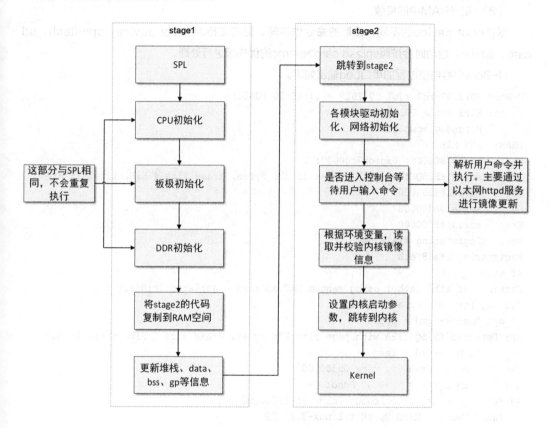

图13-2　U-Boot引导过程

每个stage都包含了需要执行的主要函数序列，即init_sequence_f和init_sequence_r。由于存在SPL，U-Boot的stage1其实并没有什么实质性的功能，整个U-Boot都运行在DDR上。在进入stage2前，U-Boot会进行代码段的重定向，由于U-Boot本身与位置无关，因此只需要同时更新堆栈信息、全局变量表等，重定向后的代码就依旧可以正常执行。按照U-Boot的思想，这个stage2运行在RAM中，因此其运行速度要比stage1快。

init_sequence_r主要进行各个模块驱动的初始化、网络的初始化和其他准备工作，比如环境变量的初始化等。在准备结束后，U-Boot最终会进入main_loop，进行控制台的初始化。此时U-Boot提供了一个时长3s的倒计时，如果不在此时间内从控制台（默认串口）进行输入，则会进入自动启动流程，根据预设的环境变量参数，进行引导启动；如果有输入，就可以停下来采用控制台的输入命令，与U-Boot进行交互。此时U-Boot会根据输入命令的情况进行解析并执行。U-Boot支持一些常用的命令，可以在U-Boot命令行中通过help命令查看，其中有一些常用的命令：boot可以让U-Boot继续执行默认命令，reset可以让芯片复位重启，sf与mmc分别对应spi-flash与mmc的操作命令。

U-Boot引导Kernel也是通过控制台命令进行的，默认的命令bootcmd保存在default env中，可以在U-Boot源代码的include/configs/sfax8.h中进行配置。

目前这个过程分为两步。

（1）将镜像从boot device中读到RAM中。

（2）引导RAM中的镜像。

根据boot device的不同，读取的命令也不同。目前支持3种boot device：spi-flash、sd card、emmc。启动时会按照spi>sd card>emmc的优先级进行处理。

U-Boot引导内核过程的串口Log输出如下。

```
U-Boot 2016.07-rc2 (Jul 09 2019 - 14:46:10 +0800)
Board: MIPS sfa18 FULLMASK P20B
       Watchdog enabled
DRAM:  256 MiB
MMC:   emmc@7800000: 0sdio@7c00000: 1
SF: Detected EN25QH128A with page size 256 Bytes, erase size 4 KiB, total 16 MiB
In:    serial@8300000
Out:   serial@8300000
Err:   serial@8300000
Net:   Registering sfa18 net
Registering sfa18 eth
sf_eth0
Warning: sf_eth0 (eth0) using random MAC address - 4a:0a:ab:7c:96:2f
Hit any key to stop autoboot:  0
do_spi_flash----cmd = probe
SF: Detected EN25QH128A with page size 256 Bytes, erase size 4 KiB, total 16 MiB
do_spi_flash----cmd = read
device 0 offset 0xa0000, size 0x300000
SF: 3145728 bytes @ 0xa0000 Read: OK
## Booting kernel from Legacy Image at 81000000 ...
   Image Name:   MIPS OpenWrt Linux-3.18.29
   Image Type:   MIPS Linux Kernel Image (lzma compressed)
   Data Size:    1680062 Bytes = 1.6 MiB
   Load Address: 80100000
   Entry Point:  80105360
   Verifying Checksum ... OK
   Uncompressing Kernel Image ... OK
All resets are held!
```

DF1A使用的是SPI Flash，当U-Boot启动时，如果没有检测到用户数据，系统会在倒计时结束时自动通过sf命令识别、读取SPI Flash内核分区进行引导。

SPI Flash的识别、读取、擦写等操作通过uboot sf命令实现。在U-Boot命令行中使用sf help查看。

从输出Log中可以看出，U-Boot首先通过sf的probe命令识别出SPI Flash的芯片，然后通过sf read命令读取内核所在分区。根据当前的分区情况，U-Boot从Flash的0xa0000位置加载内核镜像，并读取到RAM地址0x81000000。sf read的最后一个参数是读取的长度，目前默认是

0x300000，即10MB。

```
do_spi_flash----cmd = probe
SF: Detected EN25QH128A with page size 256 Bytes, erase size 4 KiB, total 16 MiB
do_spi_flash----cmd = read
device 0 offset 0xa0000, size 0x300000
SF: 3145728 bytes @ 0xa0000 Read: OK
```

完成上述过程后，就实现了内核的加载和引导。

13.1.3 Kernel启动

U-Boot读取内核所在的Flash分区到内存，调用内核固件中自带的解压代码，进行内核解压，并传递内核参数，然后调用真正的内核入口并运行。

以下是内核解压启动的串口输出。

```
## Booting kernel from Legacy Image at 81000000 ...
   Image Name:   MIPS OpenWrt Linux-3.18.29
   Image Type:   MIPS Linux Kernel Image (lzma compressed)
   Data Size:    1680062 Bytes = 1.6 MiB
   Load Address: 80100000
   Entry Point:  80105360
   Verifying Checksum ... OK
   Uncompressing Kernel Image ... OK
All resets are held!
```

从启动Log中可以看到，内核采用的是LZMA压缩算法，将压缩的内核镜像加载到内存中的0x80100000地址（Load Address），然后执行解压代码的第一条指令0x80105360（Entry Point），对内核进行解压。

内核被解压后，按以下两个阶段实现启动过程。

● 第一阶段是内核解压完成并出现"Uncompressing Kernel Image… OK,All resets are held!"之后的阶段。这部分代码实现在内核源代码linux-3.18.29/arch/mips/kernel的head.S中，该文件中的汇编代码通过查找处理器内核类型和机器码类型，来调用相应的初始化函数、建立页表，最后跳转到start_kernel()函数开始内核的初始化工作。

● 内核启动的第二阶段从start_kernel()函数开始。start_kernel()是所有Linux平台进入系统内核初始化后的入口函数，它主要完成剩余的与硬件平台相关的初始化工作，在进行一系列与内核相关的初始化后，扫描mtd的rootfs分区，识别出SquashFS分区（这部分既可以根据内核启动参数的建议"rootfstype=squashfs,jffs2"进行识别，也可以根据rootfs分区中的squashfs头信息进行识别），并将SquashFS文件系统挂载到根分区，根据内核启动参数rdinit=/sbin/init，调用第一个用户进程/sbin/init，加载文件系统中的内核模块，并等待用户进程的执行，直到完成整个系统的启动。

以下是内核启动过程中的一些Log输出内容。

```
[    0.000000] Linux version 3.18.29 (zhengqw@Ironman) (gcc version 4.8.3 (OpenWrt/
Linaro GCC 4.8-2014.04 unknown) ) #1 SMP PREEMPT Wed Jul 17 17:17:22 CST 2019
[    0.000000] memsize not set in YAMON, set to default (256MB)
[    0.000000] arcs_cmdline=memsize=256M
[    0.000000] CPU0 revision is: 5301a128 (MIPS interAptiv (multi))
[    0.000000] FPU revision is: 0173a000
```

```
[    0.000000] MIPS: machine is siflower,sf16a18-soc
[    0.000000] Determined physical RAM map:
[    0.000000]  memory: 01f00000 @ 00000000 (usable)
[    0.000000]  memory: 00100000 @ 01f00000 (reserved)
[    0.000000]  memory: 0e000000 @ 02000000 (usable)
[    0.000000] Initrd not found or empty - disabling initrd
[    0.000000] Zone ranges:
[    0.000000]   Normal   [mem 0x00000000-0x0fffffff]
[    0.000000] Movable zone start for each node
[    0.000000] Early memory node ranges
[    0.000000]   node   0: [mem 0x00000000-0x0fffffff]
[    0.000000] Initmem setup node 0 [mem 0x00000000-0x0fffffff]
[    0.000000] VPE topology {2,2} total 4
[    0.000000] Primary instruction cache 16kB, VIPT, 4-way, linesize 32 bytes.
[    0.000000] Primary data cache 16kB, 4-way, VIPT, no aliases, linesize 32 bytes
[    0.000000] MIPS secondary cache 256kB, 8-way, linesize 32 bytes.
[    0.000000] PERCPU: Embedded 9 pages/cpu @81225000 s7168 r8192 d21504 u36864
[    0.000000] Built 1 zonelists in Zone order, mobility grouping on.  Total pages: 65024
[    0.000000] Kernel command line: memsize=256M console=ttyS0,115200n8 rootfstype=squashfs,jffs2 rdinit=/sbin/init
...
[    1.181854] sfax8-spi 18200000.spi: sfax8_spi_probe...
[    1.188947] m25p80 spi0.0: found en25qh128, expected w25q128
[    1.194719] m25p80 spi0.0: en25qh128 (16384 Kbytes)
[    1.200111] 5 ofpart partitions found on MTD device spi0.0
[    1.205650] Creating 5 MTD partitions on "spi0.0":
[    1.210479] 0x000000000000-0x000000020000 : "spl-loader"
[    1.217396] 0x000000020000-0x000000080000 : "u-boot"
[    1.223940] 0x000000080000-0x000000090000 : "u-boot-env"
[    1.230729] 0x000000090000-0x0000000a0000 : "factory"
[    1.237357] 0x0000000a0000-0x000001000000 : "firmware"
[    1.296853] 2 uimage-fw partitions found on MTD device firmware
[    1.302858] 0x0000000a0000-0x00000023a2fe : "kernel"
[    1.309466] 0x00000023a2fe-0x000001000000 : "rootfs"
[    1.316065] mtd: device 6 (rootfs) set to be root filesystem
[    1.321919] 1 squashfs-split partitions found on MTD device rootfs
[    1.328169] 0x000000930000-0x000001000000 : "rootfs_data"
[    1.336266] sfax8-spi 18200000.spi: SFAx8 SPI Controller at 0x18200000 irq 231
[    1.344018] sfax8_factory_read_probe...
[    1.348109] macaddr is 10 16 88 14 3c fb
[    1.352141] sn is ff ff ff ff ff ff ff ff ff ff ff ff ff ff ff ff
[    1.358372] sn_flag is 0xff
...
[    3.055964] VFS: Mounted root (squashfs filesystem) readonly on device 31:6.
```

内核对init进程的调用，会根据U-Boot传递给内核的参数和内核内部执行流程进行识别。内核源代码linux-3.18.29/init/main.c依据下面的逻辑调用init进程。

（1）如果内核参数有"rdinit=XXX"，则赋值给ramdisk_execute_command，并执行外部程序。

（2）如果内核参数有"init=XXX"，则赋值给execute_command，并执行外部程序。

（3）如果没指定上述内核参数，则依次尝试执行"/etc/preinit"、"/sbin/init"、"/etc/init"、"/bin/init"、"/bin/sh"。

DF1A的U-Boot传递了内核参数rdinit=/sbin/init，所以会执行/sbin/init。/sbin/init的处理由一系列过程组成，处理期间会加载文件系统中的内核模块，执行一系列的初始化脚本，开启各种应用服务等。

13.1.4 init执行过程

/sbin/init是由procd软件包提供的程序。在OpenWrt系统的早期版本或其他Linux嵌入式版本中，init通常由BusyBox提供。

procd在init引导过程中扮演了非常重要的角色。procd具有以下特点。

- procd取代BusyBox的initd、klogd、syslogd、watchdog等功能，可进行系统init。
- procd作为父进程，可管理和监控子进程的状态。
- 旧版本的OpenWrt系统采用iniscript启动进程，新版本采用procd启动进程。
- procd可作为Hotplug daemon。

/sbin/init代码由procd/initd/init.c实现，其主要执行流程如下。

1. 注册系统信号。

（1）SIGUSR1、SIGUSR2信号使用RB_POWER_OFF事件关闭系统。

（2）SIGTERM信号使用RB_AUTOBOOT事件重启系统。

2. 调用early()(initd/early.c)，为init执行做前期准备。

（1）调用early_mounts()挂载/proc、/sysfs、/cgroup、/tmpfs、/devpts文件系统和创建软连接。

（2）调用early_dev()创建设备节点和/dev/null文件节点。

（3）初始化控制台/dev/console。

（4）在tmp目录下建立run、lock、state目录。

（5）调用early_env()设置PATH环境变量为"/usr/sbin:/usr/bin:/sbin:/bin"。

（6）在控制台输出"Console is alive"。

3. 根据/proc/cmdline内容init_debug=([0-9]+)，设置debug级别。

4. 初始化watchdog。

（1）初始化内核看门狗(/dev/watchdog)。每5s看门狗ping一次，看门狗超时时间为30s。如果内核在30s内没有收到任何数据，将重启系统。系统调用uloop定时器，设置以5s为周期向/dev/wathdog设备写一些数据通知内核，表示进程在正常工作。

（2）在控制台输出"- watchdog -"。

5. 创建子进程，执行/sbin/kmodloader，加载/etc/modules-boot.d/目录中文件描述的内核模块。这里的模块加载属于较早的流程，后续流程中还会加载一些内核模块。该子进程首先在/etc/modules-boot.d/遍历所有的文件内容，根据文件名的ASCII码排序，然后根据文件内容，加载.ko模块。开发板中的/etc/modules-boot.d/内容如图13-3所示。

```
DF1A:$ ls /etc/modules-boot.d/
02-crypto-hash      20-lib-crc16            30-fs-ext4          31-sf-ts            55-sf_switch
09-crypto-aead      30-button-hotplug       31-reset-button     40-scsi-core        60-sf_gmac
```

图13-3 /etc/modules-boot.d/包含的模块文件

以40-scsi-core描述的模块的加载优先级为例，其优先级顺序为scsi_mod>sd_mod，会依次查找、加载scsi_mod.ko和sd_mod.ko，如图13-4所示。

```
DF1A:$ cat /etc/modules-boot.d/40-scsi-core
scsi_mod
sd_mod
```

图13-4 40-scsi-core模块文件包含的模块

6. 执行preinit()，在内部完成init的主要流程。

（1）在控制台输出"- preinit -"。

（2）创建子进程执行/sbin/procd -h /etc/hotplug-preinit.json，这是第一次运行/sbin/procd，主要是对uevent的处理进行监听，并设置进程执行完成后调用的回调函数。

（3）设置PREINIT环境变量的值为1。

（4）创建子进程执行/etc/preinit脚本，设置该脚本执行结束后的回调函数spawn_procd()。

（5）执行回调spawn_procd()函数，第二次运行/sbin/procd。

以上的流程可以体现出/sbin/init是一个神奇的程序，既复杂又优美。经过一系列巧妙的操作后，和启动相关的线索终于显现。它启动了两个子进程执行系统启动。

（1）子进程1，执行/sbin/procd -h /etc/hotplug-preinit.json，这是带-h参数运行的procd程序，作为Hotplug daemon，这是/sbin/procd的第一次启动。

（2）子进程2，执行/etc/preinit脚本，这个脚本会调用一系列脚本完成preinit操作。

（3）主进程在子进程2执行完/etc/preinit脚本后，第二次启动/sbin/procd。

（4）/sbin/procd的第二次运行，会根据procd内部状态机和接收的信号来执行相关流程，其中包含我们熟知的/etc/inittab和/etc/rcS流程。此时/sbin/procd替换/sbin/init进程成为新的1号进程，完成开机的初始化和各种用户服务的启动。

下面让我们抽茧剥丝，沿着这些线索，拨开启动过程的神秘面纱。

/sbin/procd的第一次启动

/sbin/init第一次运行procd时，启动参数为"/sbin/procd -h /etc/hotplug-preinit.json"，这个过程主要是创建Netlink事件，对内核uevent事件进行处理。

/etc/hotplug-preinit.json的内容如下。

```
DF1A:$ cat /etc/hotplug-preinit.json
[
    [ "case", "ACTION", {
        "add": [
            [ "if",
                [ "has", "FIRMWARE" ],
                [
                    [ "exec", "/sbin/hotplug-call", "%SUBSYSTEM%" ],
                    [ "load-firmware", "/lib/firmware" ],
```

```
                                [ "return" ]
                            ]
                        ],
                    ],
                }, ],
                [ "if",
                    [ "and",
                        [ "eq", "SUBSYSTEM", "button" ],
                    ],
                    [ "exec", "/etc/rc.button/failsafe" ]
                ],
        ]
```

/etc/hotplug-preinit.json主要描述了当uevent事件的ACTION为add类型时,加载固件和故障保护模式(failsafe)的Hotplug处理。

1. 如果事件中包含$FIRMWARE变量,则按下列步骤(1)、步骤(2)执行。

(1)执行/sbin/hotplug-call,识别uevent中的$SUBSYSTEM变量,运行/etc/hotplug.d/下对应目录中的脚本,包含block、iface、net、ntp几类子系统。/etc/hotplug.d包含的目录和脚本如图13-5所示。

```
DF1A:$ cat /etc/modules-boot.d/40-scsi-core
scsi_mod
sd_mod
DF1A:$ ls /etc/hotplug.d/block/
10-mount
DF1A:$ ls /etc/hotplug.d/iface/
00-netstate        20-firewall        30-relay           60-speedlimit      80-updateserverip
15-teql            25-dnsmasq         50-miniupnpd       70-subservice      95-ddns
DF1A:$ ls /etc/hotplug.d/net/
00-sysctl   01-wlanctl
DF1A:$ ls /etc/hotplug.d/ntp/
sync_acl
```

图13-5 /etc/hotplug.d包含的目录和脚本

(2)加载/lib/firmware下的对应升级固件完成模块的固件升级,如图13-6所示。

```
DF1A:$ ls /lib/firmware/
agcram.bin              rf_default_reg.bin      rwnx_aetnensis.ini      sf_rf_expa_config.ini
hdm_hwdiag.txt          rf_pmem.bin             rwnx_settings.ini       tx_adjust_gain_table.bin
mac_hwdiag.txt          rf_xdma_reg.bin         sf1688_hb_smac.bin
nxtop_hwdiag.txt        riu_hwdiag.txt          sf1688_lb_smac.bin
```

图13-6 /lib/firmware下的升级固件

2. 判断uevent事件,如果$SUBSYSTEM为button,则运行/etc/rc.button/failsafe脚本。

```
[ "${TYPE}" = "switch" ] || echo ${BUTTON} > /tmp/failsafe_button
```

该脚本判断$TYPE不等于"switch"时,创建${BUTTON},写入/tmp/failsafe_button文件,该文件会在/lib/preinit/30_failsafe_wait文件中使用。

执行/etc/preinit

正如"执行"这个词表达的那样,preinit肩负着系统初始化前的重担。preinit脚本的内容如下。

```
#!/bin/sh
# Copyright (C) 2006 OpenWrt.org
```

```
# Copyright (C) 2010 Vertical Communications
[ -z "$PREINIT" ] && exec /sbin/init
export PATH=/usr/sbin:/usr/bin:/sbin:/bin
pi_ifname=
pi_ip=192.168.1.1
pi_broadcast=192.168.1.255
pi_netmask=255.255.255.0
fs_failsafe_ifname=
fs_failsafe_ip=192.168.1.1
fs_failsafe_broadcast=192.168.1.255
fs_failsafe_netmask=255.255.255.0
fs_failsafe_wait_timeout=2
pi_suppress_stderr="y"
pi_init_suppress_stderr="y"
pi_init_path="/usr/sbin:/usr/bin:/sbin:/bin"
pi_init_cmd="/sbin/init"
. /lib/functions.sh
. /lib/functions/preinit.sh
. /lib/functions/system.sh
boot_hook_init preinit_essential
boot_hook_init preinit_main
boot_hook_init failsafe
boot_hook_init initramfs
boot_hook_init preinit_mount_root
for pi_source_file in /lib/preinit/*; do
     . $pi_source_file
done
boot_run_hook preinit_essential
pi_mount_skip_next=false
pi_jffs2_mount_success=false
pi_failsafe_net_message=false
boot_run_hook preinit_main
```

它既简洁又不失优雅,却肩负着preinit的重担,它会开启一扇大门,一步步引导我们,来到它的庞大世界。

接下来我们对脚本的主要部分进行解释。

(1)[-z "$PREINIT"] && exec /sbin/init,判断PREINIT环境变量是否设置,如果为空则表示/sbin/init没有被执行过。这段语句主要用来确保/sbin/init得到执行。在之前的分析中,我们知道环境变量PREINIT的值在/etc/preinit执行前,已经被设置为1,可以在procd源代码的initd/preinit.c中看到setenv("PREINIT","1",1)。所以,不会执行exec /sbin/init。

(2)设置环境变量PATH,初始化pi_*、fs_failsafe_*等于网络相关的一些环境变量。在后文中会对此加以解释。

(3)加载了3个Shell脚本:/lib/functions.sh、/lib/functions/preinit.sh、/lib/functions/system.sh。这样在接下来的过程中就可以调用这些脚本里面的函数。

(4)调用lib/functions/preinit.sh中的boot_hook_init函数,初始化hook链。执行后会产生下列hook变量:preinit_essential_hook、preinit_main_hook、failsafe_hook、initramfs_hook、preinit_mount_root_hook。

（5）加载/lib/preinit/目录下所有的脚本。

```
02_default_set_state              50_indicate_regular_preinit
10_indicate_failsafe              70_initramfs_test
10_indicate_preinit               80_mount_root
10_sysinfo                        99_10_failsafe_login
30_failsafe_wait                  99_10_run_init
40_run_failsafe_hook              99_rf_misc
```

这些脚本的格式都差不多，主要是将需要执行的函数添加到对应的hook链中（可以简单理解为链表），最终会形成诸如xxx_hook= funcX funcY funcZ …的内容。

当所有的脚本全部加载后，最终生成的hook链结果如下。

```
hook=preinit_essential （空）
hook=failsafe func=indicate_failsafe failsafe_netlogin failsafe_shell
hook=initramfs （空）
hook=preinit_main func=define_default_set_state preinit_ip pi_indicate_preinit do_
sysinfo_generic failsafe_wait run_failsafe_hook indicate_regular_preinit initramfs_
test do_mount_root run_init load_firmware insmod_rf insmod_mac80211 insmod_smac_lb
insmod_smac_hb
hook=preinit_mount_root （空）
```

（6）执行boot_run_hook preinit_essential。由于preinit_essential为空，什么函数也不执行。

（7）设置下列变量：pi_mount_skip_next、pi_jffs2_mount_success、pi_failsafe_net_message。这些变量在后文中会解释。

（8）执行preinit_main hook链中注册的所有函数。其中do_mount_root内部会调用preinit_mount_root hook链中的函数。

hook链调用过程如图13-7所示。

图13-7　hook链调用过程

综上所述，preinit把不同阶段要执行的内容通过hook链的方式进行了划分，并运用一些技巧，把这些要执行的Shell函数巧妙地链起来。然后调用相应hook中的函数，达到初始化的目的。

以下对这些hook链执行的内容逐一介绍。

1. preinit_essentials

负责安装必要的内核文件系统，如proc，并初始化控制台。因为这些内容在调用/etc/preinit前已经在/sbin/init中完成，所以这个hook链内容为空。preinit hook链包含的执行函数及其所在文件如表13-1所示，preinit hook链执行函数的内容介绍如表13-2所示，preinit hook链执行函数间接调用的函数如表13-3所示。

表 13-1 preinit hook 链包含的执行函数及其所在文件

所在文件	执行函数
02_default_set_state	define_default_set_state
10_indicate_preinit	preinit_ip pi_indicate_preinit
10_sysinfo	do_sysinfo_generic
30_failsafe_wait	failsafe_wait
40_run_failsafe_hook	run_failsafe_hook
50_indicate_regular_preinit	indicate_regular_preinit
70_initramfs_test	initramfs_test
80_mount_root	do_mount_root
99_10_run_init	run_init
99_rf_misc	load_firmware、insmod_rf、insmod_mac80211、insmod_smac_lb、insmod_smac_hb

表 13-2 preinit hook 链执行函数

函数	描述
define_default_set_state	仅包含/etc/diag.sh
preinit_ip	初始化网络接口
pi_indicate_preinit	将消息发送到控制台、网络或led
do_sysinfo_generic	根据/proc/device-tree信息生成 /tmp/sysinfo/board_name和 /tmp/sysinfo/model
failsafe_wait	发出消息（发送到网络和控制台），指示用户可以选择进入故障保护模式，并等待配置的时间段（默认为2s）供用户选择故障保护模式
run_failsafe_hook	如果用户选择进入故障保护模式，则运行failsafe hook链，并且不返回，意味着不再运行preinit_main hook链后面的函数
indicate_regular_preinit	向网络、控制台或led发送消息，表明它是常规引导而不是故障保护引导
initramfs_test	如果存在initramfs，运行 initramfs hook链，并退出
do_mount_root	执行mount_root，完成rootfs的切换挂载和JFFS2 Overlay挂载。执行preinit_mount_root hook链，目前为空
run_init	执行由pi_init_cmd定义的命令，环境变量由pi_init_env定义，加上路径pi_init_path
load_firmware	加载sfwifi相关startcore模块。这个模块启动lmac模块。lmac0是2.4GHz模块，lmac1是5GHz模块
insmod_rf	加载sfwifi相关sf16a18_rf模块，即rf模块
insmod_mac80211	加载sfwifi相关mac80211模块。内核为softmac提供的驱动程序
insmod_smac_lb	加载sfwifi相关smac_lb。低频softmac驱动
insmod_smac_hb	加载sfwifi相关smac_hb。高频softmac驱动

表 13-3 preinit hook 链执行函数间接调用的函数

函数	描述
preinit_ip_deconfig	取消配置preinit网络接口
preinit_net_echo	在preinit网络接口上发出消息
mount_root	fstools包提供的rootfs的切换、挂载和overlay实现
fs_wait_for_key	等待reset按钮、Crtl-C或Enter键被按下，或等待超时

2. failsafe

为进入failsafe做准备，并进入failsafe。failsafe hook链包含的执行函数及其所在文件如表13-4所示，failsafe hook链执行函数的内容介绍如表13-5所示，failsafe hook链执行函数间接调用的函数如表13-6所示。

表 13-4 failsafe hook 链包含的执行函数及其所在文件

文件	描述
10_indicate_failsafe	执行indicate_failsafe、indicate_failsafe_led
99_10_failsafe_login	执行failsafe_netlogin、failsafe_shell

表 13-5 failsafe hook 链执行函数

函数	描述
indicate_failsafe	向网络、控制台或led发送消息和状态，指示设备现在处于故障保护模式
failsafe_netlogin	如果执行telnet，则启动telnet守护进程以允许telnet登录定义的网络接口
failsafe_shell	如果存在Shell，则启动Shell以通过串行控制台访问

表 13-6 failsafe hook 链执行函数间接调用的函数

函数	描述
indicate_failsafe_led	设置led状态，表示进入failsafe模式

3. Initramfs

实现initramfs相关处理。如果存在initramfs，则执行initramfs链内容。目前没有处理。

4. preinit_mount_root

挂载根文件系统。目前没有处理，而是放到preinit_main链的do_mount_root中处理。

/etc/preinit执行过程中涉及的变量如表13-7所示。至此，/etc/preinit流程执行完毕。如果要了解更多的细节，可以进入系统查看相关的源代码包。

表 13-7 /etc/preinit 执行过程中涉及的变量

变量	描述
pi_ifname	在preinit期间用于发出网络消息的网络接口的设备名（故障保护模式除外）
pi_ip	preinit期间设置的网络的IP地址
pi_broadcast	preinit期间设置的网络的广播地址
pi_netmask	preinit期间设置的网络的子网掩码
fs_failsafe_wait_timeout	允许用户选择进入故障保护模式时暂停的时间。默认值为2s
pi_suppress_stderr	如果为"y"，则在preinit期间抑制标准错误（stderr，文件描述符2）的输出。默认是"y"
pi_init_suppress_stderr	如果pi_suppress_stderr的值不是"y"（输出stderr），则当init命令运行时不向stderr输出信息。对终端设备（如tts / 0、ttyS0、tty1、pts / 0等）有效，对伪终端设备（如pty0、pty1等）无效。默认值是"y"
pi_init_path	被init命令调用的其他命令的值搜索路径。默认为/bin:/sbin:/usr/bin:/usr/sbin
pi_init_cmd	init命令的完整路径。默认值为/sbin/init
pi_preinit_no_failsafe_netmsg	抑制netmsg，表示可以进入故障保护模式
pi_preinit_net_messages	如果启用，则显示更多网络消息，而不仅仅是可以进入故障保护模式的消息
pi_mount_skip_next	在调用preinit_mount_root hook期间，跳过大多数步骤，通常由前面的步骤设置
pi_jffs2_mount_success	在调用preinit_mount_root hook期间，由mount尝试后的步骤使用，以确定应采取的操作

/sbin/procd的第二次启动

/etc/preinit执行完毕后，会首先发送SIGKILL信号给前面用"/sbin/procd -h /etc/hotplug-preinit.json"命令启动的进程，终止这个进程后，将会第二次启动procd进程。有了/etc/preinit的前期工作，procd的第二次启动就简单多了，它会按部就班地完成自己的使命。

procd有自己的作息和处理规则，在进程执行过程中按规则切换变化。它有7个状态（状态的区别取决于procd版本的不同），分别为STATE_NONE、STATE_EARLY、STATE_UBUS、STATE_INIT、STATE_RUNNING、STATE_SHUTDOWN、STATE_HALT。当前的状态保存在一个名称为state的全局变量中，这6个状态根据序号和外部信号的触发实现变化，在内部通过procd_state_next()函数实现切换的控制。

在state.c中定义这些状态如下。

```
enum {
        STATE_NONE = 0,
        STATE_EARLY,
        STATE_UBUS,
        STATE_INIT,
        STATE_RUNNING,
        STATE_SHUTDOWN,
        STATE_HALT,
        __STATE_MAX,
};
static int state = STATE_NONE;
```

● STATE_NONE

初始状态。

● STATE_EARLY

第一个执行的状态。其执行步骤如下。

（1）在控制台输出"- early -"。

（2）初始化看门狗。

（3）根据"/etc/hotplug.json"规则监听Hotplug。

（4）procd_coldplug()函数处理，把/dev挂载到tmpfs中，创建udevtrigger进程，产生冷插拔事件，以便让Hotplug监听进行处理。

（5）udevstrigger进程处理完成后回调procd_state_next()函数，把状态从STATE_EARLY转变为STATE_UBUS。

● STATE_UBUS

从STATE_EARLY状态切换而来，注册ubus服务。其执行步骤如下。

（1）重新启用看门狗。

（2）在控制台输出"- ubus -"。

（3）连接ubusd。此时实际上ubusd并不存在，所以procd_connect_ubus()函数使用了定时器（1s）进行重连，而uloop_run()在初始化工作完成后才真正运行。当成功连接上ubusd后，将注册ubus service对象、system对象、watch_event对象，如图13-8所示（关于ubus的内容，将在后续章节详细介绍）。

```
DF1A:$ ubus -v list system
'system' @d04f04b8
        "board":{}
        "info":{}
        "upgrade":{}
        "watchdog":{"frequency":"Integer","timeout":"Integer","stop":"Boolean"}
        "signal":{"pid":"Integer","signum":"Integer"}
        "nandupgrade":{"path":"String"}
DF1A:$ ubus -v list service
'service' @a217a9f3
        "set":{"name":"String","script":"String","instances":"Table","triggers":"Array","validate":"Array"}
        "add":{"name":"String","script":"String","instances":"Table","triggers":"Array","validate":"Array"}
        "list":{"name":"String","verbose":"Boolean"}
        "delete":{"name":"String","instance":"String"}
        "update_start":{"name":"String"}
        "update_complete":{"name":"String"}
        "event":{"type":"String","data":"Table"}
        "validate":{"package":"String","type":"String","service":"String"}
        "get_data":{"name":"String","instance":"String","type":"String"}
```

图13-8　procd注册的ubus服务

（4）初始化services（服务）和validators（服务验证器）的全局AVL tree。

（5）把ubusd服务加入services管理对象中（service_start_early）。

● STATE_INIT

其执行步骤如下。

（1）在控制台输出"- init-"。

（2）根据/etc/inittab的内容把cmd、handler对应关系加入全局链表actions中。

（3）执行/etc/inittab。顺序执行respawn、askconsole、askfirst、sysinit命令。

（4）STATE_INIT转变为STATE_RUNNING。

● STATE_RUNNING

在控制台输出"- init complete -"。

● STATE_SHUTDOWN

由外部触发，当系统收到触发信号时执行。

源代码signal.c，处理接收到的信号。

```c
static void signal_shutdown(int signal, siginfo_t *siginfo, void *data)
{
        int event = 0;
        char *msg = NULL;
        switch(signal) {
        case SIGINT:
        case SIGTERM:
                event = RB_AUTOBOOT;
                msg = "reboot";
                break;
        case SIGUSR1:
        case SIGUSR2:
                event = RB_POWER_OFF;
                msg = "poweroff";
                break;
        }
        DEBUG(1, "Triggering %s\n", msg);
        if (event)
                procd_shutdown(event);
}
```

源代码state.c设置状态如下。

```c
void procd_shutdown(int event)
{
        if (state >= STATE_SHUTDOWN)
                return;
        DEBUG(2, "Shutting down system with event %x\n", event);
        reboot_event = event;
        state = STATE_SHUTDOWN;
        state_enter();
}
```

其执行步骤如下。

（1）在控制台输出"- shutdown -"。

（2）执行/etc/inittab的shutdown命令。

（3）sync同步缓存到磁盘。

● STATE_HALT

其执行步骤如下。

（1）在控制台输出"- SIGTERM processes -"。

（2）给系统中的所有进程发送SIGTERM信号。

（3）根据reboot_event，执行重启（在控制台输出"- reboot -"）或关机（在控制台输出"- power down -"）流程。

（4）在控制台输出"- SIGKILL processes -"。

（5）给系统中的所有进程发送SIGKILL信号。

（6）执行进程安全退出流程，重启。

（7）等待重启完成。

procd按照STATE_NONE→STATE_EARLY→STATE_UBUS→STATE_INIT→STATE_RUNNING的顺序运行，取代init进程成为新的1号进程。如果收到外部信号，则执行STATE_HALT或STATE_SHUTDOWN相关动作。通常情况下，procd会按照定义的ubus服务对外提供服务。

13.1.5 rcS的执行

在procd的STATE_INIT阶段执行过程中，有一个非常重要的内容，其中包含所有的精华。对这一部分内容需要单独着重说明。

在这个阶段，procd会根据/etc/inittab的内容，把action、process 的对应关系加入全局链表init_actions中。当执行procd_inittab_run("xxx") 时，会调用链表中对应handlers的callback，按顺序执行respawn、askconsole、askfirst、sysinit动作，在关闭和重启系统时会执行shutdown动作。/etc/inittab的内容如图13-9所示。

```
DF1A:$ cat /etc/inittab
::sysinit:/etc/init.d/rcS S boot
::shutdown:/etc/init.d/rcS K shutdown
::askconsole:/bin/ash --login
```

图13-9 /etc/inittab的内容

/etc/inittab文件的语法格式为：

```
id:runlevels:action:process
```

- id：用来定义在inittab文件中唯一的条目编号，长度为1～4个字符。
- runlevels：列出的运行级别为空则代表所有级别。
- action：要执行的动作。
- process：要执行的程序。

respawn动作没有内容。执行askconsole动作，即运行/bin/ash --login，如果用户通过串口或者Telnet登录，按下Enter键后，会显示登录信息，如图13-10所示。

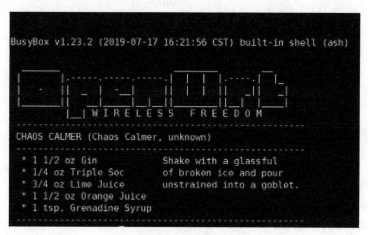

图13-10　登录信息

askfirst动作也没有内容。如果同时有askfirst和respawn动作，它们的执行动作也基本上是相同的，只是在运行前会提示"Please press Enter to activate this console."。

重要的事情一般放在后面说。sysinit动作执行/etc/init.d/rcS S boot的流程，其实就是调用/etc/init.d/rcS脚本，并传递两个参数：$1=S、$2=boot。但是我们的开发板/etc/init.d中并没有rcS脚本，可谓是"山重水复疑无路"。但是我们通过查看procd源代码inittab.c，就会发现即使没有/etc/init.d/rcS脚本，"sysinit"和"shutdown"也会调用runrc()函数来执行后面的动作，原来rcS只是个"门面"，后面的内容才是进程中要执行的内容，可谓是"柳暗花明又一村"。inittab对应的源代码handler如图13-11所示。

```
static struct init_handler handlers[] = {
    {
        .name = "sysinit",
        .cb = runrc,
    }, {
        .name = "shutdown",
        .cb = runrc,
    }, {
        .name = "askfirst",
        .cb = askfirst,
        .multi = 1,
    }, {
        .name = "askconsole",
        .cb = askconsole,
        .multi = 1,
    }, {
        .name = "respawn",
        .cb = rcrespawn,
        .multi = 1,
    }
};
```

图13-11 inittab对应的源代码handler

runrc()函数的执行流程如图13-12所示。

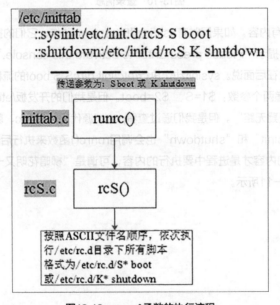

图13-12 runrc()函数的执行流程

在sysinit动作的执行过程中,最终会执行 /etc/rc.d目录下以S开头的所有脚本,如图13-13所示,并传递参数boot,默认执行脚本中的boot()函数。这些脚本都是/etc/init.d下对应脚本的软连接,可以通过/etc/init.d/xxx stop/start/restart/reload/boot/shutdown进行调用。

```
/etc/rc.d/S00sysfixtime -> ../init.d/sysfixtime
/etc/rc.d/S10boot -> ../init.d/boot
/etc/rc.d/S10system -> ../init.d/system
/etc/rc.d/S11sysctl -> ../init.d/system
/etc/rc.d/S12log -> ../init.d/log
/etc/rc.d/S12rpcd -> ../init.d/rpcd
/etc/rc.d/S13ts-init -> ../init.d/ts-init
/etc/rc.d/S14acl -> ../init.d/acl
/etc/rc.d/S19network -> ../init.d/network
/etc/rc.d/S20firewall -> ../init.d/firewall
/etc/rc.d/S35odhcpd -> ../init.d/odhcpd
/etc/rc.d/S40fstab -> ../init.d/fstab
/etc/rc.d/S50cron -> ../init.d/cron
/etc/rc.d/S50dropbear -> ../init.d/dropbear
/etc/rc.d/S50telnet -> ../init.d/telnet
/etc/rc.d/S50uhttpd -> ../init.d/uhttpd
/etc/rc.d/S51init_pctl -> ../init.d/init_pctl
/etc/rc.d/S60dnsmasq -> ../init.d/dnsmasq
/etc/rc.d/S60samba -> ../init.d/samba
/etc/rc.d/S61syncservice -> ../init.d/syncservice
/etc/rc.d/S62subservice -> ../init.d/subservice
/etc/rc.d/S80relayd -> ../init.d/relayd
/etc/rc.d/S90advanced_wifi -> ../init.d/advanced_wifi
/etc/rc.d/S94miniupnpd -> ../init.d/miniupnpd
/etc/rc.d/S95ddns -> ../init.d/ddns
/etc/rc.d/S95done -> ../init.d/done
/etc/rc.d/S96led -> ../init.d/led
/etc/rc.d/S96set_default_cpufreq -> ../init.d/set_defau
/etc/rc.d/S98sysntpd -> ../init.d/sysntpd
/etc/rc.d/S99device_listen -> ../init.d/device_listen
```

图13-13 /etc/rc.d 下以S开头的脚本

在shutdown动作的执行过程中，最终会执行/etc/rc.d目录下以K开头的所有脚本，如图13-14所示，并传递参数shutdown，默认执行脚本的shutdown()函数。

```
/etc/rc.d/K10ddns -> ../init.d/ddns
/etc/rc.d/K15miniupnpd -> ../init.d/miniupnpd
/etc/rc.d/K50dropbear -> ../init.d/dropbear
/etc/rc.d/K85odhcpd -> ../init.d/odhcpd
/etc/rc.d/K89log -> ../init.d/log
/etc/rc.d/K90network -> ../init.d/network
/etc/rc.d/K98boot -> ../init.d/boot
/etc/rc.d/K99umount -> ../init.d/umount
```

图13-14 /etc/rc.d 下以K开头的脚本

这些脚本执行的操作主要包括系统初始化、时区、ntp、log、network、ddns、dhcp、web服务等软件包相关启动和停止处理。启动时，按照前文介绍的先后次序执行，最终完成系统的启动。具体的启动过程在此不详细介绍，有意了解详情者可以对照脚本自己进行分析。

唯一值得注意的地方，就是有些脚本中会采用procd方式启动进程，这是一种新的启动进程和服务的方式。采用procd启动方式，可以对进程进行监控和重启。这种新的启动方式会逐渐替换原有的命令行启动。更多的支持请参考procd Shell API (/lib/fucntions/procd.sh)，样例模板如下。

```
#!/bin/sh /etc/rc.common
# Copyright (C) 2008 OpenWrt.org
START=98
#执行的顺序,按照字符串顺序排序而不是按数字排序
USE_PROCD=1
#使用procd启动
EXEC_BIN="/usr/bin/A"
#start_service()函数必须要重新定义
start_service() {
    procd_open_instance
    #创建一个实例, 在procd方式中,一个应用程序内可以包含多个实例
    #ubus call service list可以查看实例
    procd_set_param respawn
    #定义respawn参数,告知procd当A程序退出后尝试进行重启
    procd_set_param command "$EXEC_BIN"
    # EXEC_BIN执行的命令是"/usr/bin/A",若后面有参数可以直接在后面加上
    procd_close_instance
    #关闭实例
}
#stop_service重新定义,退出服务器后需要做的操作
stop_service() {
    rm -f /var/run/binloader.pid
}
restart() {
    stop
    start
}
```

13.1.6 从启动log看系统启动过程

在本节中再回顾一下系统串口启动log,梳理系统启动的整个执行流程。系统启动的log片段如下。

SPL阶段

```
Booting...
eth check connect ret=0 cost time=2000 ms
IROM DONE!
SiFlower SFAX8 Bootloader (Jul  9 2019 - 14:46:03)
ddr3 nt5cc128m16ip init start
MEM_PHY_CLK_DIV = 0x3
DDR training success
now ddr frequency is 400MHz!!!
ddr test
DR1BW a0000000 OK
DR1BR OK
DR2BW a0000000 OK
DR2BR OK
DR4BW a0000000 OK
DR4BR OK
DR8BW a0000000 OK
```

```
DR8BR OK
Boot from spi-flash
U-image: U-Boot 2016.07-rc2 for sfa18_p2, size is 310044
loaded - jumping to U-Boot 0xa0000000...
```

U-Boot阶段

```
U-Boot 2016.07-rc2 (Jul 09 2019 - 14:46:10 +0800)
Board: MIPS sfa18 FULLMASK P20B
       Watchdog enabled
DRAM:  256 MiB
MMC:   emmc@7800000: 0sdio@7c00000: 1
SF: Detected EN25QH128A with page size 256 Bytes, erase size 4 KiB, total 16 MiB
In:    serial@8300000
Out:   serial@8300000
Err:   serial@8300000
Net:   Registering sfa18 net
Registering sfa18 eth
sf_eth0
Warning: sf_eth0 (eth0) using random MAC address - 4a:0a:ab:7c:96:2f
Hit any key to stop autoboot:  0
do_spi_flash----cmd = probe
SF: Detected EN25QH128A with page size 256 Bytes, erase size 4 KiB, total 16 MiB
do_spi_flash----cmd = read
device 0 offset 0xa0000, size 0x300000
SF: 3145728 bytes @ 0xa0000 Read: OK
```

Kernel解压

```
## Booting Kernel from Legacy Image at 81000000 ...
   Image Name:   MIPS OpenWrt Linux-3.18.29
   Image Type:   MIPS Linux Kernel Image (lzma compressed)
   Data Size:    1680062 Bytes = 1.6 MiB
   Load Address: 80100000
   Entry Point:  80105360
   Verifying Checksum ... OK
   Uncompressing Kernel Image ... OK
All resets are held!
```

Kernel启动

```
[    0.000000] Linux version 3.18.29 (zhengqw@Ironman) (gcc version 4.8.3 (OpenWrt/
Linaro GCC 4.8-2014.04 unknown) ) #1 SMP PREEMPT Wed Jul 17 17:17:22 CST 2019
[    0.000000] memsize not set in YAMON, set to default (256MB)
[    0.000000] arcs_cmdline=memsize=256M
[    0.000000] CPU0 revision is: 5301a128 (MIPS interAptiv (multi))
[    0.000000] FPU revision is: 0173a000
[    0.000000] MIPS: machine is siflower,sf16a18-soc
[    0.000000] Determined physical RAM map:
[    0.000000]  memory: 01f00000 @ 00000000 (usable)
[    0.000000]  memory: 00100000 @ 01f00000 (reserved)
[    0.000000]  memory: 0e000000 @ 02000000 (usable)
[    0.000000] Initrd not found or empty - disabling initrd
```

```
[    0.000000] Zone ranges:
[    0.000000]   Normal   [mem 0x00000000-0x0fffffff]
[    0.000000] Movable zone start for each node
[    0.000000] Early memory node ranges
[    0.000000]   node   0: [mem 0x00000000-0x0fffffff]
[    0.000000] Initmem setup node 0 [mem 0x00000000-0x0fffffff]
[    0.000000] VPE topology {2,2} total 4
[    0.000000] Primary instruction cache 16kB, VIPT, 4-way, linesize 32 bytes.
[    0.000000] Primary data cache 16kB, 4-way, VIPT, no aliases, linesize 32 bytes
[    0.000000] MIPS secondary cache 256kB, 8-way, linesize 32 bytes.
[    0.000000] PERCPU: Embedded 9 pages/cpu @81225000 s7168 r8192 d21504 u36864
[    0.000000] Built 1 zonelists in Zone order, mobility grouping on.  Total pages: 65024
[    0.000000] Kernel command line: memsize=256M console=ttyS0,115200n8 rootfstype=squashfs,jffs2 rdinit=/sbin/init...
```

内核和硬件驱动初始化

```
[    0.000000] GIC frequency 672.00 MHz
[    0.000017] sched_clock: 16 bits at 1980kHz, resolution 505ns, wraps every 33098484ns
[    0.000149] sched_clock: 32 bits at 198MHz, resolution 5ns, wraps every 21691754490ns
[    0.000837] Calibrating delay loop... 447.28 BogoMIPS (lpj=2236416)
[    0.050529] pid_max: default: 32768 minimum: 301
[    0.051040] Mount-cache hash table entries: 1024 (order: 0, 4096 bytes)
[    0.051061] Mountpoint-cache hash table entries: 1024 (order: 0, 4096 bytes)
[    0.111411] Primary instruction cache 16kB, VIPT, 4-way, linesize 32 bytes.
[    0.111431] Primary data cache 16kB, 4-way, VIPT, no aliases, linesize 32 bytes
[    0.111448] MIPS secondary cache 256kB, 8-way, linesize 32 bytes.
[    0.111666] CPU1 revision is: 5301a128 (MIPS interAptiv (multi))
...
```

SPI分区识别

```
[    1.181731] sfax8-spi 18200000.spi: sfax8_spi_probe...
[    1.188811] m25p80 spi0.0: found en25qh128, expected w25q128
[    1.194577] m25p80 spi0.0: en25qh128 (16384 Kbytes)
[    1.199962] 5 ofpart partitions found on MTD device spi0.0
[    1.205500] Creating 5 MTD partitions on "spi0.0":
[    1.210321] 0x000000000000-0x000000020000 : "spl-loader"
[    1.217255] 0x000000020000-0x000000080000 : "u-boot"
[    1.223767] 0x000000080000-0x000000090000 : "u-boot-env"
[    1.230565] 0x000000090000-0x0000000a0000 : "factory"
[    1.237176] 0x0000000a0000-0x000001000000 : "firmware"
[    1.296775] 2 uimage-fw partitions found on MTD device firmware
[    1.302749] 0x0000000a0000-0x00000023a2fe : "kernel"
[    1.309402] 0x00000023a2fe-0x000001000000 : "rootfs"
[    1.315972] mtd: device 6 (rootfs) set to be root filesystem
[    1.321823] 1 squashfs-split partitions found on MTD device rootfs
[    1.328070] 0x000000930000-0x000001000000 : "rootfs_data"
...
```

SquashFS文件系统挂载

```
[    3.055903] VFS: Mounted root (squashfs filesystem) readonly on device 31:6.
[    3.063916] Freeing unused kernel memory: 252K (805b1000 - 805f0000)
[    3.190090] random: nonblocking pool is initialized
```

/sbin/init执行和procd第一次执行

```
[    4.480983] init: Console is alive
[    6.628134] Button Hotplug driver version 0.4.1
...
```

/etc/preinit执行

```
[    7.486726] init: - preinit -
Press the [f] key and hit [enter] to enter failsafe mode
Press the [1], [2], [3] or [4] key and hit [enter] to select the debug level
[   11.137449] mount_root: loading kmods from internal overlay
[   12.604405] jffs2: notice: (490) jffs2_build_xattr_subsystem: complete building
xattr subsystem, 2 of xdatum (0 unchecked, 2 orphan) and 2 of xref (0 dead, 2
orphan) found.
[   12.621817] block: attempting to load /tmp/jffs_cfg/upper/etc/config/fstab
[   12.644321] block: extroot: not configured
[   13.501797] jffs2: notice: (487) jffs2_build_xattr_subsystem: complete building
xattr subsystem, 2 of xdatum (0 unchecked, 2 orphan) and 2 of xref (0 dead, 2
orphan) found.
[   13.521024] mount_root: loading kmods from internal overlay
[   13.926761] block: attempting to load /tmp/jffs_cfg/upper/etc/config/fstab
[   13.941221] block: extroot: not configured
[   13.994951] mount_root: switching to jffs2 overlay
[   14.054081] startcore init fill all memory!
[   14.089117] sf_wifi_rf_probe
...
```

/sbin/procd第二次执行

```
[   17.032947] procd: - early -
[   17.914516] procd: - ubus -
[   18.919860] procd: - init -
Please press Enter to activate this console.
BusyBox v1.23.2 (2019-07-17 16:21:56 CST) built-in shell (ash)
```

```
  _____                     _____        __
 |       |.-----.-----.-----.|  |  |  |.----.|  |_
 |   -   ||  _  |  -__|     ||  |  |  ||   _||   _|
 |_____||   __|_____|__|__||_____||__|  |____|
          |__| W I R E L E S S   F R E E D O M
 -----------------------------------------------------
 CHAOS CALMER (Chaos Calmer, unknown)
 -----------------------------------------------------
  * 1 1/2 oz Gin            Shake with a glassful
  * 1/4 oz Triple Sec       of broken ice and pour
  * 3/4 oz Lime Juice       unstrained into a goblet.
  * 1 1/2 oz Orange Juice
```

```
* 1 tsp. Grenadine Syrup
--------------------------------------------------
```

将以上的log内容与前文中的启动过程介绍相对照,或许会有"众里寻他千百度,蓦然回首,那人却在,灯火阑珊处"的感觉。

13.2 ubus总线原理

ubus即Micro Bus,是一个开源的进程间的通信系统,它提供了一个通用的框架和解决方案,使进程间通信的实现变得非常简单和高效。ubus作为OpenWrt系统消息和RPC通信的总线,在OpenWrt下有着广泛的应用,很多关键的程序都支持ubus通信,如rpcd、procd、Netifd等。

在一些桌面系统的发行版(如Ubuntu)中有个消息总线叫dBus,这个消息总线主要应用于桌面系统,比较庞大。与dBus相比,ubus是一个"微总线"结构,更简洁易用,更适用于嵌入式系统。但ubus也有一些缺点,更适用于少量、低频次数据的传输,对多线程的支持不是很好。

13.2.1 ubus的基本概念

ubus实现的基础是UNIX Domain Socket(UNIX域套接字)。UNIX Domain Socket是IPC(Inter-Process Communication,进程间通信)方式之一,也叫作UDS或本地Socket,用于实现同一主机上的进程间通信,相对于网络通信的Internet Domain Socket(Internet域套接字),更为高效和可靠。

ubus在实现Unix Domain Socket基本通信上,对具体通信的实现和传输数据进行了封装。

ubus daemon

ubus daemon(ubus守护进程)是指ubus在操作系统中提供服务的进程,也就是Server,分为两类,一类由系统创建,随操作系统启动而启动,叫作system ubus daemon,即ubusd(见图13-15);另一类由用户自行创建,叫作session ubus daemon。session ubus daemon与system ubus daemon相互独立,互不影响。

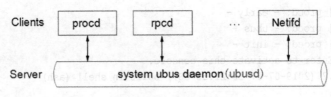

图13-15 ubusd示意图

path

path是一个可用的Linux路径名,即文件系统的路径名,是一个字符串。ubus的通信机制,是基于socket实现的进程间通信,所以这里面引入socket所使用的"路径"的概念。建立socket时需要使用一个指定的path作为socket所在的地址,而这个socket的使用者也需要连接到这个地址。ubus的执行基于socket,所以ubus也通过path来识别不同的ubus daemon。而ubus的使用者也通过path来连接到ubus daemon。

ubusd的path为默认的/var/run/ubus.sock。当连接到ubusd时,可以不指定path,系统会

设置默认值。用户在使用ubus时，可以自行启动一个ubus daemon，即通过自定义一个path，来启动（session ubus daemon）。path的定义也可以作为一个选项，写入config配置文件里。总之，在OpenWrt中建立ubus连接时，一定要指定一个path，这个path决定了要使用哪一个ubus daemon。除非使用system ubus daemon，才可以不指定path。

object

object（对象）是抽象的概念，是ubus传递消息的起点和终点。在一个Client进程上可以定义多个object。ubus上的消息，总是从一个object发送到另一个object，所以，在ubus daemon上，object既是ubus消息发送的源，也是消息发送的目的地。

图13-16示意了定义两个client，从A.object1向B.object2发送消息。

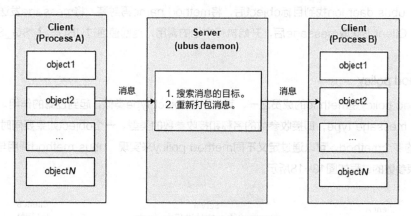

图13-16 ubus object之间通信

我们通常使用ubusd作为ubus通信的Server，ubus Client通过ubusd相互通信。Client和Server各维护一个object结构。Client端ubus上下文（ubus context）通过AVL树（平衡二叉树）数据结构维护object；Server端也通过AVL树数据结构维护object。当Client定义的object需要注册到Server时，通过发送UBUS_MSG_ADD_OBJECT请求通知Server，Server将object信息添加到自己的object结构中。添加成功后，Server将生成的object id返回给Client，Client将object以AVL树数据结构的形式添加到ubus context中。object id分配的过程如图13-17所示。

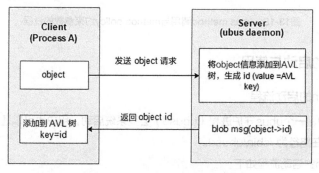

图13-17 ubus object id分配示意图

object type

object type是object的类型定义,与object是一对多的关系。object type是object的成员之一。object type包含的内容有:object type name、method、object type id。多个object可以使用同一个object type。object type存在的意义,是起到过滤的作用。一个object如果要同时与其他多个object通信,就可以通过object type来实现。

method

method是object的成员之一。method用于实现具体的功能。每个object可以有多个method。method包括method name、method handler函数、method policy。例如:Client A通过向ubus daemon发送UBUS_MSG_INVOKE消息,将要调用的Client B的method name传递给ubus daemon,ubus daemon找到目标object后,将method name再封装,将message发送到Client B进程上,Client B获取message后,开始执行本地的调用,返回值通过UBUS_MSG_STATUS消息传递。

method policy

method policy是method的成员之一,其作用与object type类似,起到过滤的作用。policy包括name、message type,即接收参数的名称和接收参数的类型。一个object如果要同时调用另一个object的多个method,可以通过定义不同method policy来实现。ubus method调用与method policy约束参数的过程如图13-18所示。

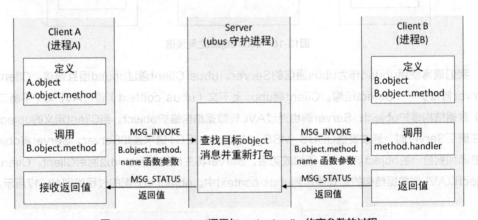

图13-18　ubus method调用与method policy约束参数的过程

13.2.2　ubus的启动与使用

ubus daemon的启动流程

在OpenWrt中,一个ubus守护进程(ubusd)会随系统启动。用户也可以通过命令、配置文件来启动它。进程所在路径为 /sbin/ubusd。

ubus daemon的启动流程如下。

(1)注册信号:向系统内核申请处理方法为SIGPIPE的signal资源,使内核可以通过signal通知ubusd进程异步事件的发生。这个是给OpenWrt的Linux内核管理ubusd进程使用的,用户无须关注。

（2）初始化loop：调用uloop_init()，内部通过调用系统内核函数kqueue()，申请内核事件队列资源。

（3）初始化socket：指定path，向内核申请socket服务，获得一个socket的文件描述符。ubus中连接的建立、数据的传输，都将通过此socket进行。

（4）注册事件：调用uloop_fd_add()，通过调用kevent，向之前申请的事件队列中注册EVFILT_READ类型的事件，此事件用于监听socket，当socket缓冲区中有数据时，此事件将得到通知。

（5）启动事件循环：调用uloop_run()，设置定时器，周期性地判断check事件队列中是否有要处理的事件。

ubus的使用流程

1. 与ubus daemon建立连接

（1）定义object信息，如method、object type、object。

（2）使用ubus_connect()函数，通过传入ubus daemon的socket path(/var/run/ubus.sock)，与ubus daemon建立连接，成功后会获得一个ubus上下文。

（3）调用ubus_add_uloop()函数，向内核注册事件。

（4）调用ubus_add_object()函数，将自定义的object信息，诸如object type、method等，通过消息（message）添加到ubus daemon上，并根据ubus daemon添加的结果，赋值到本地的ubus上下文中。

2. 进行method调用

（1）当"调用者"和"被调用者"都注册到ubus daemon上后，"调用者"要根据"被调用者"的"设计"，明确"被调用者"的object name和method name。

（2）"调用者"根据"被调用者"的object name，获取"被调用者"object在ubus daemon上对应的id。

（3）根据object id，通过调用ubus_invoke()函数，将method name等信息通过ubus内部消息发送到"被调用者"进程上，完成调用。调用结束后的返回值，也通过同样的发送流程，返回"调用者"进程。

（4）调用uloop_run，根据timeout等待时间值，等待接收返回值事件。

3. 接收消息调用

（1）接收定义的method，包括method name、handler function等。

（2）接收定义的object，并将method加入object中。

（3）启动连接ubus daemon的流程。ubus会在ubus daemon上统一记录和管理object信息。

（4）本地进程启动uloop run，等待接收消息，收到method调用的消息后会在本地结构体中找到对应注册的method，并执行method.handler()函数。

ubus流程结构如图13-19所示。

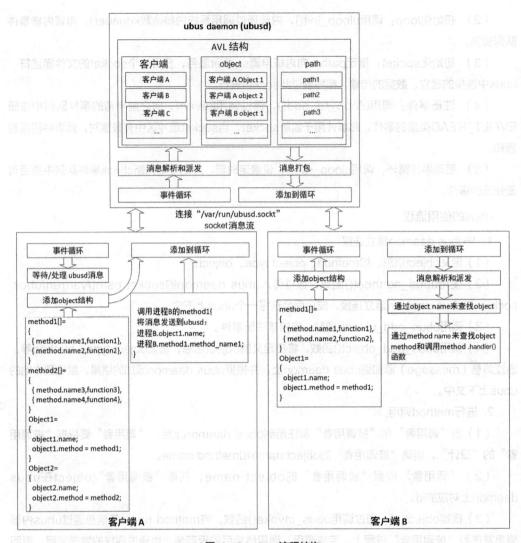

图13-19 ubus 流程结构

13.2.3 ubus的基本概念

ubus是消息总线的一种,采用socket原理,实现进程与进程之间的通信。在OpenWrt中,ubus的架构分4层,基本架构如图13-20所示。

(1)函数库libubus:ubus提供的接口函数库,应用程序通过对API函数的调用,完成与其他应用程序间的呼叫和交互消息。

(2)ubus daemon:总线守护进程可以同时与多个应用程序相连,并把消息从一个应用程序转到另一个或多个应用程序中。

(3)libubox:公共函数、结构体综合库,其中包含了AVL树结构、链表结构等ubus实现时需要使用的资源。

(4)Kernel/socket:ubus的实现需要获得内核消息的队列,并且ubus的底层通信使用的是socket,也需要获取socket资源。

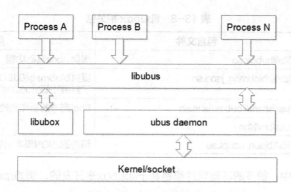

图13-20 ubus 软件架构

ubus是在libubox库基础上实现的。libubox是ubus实现功能的基础软件包，是OpenWrt下的一个综合库，提供以下几个方面的功能实现。

● event Loop

事件循环处理，包含文件描述符事件、定时器事件、进程事件，通过epoll实现。

● 任务队列

任务队列通过event loop定时器实现，把定时器超时时间设置为1，通过uloop事件循环来处理，定时器就会处理任务队列中的任务。

● AVL树和链表

用于快速查找实现平衡二叉树（AVL树）和各种链表结构（包括vlist、safe_list、key/value存储kvlist）。

● BLOB（Binary Large Objects）message

BLOB是一个以二进制形式保存数据的message结构。依靠ubus的进程与ubus daemon间使用BLOB message传输，可以提高数据传递的效率。uloop可以实现BLOB二进制对象(blob)、BLOB消息对象(blobmsg)、BLOB JSON消息对象(blobmsg-json)。

● MD5和BASE64

实现MD5和BASE64。

● JSON(JavaScript Object Notation)

JSON是一种轻量级的数据交换格式，在ubus中执行终端命令时，某些参数可以通过JSON字符串传入，执行结果也可以采用JSON形式输出到终端。

● 其他实用库

如usock、ustream、ulog等。

libubox软件包在OpenWrt功能目录package/libs/libubox下。libubox软件包提供以下几个库文件：libubox、libblobmsg-json、jshn、libjson-script、libubox-lua，如表13-8所示。

表 13-8 libubox 相关包

包名称	包含文件	用途
libubox	/lib/libubox.so	提供 ubox基础功能
libblobmsg-json	/lib/libblobmsg_json.so	提供blobmsg和JSON转换相关功能，提供了封装和解析JSON数据的接口
jshn	/usr/share/libubox/jshn.sh	Shell脚本解析和创建JSON相关的函数
	/usr/bin/jshn	
libjson-script	/lib/libjson_script.so	精简的JSON脚本引擎

在OpenWrt系统中，很多的基础软件都是基于libubox来开发的，例如rpcd、ubox、procd、uhttpd、relayd等。

在OpenWrt系统源代码中，ubus软件包在package/system/ubus目录下。ubus软件包中包括libubus、libubus-lua、ubus、ubusd，包含的文件和用途如表13-9所示。

表 13-9 ubus 相关软件包提供的文件

包名称	包含文件	用途
libubus	/lib/libubus.so	提供ubus客户端库。C程序客户端需要使用
libubus-lua	/usr/lib/lua/ubus.so	提供ubus Lua的绑定库。Lua程序客户端需要使用
ubus	/bin/ubus	ubus命令行程序。Shell客户端调用
ubusd	/sbin/ubusd	ubus daemon

13.2.4 ubus命令

我们可以在Shell中使用ubus命令连接到ubus总线，进行ubus调用。在OpenWrt中，可通过ubus help查看ubus支持的命令。

```
DF1A:~$ ubus help
Usage: ubus [<options>] <command> [arguments...]
Options:
 -s <socket>:          Set the unix domain socket to connect to
 -t <timeout>:         Set the timeout (in seconds) for a command to complete
 -S:                   Use simplified output (for scripts)
 -v:                   More verbose output
Commands:
 - list [<path>]                       List objects
 - call <path> <method> [<message>]    Call an object method
 - listen [<path>...]                  Listen for events
 - send <type> [<message>]             Send an event
 - wait_for <object> [<object>...]     Wait for multiple objects to appear on ubus
```

通过ps命令查看我们的开发板，只有system ubus daemon启动，即ubusd，所以接下来的ubus命令都是连接到默认的ubus daemon。如果想连接到其他ubus daemon上，通过-s命令指定daemon path即可。后续的操作中如果没有特殊说明，都是连接到默认的system ubus daemon（ubusd）上。

ubus list

命令含义：显示所有连接到ubus daemon的object。

```
ubus [-s <socket> | -t <timeout> | -S| -v ] list <path>
```

ubus list的参数如表13-10所示。

表13-10 ubus list 的参数

参数名	说明
-s \<socket\>	Option选项，可选参数，用于指定ubus daemon。 不输入此参数时，用于选择默认的ubus daemon。 输入此参数时，用于选择socket所指定的ubus daemon。 socket为字符串，长度为1~107
-t \<timeout\>	Option选项，可选参数，设置命令执行的超时时间，单位为秒
-v	显示object更详细的信息，例如method及参数等
-S	Option选项，可选参数，简单输出信息
list	必选参数，显示连接到默认的ubus daemon上的object
\<path\>	可选参数，object name，字符串

ubus list命令的一些示例如下。

1. 查看默认ubus daemon上的object名称。

```
DF1A:~$ ubus list
dhcp
file
hostapd.wlan0
hostapd.wlan1
iwinfo
lepton.network
log
network
network.device
network.interface
network.interface.cfg094d8f
network.interface.guest
network.interface.lan
network.interface.lease
network.interface.loopback
network.interface.wan
network.interface.wan6
network.wireless
service
session
system
uci
```

2. 详细列出默认ubus daemon上的object名称、提供的method及参数。

```
DF1A:~$ ubus -v list
'dhcp' @80692849
    "ipv4leases":{}
    "ipv6leases":{}
'file' @eba9285e
```

```
      "read":{"path":"String","base64":"Boolean"}
      "write":{"path":"String","data":"String","append":"Boolean","mode":"Integer"
,"base64":"Boolean"}
      "list":{"path":"String"}
      "stat":{"path":"String"}
      "md5":{"path":"String"}
      "exec":{"command":"String","params":"Array","env":"Table"}
'hostapd.wlan0' @782f7ed5
      "get_clients":{}
      "del_client":{"addr":"String","reason":"Integer","deauth":"Boolean","ban_
time":"Integer"}
      "list_bans":{}
      "wps_start":{}
      "wps_cancel":{}
      "deauth":{}
      "update_beacon":{}
      "switch_chan":{"freq":"Integer","bcn_count":"Integer"}
      "set_vendor_elements":{"vendor_elements":"String"}
      "update_params":{"hidden":"Integer"}
      "show_sta_whitelist":{}
      "add_sta_whitelist":{"sta_macaddr_list":"Table"}
      "del_sta_whitelist":{"sta_macaddr_list":"Table"}
'hostapd.wlan1' @6c0dbdf0
      "get_clients":{}
      "del_client":{"addr":"String","reason":"Integer","deauth":"Boolean","ban_
time":"Integer"}
      "list_bans":{}
      "wps_start":{}
      "wps_cancel":{}
      "deauth":{}
      "update_beacon":{}
      "switch_chan":{"freq":"Integer","bcn_count":"Integer"}
      "set_vendor_elements":{"vendor_elements":"String"}
      "update_params":{"hidden":"Integer"}
      "show_sta_whitelist":{}
      "add_sta_whitelist":{"sta_macaddr_list":"Table"}
      "del_sta_whitelist":{"sta_macaddr_list":"Table"}
...
```
#由于显示的内容较多,仅列出一部分

3. 仅查看network.interface.lan 提供的method及参数。

```
DF1A:~$ ubus -v list network.interface.lan
'network.interface.lan' @39dc1195
      "up":{}
      "down":{}
      "status":{}
      "prepare":{}
      "dump":{}
      "add_device":{"name":"String","link-ext":"Boolean"}
      "remove_device":{"name":"String","link-ext":"Boolean"}
      "notify_proto":{}
```

```
"remove":{}
"set_data":{}
```

ubus call

命令含义:调用指定object的method,message参数必须为JSON格式。

```
ubus [-s <socket> | -t <timeout> | -v | -S] call <path> <method> [<message>]
```

ubus call的参数如表13-11所示。

表13-11 ubus call 的参数

参数名	说明
-s \<socket\> -t \<timeout\> -v -S	与ubus list同名参数的用法相同
call	必选参数,对指定object的method产生调用
\<path\>	必选参数,object name,字符串
\<method\>	必选参数,method name,字符串
\<message\>	可选参数,method handler函数的参数。 必须是JSON字符串类型,传入参数的关键字字符串和参数值

通过ubus call可以调用连接到ubus总线上的其他进程,从而实现进程间的通信和命令传递。ubus call命令的一些示例如下。

1. 调用system的无参数board方法,获取板子信息。

```
DF1A:~$ ubus call system board
{
        "kernel": "3.18.29",
        "hostname": "SiWiFi8867",
        "system": "MIPS sf16a18",
        "model": "sf16a18-df1a",
        "release": {
                "distribution": "SiWiFi",
                "version": "Chaos Calmer",
                "revision": "unknown",
                "codename": "sf16a18-df1a",
                "target": "siflower\/sf16a18-fullmask",
                "description": "SiWiFi SF16A18-DF1A openwrt_master_df1a_fullmask_rel_"
        }
}
```

2. 调用network.device的status方法,查看eth0.1接口的信息。

```
DF1A:~$ ubus call network.device status '{"name":"eth0.1"}'
{
      "external": false,
      "present": true,
      "type": "Network device",
      "up": true,
      "carrier": false,
      "link-advertising": [
```

```
            "10H",
            "10F",
            "100H",
            "100F",
            "1000F"
    ],
    "link-supported": [
            "10H",
            "10F",
            "100H",
            "100F",
            "1000F"
    ],
    "speed": "10H",
    "mtu": 1500,
    "mtu6": 1500,
    "macaddr": "10:16:88:14:3d:04",
    "txqueuelen": 1000,
    "ipv6": true,
    "promisc": false,
    "rpfilter": 0,
    "acceptlocal": false,
    "igmpversion": 0,
    "mldversion": 0,
    "neigh4reachabletime": 30000,
    "neigh6reachabletime": 30000,
    "dadtransmits": 1,
    "statistics": {
            "collisions": 0,
            "rx_frame_errors": 0,
            "tx_compressed": 0,
            "multicast": 0,
            "rx_length_errors": 0,
            "tx_dropped": 0,
            "rx_bytes": 0,
            "rx_missed_errors": 0,
            "tx_errors": 0,
            "rx_compressed": 0,
            "rx_over_errors": 0,
            "tx_fifo_errors": 0,
            "rx_crc_errors": 0,
            "rx_packets": 0,
            "tx_heartbeat_errors": 0,
            "rx_dropped": 0,
            "tx_aborted_errors": 0,
            "tx_packets": 8,
            "rx_errors": 0,
            "tx_bytes": 1179,
            "tx_window_errors": 0,
            "rx_fifo_errors": 0,
```

```
            "tx_carrier_errors": 0
        }
}
```

ubus listen

命令含义：建立一个监听socket，并将收到的事件打印至终端。

```
ubus [-s <socket> | -t <timeout> | -v | -S] listen [<path>...]
```

ubus listen的参数如表13-12所示。

表13-12　ubus listen 的参数

参数名	说明
-s <socket> -t <timeout> -v -S	与ubus list同名参数的用法相同
listen	必选参数，监听收到的事件
<path>	可选参数，object name，字符串，只监听指定object发送过来的事件，可以同时指定多个

ubus listen配合ubus send使用，由ubus send发出的事件可以被ubus listen接收到。ubus listen支持下列的事件过滤条件。

● 如果listen后面没有指定事件参数，则接收全部事件。

● listen后面可以带多个事件。

● listen后面支持正则表达式，用于过滤事件。

ubus listen命令的一些示例。

1. 监听指定事件。

```
#启动一个终端A用于监听eventA。
DF1A:~$ ubus listen eventA
#也可以在命令后加上&符，使进程在后台运行
#启动另一个终端B用于发送eventA，并携带JSON消息
DF1A:~$ ubus send eventA '{"active":"darts"}'
#我们会在终端A侧接收到JSON消息,包含event名称和消息内容
DF1A:~$ ubus listen eventA
{ "eventA": {"active":"darts"} }
```

2. 监听多个事件。

```
#启动终端用于监听eventA、eventB,该进程在后台运行
DF1A:~$ ubus listen eventA eventB &
#发送eventA、eventB、eventC(未监听)
DF1A:~$ ubus send eventA '{"active":"darts"}'
DF1A:~$ ubus send eventB '{"active":"cats"}'
DF1A:~$ ubus send eventC '{"active":"dogs"}'
#我们会收到2条JSON消息,包含event名称和消息内容,未收到eventC
{ "eventA": {"active":"darts"} }
{ "eventB": {"active":"cats"} }
#结束后台进程
DF1A:~$ killall ubus
```

3. 监听全部事件。

```
#启动一个终端A用于监听所有事件
DF1A:~$ ubus listen
#启动另一个终端B用于发送eventA、eventB、eventC
DF1A:~$ ubus send eventA '{"active":"darts"}'
DF1A:~$ ubus send eventB '{"active":"cats"}'
DF1A:~$ ubus send eventC '{"active":"dogs"}'
#我们会在终端A侧接收所有JSON消息,包含事件名称和消息内容
DF1A:~$ ubus listen
{ "eventA": {"active":"darts"} }
{ "eventB": {"active":"cats"} }
{ "eventC": {"active":"dogs"} }
```

ubus send

命令含义：发送一个事件通知，可携带JSON格式消息。

```
ubus [-s <socket> | -t <timeout> | -v | -S] send <type> [<message>]
```

ubus send的参数如表13-13所示。

表 13-13　ubus send 的参数

参数名	说明
-s \<socket\>	与ubus list同名参数的用法相同
-t \<timeout\>	
-v	
-S	
send	必选参数，发送一个事件通知
\<type\>	必选参数，指定要发送的目标object的object type name，字符串
\<message\>	可选参数，JSON字符串，指定要发送的消息内容

ubus wait_for

命令含义：等待一个或多个object出现在ubus中。

```
ubus [-s <socket> | -t <timeout> | -v | -S] wait_for <object> [<object>...]
```

ubus wait_for的参数如表13-14所示。

表 13-14　ubus wait_for 的参数

参数名	说明
-s \<socket\>	与ubus list同名参数的用法相同
-t \<timeout\>	
-v	
-S	
wait_for	必选参数，等待指定的一个或多个object出现在ubus中
\<object\>	必选参数，指定object名称，字符串

ubus wait_for命令示例如下。

```
#等待object "hello"和"foo"出现在ubus中
DF1A:$ ubus -t 30 wait_for hello foo
```

如果hello和foo都出现在ubus中，则命令直接返回，如果其中任意一个没有出现，则命令等待30秒后返回。

13.2.5 使用Lua语言实战ubus

接下来通过实例学习如何使用Lua语言进行ubus通信。

测试Lua客户端之前,需要在开发板中安装相关的依赖软件包。

```
DF1A:$ opkg update
DF1A:$ opkg install libubox libblobmsg-json jshn libjson-script libubox-lua
```

创建测试代码test.lua。

```lua
#!/usr/bin/env lua
require "ubus"
require "uloop"
uloop.init()
local conn = ubus.connect()
if not conn then
        error("Failed to connect to ubus")
end
local my_method = {
        demo = {
                func_null = {
                        function(req,msg)
                                print("Call to function null");
                        end,{}
                },
                func_hello = {
                        function(req,msg)
                                print("Call to function hello " .. "id:" .. msg.id .. " message:" .. msg.msg)
                                conn:reply(req, {id=msg.id,message=msg.msg,student=msg.student,skill=msg.skill,kownledge=msg.kownledge});
                        end, {id = ubus.INT32, msg = ubus.STRING, student=ubus.INT8,skill=ubus.ARRAY,kownledge=ubus.TABLE  }
                },
        },
        test = {
                func_foo = {
                        function(req, msg)
                                conn:reply(req, {message="foo"});
                                print("Call to function 'func_foo'")
                                for k, v in pairs(msg) do
                                        print("key=" .. k .. " value=" .. tostring(v))
                                end
                        end, {id = ubus.INT32, msg = ubus.STRING }
                },
                func_bar = {
                        function(req,msg)
                                conn:reply(req, {id=msg.id,message=msg.msg});
                                conn:reply(req, {message="bar"});
                                print("Call to function 'func_bar'")
                        end, {id = ubus.INT32, msg = ubus.STRING }
                }
```

```
        }
}
conn:add(my_method)
local my_event = {
        test_event = function(msg)
                print("Call to test event")
                for k, v in pairs(msg) do
                        print("key=" .. k .. " value=" .. tostring(v))
                end
        end
}
conn:listen(my_event)
uloop.run()
```

上述代码的说明如下。

（1）require "ubus"和require "uloop"，分别引用Lua ubus库和libubox-lua库。libubox-lua提供uloop循环。

（2）使用uloop.init()对uloop循环进行初始化。

（3）连接ubus总线。使用ubus.connect()函数，得到一个ubus实例conn，接下来通过conn实例进行ubus API调用。

（4）定义对外ubus服务的对象和方法。对象名称为"my_method"，类型是table，使用conn:add(my_method)注册到ubus总线上。my_method对象内部定义了"demo"和"test"这两个table对象，这两个对象会出现在ubus总线中。这两个对象对外提供调用方法。demo对象对外提供func_hello方法，test对象对外提供func_foo和func_bar方法。

（5）在一个对象中可以定义多个方法，每个方法包含2个参数，第一个参数为请求对象，一般在回复给调用方时使用，如conn:reply(req,XXXX)；第二个参数为调用方传递过来的参数对象。

ubus Lua method参数类型与参数如图13-21所示。

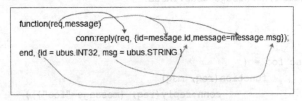

图13-21 ubus Lua method参数类型与参数示意图

（6）function有多种写法。

方法可以有参数，也可以没有参数。没有参数的方法，使用下面的格式：

```
function(req,msg)
...
end,{}
```

有参数的方法，使用下面的格式：

```
function(req,msg)
...
end,{id = ubus.INT32, msg = ubus.STRING, student=ubus.INT8,skill=ubus.ARRAY,kownledge=ubus.TABLE }
```

#{}之间的内容,根据所要传递的参数来进行定义。格式为:
{参数=ubus类型}

Lua的ubus参数类型如表13-15所示。

表13-15 Lua 的 ubus 参数类型

ubus 类型	表示类型	含义 / 示例
ubus.INT8	Boolean	布尔值，可选值为true/false。例如：student=ubus.INT8，传递参数"student":true
ubus.STRING	String	字符串。例如：msg=ubus.STRING，传递参数"msg":"foo"
ubus.INT32	Integer	整型。例如：id=ubus.INT32，传递参数"id":1
ubus.ARRAY	Array	数组。例如：skill=ubus.ARRAY，传递参数"skill":["PingPong","Yoga"]
ubus.TABLE	Table	表。例如：kownledge=ubus.TABLE，传递参数"kownledge":{"Math":"Junior","Python":"Advanced"}

（7）注册到ubus总线。使用conn:add(my_method)函数进行注册，my_method中定义的对象和方法会提供给其他连接到ubus总线上的进程使用。

（8）定义事件对象，监听名为"test_event"的事件。

（9）使用conn:listen(my_event)方法注册监听事件。

（10）使用uloop_run()进行循环监控。

对以上的代码进行测试

运行测试代码（后台，测试完需要执行killall lua）。

```
DF1A:~$ lua test.lua &
```

使用ubus命令查看ubus对象。

```
#发现对象demo和test已经被注册到ubus总线中
DF1A:~$ ubus list
demo
dhcp
file
hostapd.wlan0
hostapd.wlan1
iwinfo
lepton.network
log
network
network.device
network.interface
network.interface.cfg094d8f
network.interface.guest
network.interface.lan
network.interface.lease
network.interface.loopback
network.interface.wan
network.interface.wan6
network.wireless
nlpk-rpc-api
service
session
```

```
system
test
uci
```

查看demo和test对象支持的方法。

```
#可以看到对象的id、支持的方法列表、方法的名称和方法支持的参数都已经列出来了
DF1A:~$ ubus -v list demo
'demo' @d4088b7b
    "func_null":{}
    "func_hello":{"kownledge":"Table","id":"Integer","msg":"String","student":
"Boolean","skill":"Array"}
DF1A:~$ ubus -v list test
'test' @aa6e5fcb
    "func_foo":{"id":"Integer","msg":"String"}
    "func_bar":{"id":"Integer","msg":"String"}
```

调用对象提供的方法。

```
#使用ubus call命令调用demo对象的空方法。控制台输出"Call to function null"
DF1A:~$ ubus call demo func_null
Call to function null
```

调用demo对象的func_hello方法，传递参数和类型（对于func_foo和func_bar的相应方法，读者可以自行测试）。

```
DF1A:~$ ubus call demo func_hello \
'{"id":1,"msg":"foo","student":true,"skill":["PingPong","Yoga"],"kownledge":{"Math
":"Junior","Python":"Advanced"}}'
#以下为输出
Call to function hello id:1 message:foo
{
    "message": "foo",
    "kownledge": {
        "Math": "Junior",
        "Python": "Advanced"
    },
    "id": 1,
    "student": true,
    "skill": [
        "PingPong",
        "Yoga"
    ]
}
```

调用ubus send发送test_event。

```
DF1A:~$ ubus send test_event '{"name":"foo"}'
```

完成测试后，终端输出接收到的事件的内容。

```
Call to test event
key=name value=foo
```

13.2.6 使用C语言实战ubus

下面通过实例来学习如何使用C语言进行ubus通信，结合ubus命令（Shell）进行测试。创建一个软件包hello_cubus，提供hello_cubus可执行程序，此程序启动后连接ubus，提供hello-cubus对象和func_hello方法。软件包的创建方法已经在11.2节中详细讲解过，在接下来的示例中，直接创建软件包。创建的软件包文件结构如图13-22所示。

```
hello_cubus/
├── files
│   └── hello_cubus.init
├── Makefile
└── src
    ├── CMakeLists.txt
    └── hello_cubus.c
```

图13-22　hello_cubus软件包文件结构

创建软件包目录。

```
HOST:chaos_calmer_15_05_1$ mkdir -p package/df1a/hello_cubus
```

创建Makefile文件，src、files目录。

```
HOST:chaos_calmer_15_05_1$ cd package/df1a/hello_cubus
HOST:chaos_calmer_15_05_1$ touch Makefile
HOST:chaos_calmer_15_05_1$ mkdir src files
```

创建Makefile文件。

```
#
# Copyright (C) 2019 Freeirs
#
# This is free software, licensed under the GNU General Public License v2.
# See /LICENSE for more information.
#
include $(TOPDIR)/rules.mk
PKG_NAME:=hello_cubus
PKG_VERSION:=1.0
PKG_BUILD_DIR:=$(BUILD_DIR)/$(PKG_NAME)-$(PKG_VERSION)
PKG_LICENSE:=GPL-2.0
PKG_LICENSE_FILES:=
PKG_BUILD_PARALLEL:=1
PKG_MAINTAINER:=Zheng QiWen <zhengqwmail@gmail.com>
include $(INCLUDE_DIR)/package.mk
include $(INCLUDE_DIR)/cmake.mk
define Package/$(PKG_NAME)
  SECTION:=example
  CATEGORY:=DF1A Packages
  DEPENDS:= +libubus +ubus +ubusd +jshn +libubox
  TITLE:=df1a ubus package example
endef
define Package/$(PKG_NAME)/description
    df1a ubus example
```

```
endef
define Build/Prepare
    rm -rf $(PKG_BUILD_DIR)
    mkdir -p $(PKG_BUILD_DIR)
    $(CP) ./src/* $(PKG_BUILD_DIR)/
endef
define Package/$(PKG_NAME)/install
    $(INSTALL_DIR)  $(1)/usr/bin
    $(INSTALL_BIN)  $(PKG_INSTALL_DIR)/usr/sbin/hello_cubus $(1)/usr/bin
    $(INSTALL_DIR)  $(1)/etc/init.d
    $(INSTALL_BIN)  ./files/hello_cubus.init $(1)/etc/init.d/hello_cubus
endef
$(eval $(call BuildPackage,$(PKG_NAME)))
```

创建files/hello_cubus.init文件。

```
#!/bin/sh /etc/rc.common
START=99
STOP=99
USE_PROCD=1
PROG=/usr/bin/hello_cubus
start_service() {
    procd_open_instance
    procd_set_param command "$PROG"
    procd_set_param respawn
    procd_set_param stdout 1
    procd_set_param stderr 1
    procd_close_instance
}
reload_service() {
    restart
}
```

创建src/hello_cubus.c文件。

```c
#include <stdio.h>
#include <stdlib.h>
#include <string.h>
#include <getopt.h>
#include <signal.h>
#include <stdarg.h>
#include <syslog.h>
#include <libubox/ulog.h>
#include <libubox/uloop.h>
#include <libubox/utils.h>
#include <libubox/list.h>
#include <libubox/usock.h>
#include <libubus.h>
enum {
    HELLO_ID,
    HELLO_MSG,
    HELLO_STUDENT,
    HELLO_SKILL,
```

```c
        HELLO_KOWNLEDGE,
        __HELLO_MAX,
};
static struct ubus_context *ctx;
static struct blob_buf buf;
static int api_func_hello(struct ubus_context *ctx, struct ubus_object *obj,
        struct ubus_request_data *req, const char *method,
        struct blob_attr *msg);
static const struct blobmsg_policy hello_policy[__HELLO_MAX] = {
        [HELLO_ID] = { .name = "id", .type = BLOBMSG_TYPE_INT32 },
        [HELLO_MSG] = { .name = "msg", .type = BLOBMSG_TYPE_STRING },
        [HELLO_STUDENT] = { .name = "student", .type = BLOBMSG_TYPE_INT8 },
        [HELLO_SKILL] = { .name = "skill", .type = BLOBMSG_TYPE_ARRAY },
        [HELLO_KOWNLEDGE] = { .name = "kownledge", .type = BLOBMSG_TYPE_TABLE },
};
static const struct ubus_method api_methods[] = {
        UBUS_METHOD("func_hello", api_func_hello, hello_policy),
};
static struct ubus_object_type api_type =
        UBUS_OBJECT_TYPE("hello-cubus", api_methods);
static struct ubus_object hello_obj = {
        .name = "hello-cubus",
        .type = &api_type,
        .methods = api_methods,
        .n_methods = ARRAY_SIZE(api_methods),
};
static struct ubus_event_handler hello_listener;
static void handle_signal(int signo)
{
        ULOG_INFO("handle_signal\n");
        if(!ctx){
                ubus_remove_object(ctx, &hello_obj);
                ubus_free(ctx);
        }
        uloop_end();
}
static void setup_signals(void)
{
        struct sigaction s;
        memset(&s, 0, sizeof(s));
        s.sa_handler = handle_signal;
        s.sa_flags = 0;
        sigaction(SIGINT, &s, NULL);
        sigaction(SIGTERM, &s, NULL);
        sigaction(SIGUSR1, &s, NULL);
        sigaction(SIGUSR2, &s, NULL);
        s.sa_handler = SIG_IGN;
        sigaction(SIGPIPE, &s, NULL);
}
static int api_func_hello(struct ubus_context *ctx, struct ubus_object *obj,
```

```c
                            struct ubus_request_data *req, const char *method,
                            struct blob_attr *msg)
{
    struct blob_attr *tb[__HELLO_MAX];
    void *instance;
    struct blob_attr *attr;
    blobmsg_parse(hello_policy, __HELLO_MAX, tb, blob_data(msg), blob_len(msg));
    /* id and msg are must be parameters*/
    if (!tb[HELLO_ID] || !tb[HELLO_MSG])
            return UBUS_STATUS_INVALID_ARGUMENT;
    blob_buf_init(&buf, 0);
    if(tb[HELLO_ID]) {
            blobmsg_add_u32(&buf, "id", blobmsg_get_u32(tb[HELLO_ID]));
    }
    if(tb[HELLO_MSG]) {
            blobmsg_add_string(&buf, "msg", blobmsg_get_string(tb[HELLO_MSG]));
    }
    if(tb[HELLO_STUDENT]) {
            blobmsg_add_u8(&buf, "student", blobmsg_get_u8(tb[HELLO_STUDENT]));
    }
    if(tb[HELLO_SKILL]
            && blobmsg_type(tb[HELLO_SKILL]) == BLOBMSG_TYPE_ARRAY) {
            int len = blobmsg_data_len(tb[HELLO_SKILL]);
            instance = blobmsg_open_array(&buf, "skill");
            __blob_for_each_attr(attr, blobmsg_data(tb[HELLO_SKILL]), len) {
                    blobmsg_add_blob(&buf, attr);
            }
            blobmsg_add_string(&buf, NULL, "DF1A");
            blobmsg_close_array(&buf, instance);
    }

    if(tb[HELLO_KOWNLEDGE]
            && blobmsg_type(tb[HELLO_KOWNLEDGE]) == BLOBMSG_TYPE_TABLE) {
            int len = blobmsg_data_len(tb[HELLO_KOWNLEDGE]);
            instance = blobmsg_open_table(&buf, "kownledge");
            __blob_for_each_attr(attr, blobmsg_data(tb[HELLO_KOWNLEDGE]), len) {
                    blobmsg_add_blob(&buf, attr);
            }
            blobmsg_close_table(&buf, instance);
    }
    ubus_send_reply(ctx, req, buf.head);
    return UBUS_STATUS_OK;
}
static void hello_recv_event(struct ubus_context *ctx, struct ubus_event_handler
*ev,const char *type, struct blob_attr *msg)
{
    struct blob_attr *tb[__HELLO_MAX];
    ULOG_INFO("hello_recv_event\n");
    blobmsg_parse(hello_policy, __HELLO_MAX, tb, blob_data(msg), blob_len(msg));
```

```c
    if (tb[HELLO_ID] && tb[HELLO_MSG])
        ULOG_INFO("hello_recv_event id=%d msg=%s\n",blobmsg_get_u32(tb[HELLO_
ID]),blobmsg_get_string(tb[HELLO_MSG]));
}
int main(int argc, char **argv)
{
        const char *ubus_socket = NULL;
        int ch;
        int ret = 0;
        while ((ch = getopt(argc, argv, "s:")) != -1) {
                switch (ch) {
                case 's':
                        ubus_socket = optarg;
                        break;
                default:
                        break;
                }
        }
        setup_signals();
        uloop_init();
        ctx = ubus_connect(ubus_socket);
        if (!ctx) {
                ULOG_ERR( "Failed to connect to ubus\n");
                return -1;
        }
        /* add object */
        ret = ubus_add_object(ctx, &hello_obj);
        if (ret) {
                ULOG_ERR("ubus_add_object error %s", ubus_strerror(ret));
                goto ERROR;
        }
        /*add event handler*/
        hello_listener.cb = hello_recv_event;
          ret = ubus_register_event_handler(ctx, &hello_listener, "test_event");
          if (ret) {
                ULOG_ERR("ubus_register_event_handler error %s", ubus_strerror(ret));
                goto ERROR;
        }
        ULOG_INFO("Connected to ubus, id=%08x\n", ctx->local_id);
        ubus_add_uloop(ctx);
        uloop_run();
ERROR:
        if(!ctx){
                ubus_remove_object(ctx, &hello_obj);
                ubus_free(ctx);
        }
        uloop_done();
        return ret;
}
```

编写src/CMakeLists.txt文件。

```
cmake_minimum_required(VERSION 2.6)
PROJECT(hello_cubus C)
ADD_DEFINITIONS(-Os -Wall -Werror -Wmissing-declarations)
SET(SRCS hello_cubus.c)
SET(LIBS ubox ubus json-c blobmsg_json)
ADD_EXECUTABLE(hello_cubus ${SRCS})
TARGET_LINK_LIBRARIES(hello_cubus ${LIBS})
INSTALL(TARGETS hello_cubus
        RUNTIME DESTINATION sbin
)
```

源代码说明

接下来对源代码主要框架进行说明。

1. 判断命令行参数。支持"-s"参数，后面可以跟随ubus socket文件。默认使用/var/run/ubus.sock文件，连接到ubusd上。

2. 设置信号。通过函数setup_signals()设置信号处理函数。

3. uloop_init循环初始化。

4. 通过ubus_connect(ubus_socket)连接到ubus总线，返回ubus_context上下文，后续的ubus操作都围绕这个上下文进行。

5. 通过ubus_add_object(ctx, &hello_obj)函数向ubus上下文注册hello-cubus对象。hello-cubus对象和api_methods方法绑定。

```
static struct ubus_object hello_obj = {
        .name = "hello-cubus",
        .type = &api_type,
        .methods = api_methods,
        .n_methods = ARRAY_SIZE(api_methods),
};
```

```
#api_methods定义了使用的method名称(func_hello)、method对应的处理函数(api_func_hello)、
method policy(hello_policy)
static const struct ubus_method api_methods[] = {
        UBUS_METHOD("func_hello", api_func_hello, hello_policy),
};
```

6. method policy定义了方法支持的参数名称和参数的类型，实际上是对参数的约束。

```
static const struct blobmsg_policy hello_policy[__HELLO_MAX] = {
        [HELLO_ID] = { .name = "id", .type = BLOBMSG_TYPE_INT32 },
        [HELLO_MSG] = { .name = "msg", .type = BLOBMSG_TYPE_STRING },
        [HELLO_STUDENT] = { .name = "student", .type = BLOBMSG_TYPE_INT8 },
        [HELLO_SKILL] = { .name = "skill", .type = BLOBMSG_TYPE_ARRAY },
        [HELLO_KOWNLEDGE] = { .name = "kownledge", .type = BLOBMSG_TYPE_TABLE },
};
```

```
#ubus目前支持的参数类型有以下几种(在libubox库的blobmsg.h文件中定义)
enum blobmsg_type {
```

```
        BLOBMSG_TYPE_UNSPEC,
        BLOBMSG_TYPE_ARRAY,
        BLOBMSG_TYPE_TABLE,
        BLOBMSG_TYPE_STRING,
        BLOBMSG_TYPE_INT64,
        BLOBMSG_TYPE_INT32,
        BLOBMSG_TYPE_INT16,
        BLOBMSG_TYPE_INT8,
        __BLOBMSG_TYPE_LAST,
        BLOBMSG_TYPE_LAST = __BLOBMSG_TYPE_LAST - 1,
        BLOBMSG_TYPE_BOOL = BLOBMSG_TYPE_INT8,
};
```

7. func_hello方法的实现函数为api_func_hello()，通过blobmsg_parse()函数对调用者传递的参数进行解析。func_hello()函数将收到的所有参数直接返回，仅对skill会附加一个字符串类型值"DF1A"。func_hello()函数默认支持5个参数。

（1）id：必选传递参数。

（2）msg：必选传递参数。

（3）student：可选。

（4）skill：可选。

（5）kownledge：可选。

8. 注册名为test_event的事件。目前只打印出id和msg参数。

9. 主函数调用ubus_add_uloop()函数把ctx添加到uloop循环中，然后调用uloop_run()进入消息循环。

10. ubus_free(ctx)和uloop_done()不会被执行。

编译测试

使用make menuconfig命令配置软件包。

```
#请先切换到OpenWrt的根目录后操作如下命令
HOST:chaos_calmer_15_05_1$ make menuconfig
```

配置hello_cubus软件包的过程如图13-23所示。

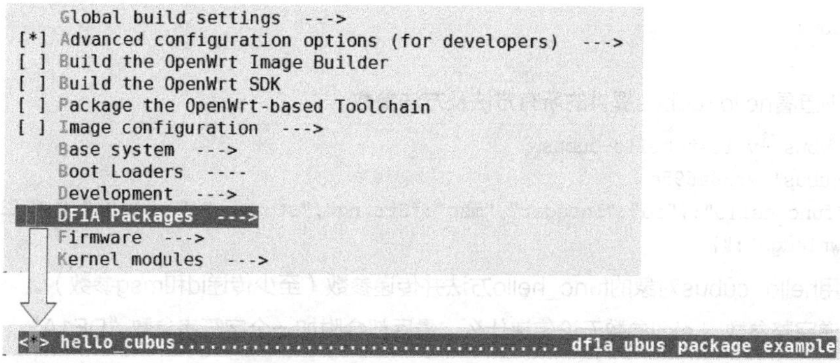

```
    Global build settings  --->
[*] Advanced configuration options (for developers)  --->
[ ] Build the OpenWrt Image Builder
[ ] Build the OpenWrt SDK
[ ] Package the OpenWrt-based Toolchain
[ ] Image configuration  --->
    Base system  --->
    Boot Loaders  ----
    Development  --->
    DF1A Packages  --->
    Firmware  --->
    Kernel modules  --->


<*> hello_cubus.................................. df1a ubus package example
```

图13-23　配置hello_cubus软件包

编译软件包

编译软件包在对OpenWrt进行固件编译时会自动完成，这里采用单独编译用于测试。

```
HOST:chaos_calmer_15_05_1$ make package/df1a/hello_cubus/{clean,compile} V=99
```

编译生成的软件包放在下列目录中。

```
HOST:chaos_calmer_15_05_1$ find ./bin -name hello_cubus*
./bin/siflower/packages/base/hello_cubus_1.0_mips_siflower.ipk
```

软件包安装测试

通过scp命令把hello_cubus_1.0_mips_siflower.ipk软件包上传到开发板的/tmp目录。假设开发板IP地址为172.16.10.126。

1. 使用SCP命令将软件包上传到开发板中。

```
HOST:chaos_calmer_15_05_1$ scp ./bin/siflower/packages/base/hello_cubus_1.0_mips_siflower.ipk admin@172.16.10.126:/tmp
admin@172.16.10.126's password:
hello_cubus_1.0_mips_siflower.ipk           100% 3962     3.9KB/s   00:00
```

2. 在开发板中安装hello_cubus软件包。

```
DF1A:~$ opkg install /tmp/hello_cubus_1.0_mips_siflower.ipk
Installing hello_cubus (1.0) to root...
Configuring hello_cubus.
```

3. 查看hello_cubus进程是否已经启动。

```
DF1A:~$ ps -www|grep hello_cubus|grep -v grep
20357 admin     1304 S    /usr/bin/hello_cubus
```

4. 用logread命令查看ubus注册信息。因为代码中使用了ULOG函数，所以可以通过logread读出打印信息。

```
DF1A:$ logread -l 1
Sun Nov 24 09:40:54 2019 daemon.info hello_cubus: Connected to ubus, id=8361a2cc
```

5. 使用ubus命令查看ubus对象。下列代码中显示对象hello-cubus已经被注册到ubus总线中。

```
DF1A:~$ ubus list
...
hello-cubus
...
```

6. 详细查看hello-cubus提供的所有方法及方法参数。

```
DF1A:~$ ubus -v list hello-cubus
'hello-cubus' @c38a695e
    "func_hello":{"id":"Integer","msg":"String","student":"Boolean","skill":"Array","kownledge":"}
```

7. 调用hello-cubus对象的func_hello方法并传递参数（至少传递id和msg参数）。

8. 传递完整参数，skill参数无论传递什么，最后都会附加一个字符串参数"DF1A"。

```
DF1A:~$ ubus call hello-cubus func_hello '{"id":1,"msg":"Hi gays!","student":false,"skill":["Python","Golang","PHP",true,32],"kownledge":{"Math":"Junior","Python":"Advanced","Science":true,"IELTS":8}}'
#以下为输出
```

```
{
    "id": 1,
    "msg": "Hi gays!",
    "student": false,
    "skill": [
        "Python",
        "Golang",
        "PHP",
        true,
        32,
        "DF1A"
    ],
    "kownledge": {
        "Math": "Junior",
        "Python": "Advanced",
        "Science": true,
        "IELTS": 8
    }
}
```

9. 发送test_event进行测试。

```
DF1A:$ ubus send test_event '{"id":1,"msg":"hello event"}'
#通过logread查看
DF1A:$ logread -l 3
Sun Nov 24 09:40:54 2019 daemon.info hello_cubus: Connected to ubus, id=8361a2cc
Sun Nov 24 09:41:33 2019 daemon.info hello_cubus: hello_recv_event
Sun Nov 24 09:41:33 2019 daemon.info hello_cubus: hello_recv_event id=1 msg=hello event
```

13.3 Netifd原理

Netifd即Network Interface Daemon，是OpenWrt系统中用于进行网络配置的守护进程，是最为基础和重要的程序之一，绝大多数网络接口设置由Netifd来处理完成。

13.3.1 Netifd软件架构

Netifd依赖于libuci库、libubus库、libnl-tiny库、libubox库等。Netifd的软件架构如图13-24所示。

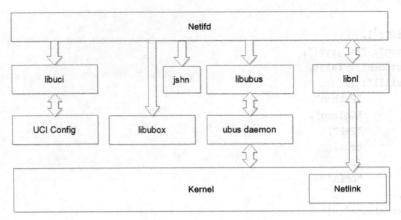

图13-24　Netifd的软件架构

接下来对图13-24中各库进行介绍。

libuci

libuci提供操作UCI配置文件的C语言库。Netifd所需网络接口的信息，如上网协议、接口、IP地址等，都放在/etc/config/network和/etc/config/wireless UCI配置文件中，Netifd通过libuci来操作和读取配置文件。

libubox

Netifd基于libubox库实现事件监听和数据结构管理。

libubus

libubus提供ubus相关操作接口API，如连接ubus总线，创建object、method等操作。默认Netifd连接到ubusd(system ubus daemon)上。Netifd默认会创建4个object（network、network.device，network.wireless、network.interface），后面会根据UCI配置文件中描述的接口类型，动态创建如network.interface.lan、network.interface.wan等。每个object都会提供接口操作的方法。通过ubus Shell命令可以查看object提供的接口方法。

Netlink

Netlink提供了一种在用户空间和内核间进行双向通信的IPC机制，能以标准套接字的扩展形式实现全双工的通信连接。Netlink广泛用于内核网络子系统，用于替代ioctl。RFC 3549对Netlink进行了详细的定义。

Netlink有如下特点。

- 双向传输，异步通信。
- 在用户空间中使用标准socket API。
- 在内核空间中使用专门的API。
- 支持多播。
- 可由内核端发起通信。
- 支持32种协议类型。

Netlink协议实现的源代码位于内核源代码目录net/netlink下。
- af_netlink.c和af_netlink.h：提供Netlink内核套接字API。
- diag.c：监视接口模块，提供用于读写有关Netlink套接字的信息。
- genetlink.c：提供新的Netlink API，使用新接口创建Netlink消息更为容易。

采用Netlink实现与内核通信的应用很多，如NETLINK_ROUTE、NETLINK_USERSOCK、NETLINK_NFLOG、NETLINK_NETFILTER、NETLINK_IP6_FW、NETLINK_KOBJECT_UEVENT、NETLINK_GENERIC等，其中网络子系统最常使用的几个有NETLINK_KOBJECT_UEVENT、NETLINK_ROUTE、NETLINK_NETFILTER。

libnl

为了简化Netlink编程，通常使用Netlink库来进行操作，推荐使用libnl API。libnl库如图13-25所示。

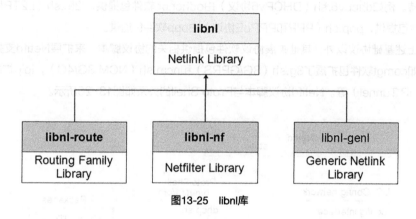

图13-25　libnl库

libnl库软件包由一系列库组成。
- Netlink Library(libnl)：libnl核心库，功能包含socket处理、发送和接收消息、消息构造和解析等。
- Routing Family Library(libnl-route)：功能包含地址、连接、邻居子网、路由、流量控制、邻居表等。
- Netfilter Libraray(libnl-nf)：功能包含连接跟踪、日志、队列。
- Generic Netlink Library(libnl-genl)：功能包含控制器API、协议族和命令注册等。

Netifd监控设备接口信息通过Netlink协议与内核通信，设置和获取相关信息。Netifd使用简化的libnl库（libnl-tiny）对Netlink消息进行处理，由Netifd源代码system-linux.c实现。

13.3.2　Netifd软件包

Netifd源代码软件包在package/config/netifd目录。生成的Netifd软件包包含以下文件。

（1）Shell脚本：/sbin/ifup、/sbin/ifdown（ifup的软连接）、/sbin/ifstatus、/sbin/devstatus。

（2）init.d启动脚本：/etc/init.d/network。

（3）Hotplug脚本：/etc/hotplug.d/iface/00-netstate。

（4）udhcpc脚本：/usr/share/udhcpc/default.script、/lib/netifd/dhcp.script。

（5）Netifd守护进程：/sbin/netifd。

（6）Netifd支持Shell库：/lib/netifd/netifd-proto.sh、/lib/network/config.sh、/lib/netifd/utils.sh、/lib/netifd/netifd-wireless.sh。

（7）Proto Shell（协议脚本）：/lib/netifd/proto/dhcp.sh（DHCP协议）。

下面对一些重要的脚本和目录进行说明。

/lib/netifd/proto目录

这是协议脚本目录，用于存放与接口支持的协议相关的Shell脚本，即"Proto Shell"。由Netifd转换生成协议对应的protocol handler。

默认Netifd软件包只支持static协议和DHCP协议，其中static协议，不是通过Proto Shell扩展实现的，是在Netifd源代码中实现的，而DHCP协议由dhcp.sh来支持。其他基础协议由相关软件包来支持。例如dhcpv6.sh（DHCPv6协议）由odhcp6c软件包提供，l2tp.sh（L2TP协议）由xl2tpd软件包提供，ppp.sh（PPP和PPPoE协议）由ppp软件包提供。

除了上述基础协议以外，其他扩展协议软件包提供相关的协议脚本，来扩展Netifd支持的接口协议。例如comgt软件包扩展了3g.sh（3G/GPRS）和ncm.sh（NCM 3G/4G），ipip扩展了ipip.sh（IP in IP Tunnel）等。Netifd协议脚本与Proto Shell的关系如图13-26所示。

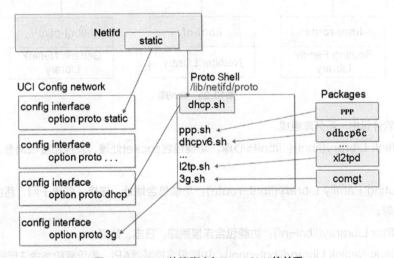

图13-26　Netifd协议脚本与Proto Shell的关系

通过ubus命令可以查看系统当前支持的协议及参数信息。

```
DF1A:$ ubus call network get_proto_handlers
```

/sbin/ifdown和/sbin/ifup脚本

这两个脚本实现对某个网络接口关闭（down）和启用（up）操作。/sbin/ifdown脚本实际上是/sbin/ifup脚本的软连接，功能由/sbin/ifup来实现。down/up操作在脚本内部通过调用Netifd注册到ubus的对象的方法来实现功能。

ifdown/ifup脚本的语法格式如下。

```
/sbin/{ifup|ifdown} [-a] [-w] [interface]
```

该脚本支持的参数如表13-16所示。

表13-16 ifup/down 脚本支持的参数

参数	含义
-a	表示对所有接口均执行相同的操作，此时将忽略interface设置。通过ubus -S list 'network.interface.*'命令来列出所有接口，然后对所有接口执行命令
-w	表示是否执行wifi up（启用无线）操作。如果有此参数，则/sbin/wifi up操作不会被执行。如果未指定，则在ifup的时候，/sbin/wifi up也会被执行
interface	指定down/up操作的目标接口。通过ubus -S list 'network.interface.*'|awk -F 'interface.' '{print $2}'命令查看都有哪些接口

禁用和启用lan接口的命令示例如下。

```
DF1A:~$ ifdown lan
DF1A:~$ ifup lan
```

Hotplug脚本

Hotplug直译就是热插拔。在OpenWrt中，由procd进程通过Netlink套接字监听NETLINK_KOBJECT_UEVENT消息，然后调用/etc/hotplug.json脚本进行消息的判断和分发。

每次网络接口启动或者关闭的时候，都会按照ASCII顺序（通常为数字前缀<nn>-<scriptname>），依次执行/etc/hotplug.d/iface/目录下的脚本。利用Hotplug脚本，可以把要在网络接口启动或关闭时执行的动作预定义到脚本里。

iface中系统传递的环境变量如表13-17所示。

表13-17 iface 环境变量

变量名称	说明
ACTION	值为ifup或ifdown
INTERFACE	网络接口名称，如wan、lan等
DEVICE	物理设备的名称，如br-lan

/etc/hotplug.d/iface/00-netstate脚本是网络接口变化时执行的脚本，由Netifd提供。当网络接口ifup启动时，会根据接口的名称和状态写入/var/state/network临时文件。整体流程如图13-27所示。

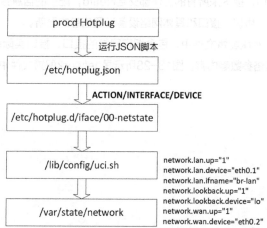

图13-27 iface Hotplug流程

udhcpc脚本

udhcpc即Micro DHCP Client，是BusyBox软件包提供的面向嵌入式系统的非常小型的DHCP客户端。udhcpc有一个"-s"参数表示接收到dhcp event时需要运行的脚本。

Netifd提供两个和udhcpc相关的脚本。

● /usr/share/udhcpc/default.script：在没有指定"-s"参数的情况下，udhcpc接收到dhcp evnet时默认运行的脚本。在BusyBox源代码中已经定义了这个脚本的宏变量。请参考busybox-1.23.2/include/config/udhcpc/default/script.h。

```
define CONFIG_UDHCPC_DEFAULT_SCRIPT "/usr/share/udhcpc/default.script"
```

● /lib/netifd/dhcp.script：这个脚本是OpenWrt系统中udhcpc启动时，通过"-s"参数指定的启动脚本。

```
DF1A:$ ps -wwww|grep udhcpc|grep -v grep
20372 admin     1568 S    udhcpc -p /var/run/udhcpc-eth1.pid -s /lib/netifd/dhcp.
script -f -t 0 -i eth1 -x hostname SiWiFic04c -C
```

/lib/netifd/dhcp.script脚本接收3个命令参数，如表13-18所示。

表13-18 dhcp.script脚本命令参数

命令参数	说明
deconfig	刚启动客户端时使用
bound	当客户端从一个没有绑定的状态添加进来，要求进行绑定时，运行脚本可以设定参数的运行环境变量，如默认网关、DNS服务器等
renew	重新启动客户端请求租约时使用

13.3.3 Netifd基本概念

Netifd是OpenWrt下专有的网络配置守护进程，下面我们就从OpenWrt下配置网络的过程入手，引入Netifd设计理念和一些基本概念。

OpenWrt中最基本的网络配置过程，可以概括为"配置网络参数，把配置参数重新加载到网络接口"。配置网络参数就是把网络的配置信息写入对应的UCI配置文件中，有线网络需要修改/etc/config/network，无线网络需要修改/etc/config/wireless，然后执行/etc/init.d/network restart或/etc/init.d/network reload，接下来所有的工作都交给Netifd，配置的信息会被Netifd解析和处理，由Netifd来完成配置参数、协议、接口和具体网络设备设置信息的更新。

仔细观察会发现network配置文件中，包含指定网络的接口、接口实际绑定的网络设备名称、接口的上网方式和其他网络参数等内容。图13-28所示是network配置文件中对lan接口的配置。

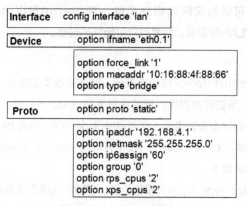

图13-28 lan接口配置

协议、接口、设备是Netifd的基本组件，Netifd的程序设计就是围绕着这几部分来展开的。Netifd引入了3个主要概念——Proto Handler、Interface、Device，它们分别对应network配置文件中的Proto、Interface、Ifname。

Proto Handler、Interface和Device之间的关系如图13-29所示。

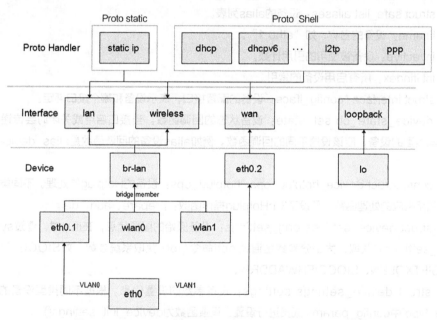

图13-29 Proto Handler、Interface和Device之间的关系示意图

Netifd根据network配置文件信息，创建接口（Interface）并指明其在实际网络中代表的设备（Device），以及其绑定的上网方式（Proto Handler），完成一个网络接口的配置并启用接口开始工作。当网络状态变化时，这3部分互相配合和通知，完成网络状态转换。

13.3.4 Netifd设备

在Netifd中，有几种设备：Network设备、Network alias设备、Bridge桥设备、MACVLAN、IP Tunnel、VLAN、VLANDEV。

表示链路层的设备可以为实际的物理设备，例如eth0可以为IP Tunnel，也可以为MACVLAN、Bridge、VLAN等设备，后者可以引用其他设备。

设备

在Netifd中，设备用struct device结构来表示，具体设备类型由struct device_type结构来表示。struct device表示设备的通用属性，具体设备类型struct device_type为设备的特有属性。创建设备时，需要指定待创建设备的类型。例如struct device_type bridge_device_type是创建Bridge桥设备时的指定结构，struct device_type simple_device_type是默认的设备类型。

struct device部分成员如下。

- const struct device_type *type：设备类型。创建不同类型设备时使用，包含设备的配置参数、创建方法、配置初始化、打印调试信息、状态、状态检测方法等。常见的设备类型有Network alias、Bridge、MACVLAN、IPTunnel、VLAN、VLANDEV。
- struct avl_node avl：设备树节点。用于把设备插入AVL设备树中，便于快速查找某个设备。设备树的根为struct avl_tree devices。
- struct safe_list users：设备的使用者列表，所有使用设备的使用者都会被添加到这个列表中。
- struct safe_list aliases：设备的alias列表。
- ifname：设备的名称，如"eth0.1"。
- int active：设备被使用的引用计数。
- int ifindex：所有启用设备的索引。
- struct interface *config_iface：设备的配置接口，表示设备和哪个接口绑定。
- device_state_cb set_state：设备状态的回调函数。当接口启用或关闭时会传递状态。对于不同类型的设备，应该设置不同的回调函数。例如alias设备的回调函数是alias_device_set_state。
- const struct device_hotplug_ops *hotplug_ops：设备的Hotplug的处理。不同类型的设备应该采用相应的处理函数。预设了3个Hotplug操作函数：prepare、add、del。
- struct device_settings orig_settings：设备原始的配置信息，在ifup时，通过system_if_get_settings()获取，大部分参数是通过ioctl命令/porc获取系统参数，如SIOCGIFMTU、SIOCGIFTXQLEN、SIOCGIFHWADDR等。
- struct device_settings settings：设备参数的设置信息。根据不同类型设备的struct device_type 中config_params 成员进行设置。设置函数为device_init_settings()。

设备使用者

创建一个设备后，如果要引用这个设备，就要注册一个设备使用者，在Netifd中用struct device_user结构来表示。设备的使用者一般是接口（Interface），也有设备之间相互引用的情况，例如bridge和bridge member的关系。

struct device_user的主要成员有以下几个。

- struct safe_list list：链表成员。设备的users成员使用，用于设备查找设备使用者。
- bool claimed：表示是否已经与设备关联。如果已关联设备，则为true，否则为false。

- struct device *dev：引用的设备对象的指针。
- void (*cb)(struct device_user *, enum device_event)：当关联设备的状态发生改变时，会通过该函数传递通知事件。设备的事件类型如下。
  - DEV_EVENT_ADD：设备已经存在，当添加一个新的device_user时，立刻产生该事件。
  - DEV_EVENT_REMOVE：设备不可用、不存在或是正在移除中，所有的使用者都应该删除引用并清理相关资源。
  - DEV_EVENT_SETUP：设备即将启用。
  - DEV_EVENT_UP：设备被成功启动。
  - DEV_EVENT_TEARDOWN：设备即将被关闭。
  - DEV_EVENT_DOWN：设备已经被关闭。
  - DEV_EVENT_UPDATE_IFNAME：更新接口名。仅用于VLAN和alias类型的设备使用者。
  - DEV_EVENT_UPDATE_IFINDEX：更新接口索引号。仅用于alias类型的设备使用者。
  - DEV_EVENT_LINK_UP：已经在此设备建立连接。仅用于Interface和alias类型的设备使用者。
  - DEV_EVENT_LINK_DOWN：链路丢失。仅用于Interface和alias类型的设备使用者。
  - DEV_EVENT_TOPO_CHANGE：拓扑变化，如增加桥设备成员。

设备与设备使用者之间的关系

在设备struct device结构中维护了一个引用计数（成员active），来控制设备的UP/DOWN状态。

- 如果设备使用者要绑定一个设备，需要调用device_claim(struct device_user user)函数，如果函数返回成功，则该设备的引用计数值加1。
- 如果设备使用者要取消绑定设备，需要调用device_release(struct device_user user)函数，如果函数返回成功，则该设备的引用计数值减1。
- 当引用计数从0变为1时，表示有一个设备使用者使用了该设备，设备就会被启用；而当引用计数从1变为0时，表示最后一个设备使用者离开，设备就会立即被关闭。

设备的引用计数如图13-30所示。

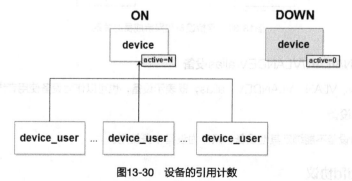

图13-30　设备的引用计数

网桥设备

网桥设备(bridge device)用struct bridge_state结构表示，其内部包含struct device dev。网桥设备本身也是一种设备。网桥设备还包含网桥成员(bridge_member)，每个网桥成员都是设备使用者，可以绑定其他设备。网桥设备通过网桥成员struct vlist_tree members与bridge_member关联。

```
struct bridge_state {
    struct device dev;
...
    struct bridge_member *primary_port;
    struct vlist_tree members;
...
};
struct bridge_member {
    struct vlist_node node;
    struct bridge_state *bst;
    struct device_user dev;
    bool present;
    char name[];
};
```

网桥设备和网桥成员的关系如图13-31所示。

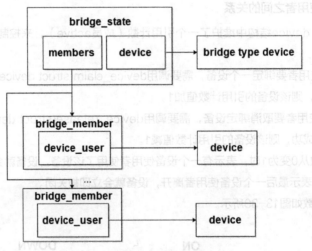

图13-31　网桥设备和网桥成员的关系

MACVLAN/VLAN/VLANDEV/alias设备

MACVLAN、VLAN、VLANDEV、alias，既表示设备，也可以作为设备使用者绑定其他设备。

IP Tunnel设备

这种类型的设备不能绑定其他设备，只作为设备本身使用。

13.3.5 Netifd协议

Proto即协议，表示具体应用到接口配置中的网络协议名称。如静态设置网络IP地址(static)、动态获取IP地址(dhcp)等。

Proto Handler

Proto Handler是Netifd对协议的表示，每个协议都有对应的struct proto_handler结构的实例。结构体的主要成员有以下几个。

● struct avl_node avl：协议的AVL节点。所有协议最终会作为一个AVL节点被插入struct avl_tree handlers协议树根节点中。

● unsigned int flags：协议处理事件的标志。

● const char *name：协议的名称，如"dhcp"。

● const struct uci_blob_param_list *config_params：协议的参数列表。这是协议参数解析后的结果。

● struct interface_proto_state *(*attach)(const struct proto_handler *h,struct interface *iface, struct blob_attr *attr)：绑定协议与接口的函数。这个函数会分配一个struct interface_proto_state结构并赋值给结构体struct interface中的proto数据成员，实现接口和协议的绑定。

Proto Shell Handler

Proto Shell Handler是解析协议脚本用的结构。解析协议脚本时，会为每个脚本分配一个struct proto_shell_handler结构，内部包含struct proto_handler proto成员。

13.3.6 Netifd接口

接口代表应用在一个或多个设备上的网络配置。如接口lan是应用在设备eth0.1上的LAN网络的配置。

接口

接口在Netifd中用struct interface结构来表示，使用时必须绑定在一个主设备(main device)和一个3层设备(L3 device)上。采用struct interface_user 结构来表示接口的使用者，指向一个接口实例。

在network配置文件中，存在一种别名技术alias，这是一种特殊的接口，用于扩展接口配置。alias设备会产生一个接口使用者，用于指明自己的parent，alias自身作为使用者的信息保存在struct interface的parent_iface成员中，而其parent则由struct interface的parent_ifname来指定。如以下程序所示，alias接口扩展了lan接口的配置。

```
config interface 'lan'
     option ifname 'eth0.1'
     option force_link '1'
     option macaddr '10:16:88:4f:88:66'
     option type 'bridge'
     option proto 'static'
     option ipaddr '192.168.4.1'
     option netmask '255.255.255.0'
     option ip6assign '60'
     option group '0'
     option rps_cpus '2'
     option xps_cpus '2'
```

```
config alias
    option interface 'lan'
    option proto 'static'
    option ipaddr '192.168.5.251'
    option netmask '255.255.255.0'
```

struct interface的主要成员有以下几个。

● struct vlist_node node：所有接口的实例都作为一个节点，被插入接口树(struct vlist_tree interfaces)中。

● struct list_head hotplug_list：处理Hotplug事件列表。

● enum interface_event hotplug_ev：具体的接口Hotplug事件类型。包含的事件有IFEV_DOWN、IFEV_UP、IFEV_UPDATE、IFEV_FREE、IFEV_RELOAD。

● enum interface_state state：接口状态。

　　■ IFS_SETUP：接口当前正在被protocol handler配置。
　　■ IFS_UP：接口启用，接口已被完全配置。
　　■ IFS_TEARDOWN：接口正在被取消配置。
　　■ IFS_DOWN：接口关闭。

● const char *name：接口的名称，与network配置文件中的config interface名称一致，如"lan"。

● const char *ifname：物理接口的名称，与option ifname名字一致，如eth0.1。

● bool available：标识接口是否可用，如果可用，表示接口随时可以启动(UP)。

● bool autostart：接口配置后，是否自动执行启动操作。如果接口状态从非活动切换到活动状态，Netifd将尝试立即启动它。如果手动将接口设置为UP，不管是否成功将设置此标志为true；如果手动将接口设置为DOWN，则强制设置此标志为false。

● const char *parent_ifname：alias中option interface包含的接口名称。仅限alias接口使用。

● struct interface_user parent_iface：alias指向的接口名称。仅限alias接口使用。

● struct device_user main_dev：接口绑定的主设备，使用时必须设置。

● struct device_user ext_dev：接口绑定的扩展设备。接口被连接时一般需要设定。

● struct device_user l3_dev：绑定的L3设备。

● struct blob_attr *config：接口配置。默认的参数有proto、ifname、auto、defaultroute、peerdns、metric、dns、dns_search、interface、ip6assign、ip6hint、ip4table、ip6table、ip6class、delegate、ip6ifaceid、force_link。完成解析network接口配置后，由对应的接口参数赋值。

● const struct proto_handler *proto_handler：指向proto handler，由network配置文件中的proto选项决定所使用的协议proto handler（根据proto名称，在协议树中查找对应的proto handler）。

● struct interface_proto_state *proto：由proto handler中的attch()函数来分配，所有内容由proto handler实例来填充，用于协议的绑定和协议变化的通知。

● struct interface_ip_settings proto_ip：更新时使用的接口的具体配置，包含多条IP、

Route、DNS等。

● struct interface_ip_settings config_ip：初始化时使用的接口的具体配置，包含多条IP、Route、DNS等。

13.3.7 Netifd处理流程

Netifd作为网络的守护进程，配置网络接口的上网协议，对网络接口进行管理，通过Netlink监听和设置内核网络接口信息，注册ubus对外接口等。主要处理过程有以下几种。

● 注册ubus对象和方法，对外提供调用接口。
● 解析协议脚本，生成协议Proto Handler，构建协议树。
● 创建和监听NetLink消息。
● 解析network和wireless配置文件。
● 根据配置文件信息，创建不同类型的设备（Device），构建设备树。
● 根据配置文件信息，创建接口（Interface），构建接口树，查找、创建接口使用者（Device user）、绑定接口，查找、绑定协议（Proto Handler）。
● 启用所有接口和无线设备。

下面对主要过程进行分析。

注册ubus对象和方法

Netifd启动后，会首先向ubusd注册object和method，相当于对外提供调用接口。Netifd主要注册了4个object，名字分别为network、network.device、network.wireless、network.interface，如图13-32所示。

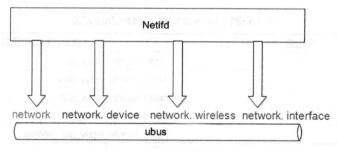

图13-32　Netifd向ubus注册4个object

● main_object

object name为network，对应的方法如表13-19所示。

表13-19　network ubus 方法

方法名	对应源代码处理函数
restart	netifd_handle_restart
reload	netifd_handle_reload
add_host_route	netifd_add_host_route
get_proto_handlers	netifd_get_proto_handlers
add_dynamic	netifd_add_dynamic

查看network支持的方法和参数,如下所示。

```
DF1A:$ ubus -v list network
'network' @346f0b01
    "restart":{}
    "reload":{}
    "add_host_route":{"target":"String","v6":"Boolean","interface":"String"}
    "get_proto_handlers":{}
    "add_dynamic":{"name":"String"}
```

● dev_object

object name为"network.device",对应的方法如表13-20所示。

表 13-20　network.device ubus 方法

方法名	对应源代码处理函数
status	netifd_dev_status
set_alias	netifd_handle_alias
set_state	netifd_handle_set_state

查看network.device支持的方法和参数,如下所示。

```
DF1A:$ ubus -v list network.device
'network.device' @5b2a2bd8
    "status":{"name":"String"}
    "set_alias":{"alias":"Array","device":"String"}
    "set_state":{"name":"String","defer":"Boolean"}
```

● wireless_object

object name为"network.wireless",对应的方法如表13-21所示。

表 13-21　network.wireless ubus 方法

方法名	对应源代码处理函数
up	netifd_handle_wdev_up
down	netifd_handle_wdev_down
status	netifd_handle_wdev_status
notify	netifd_handle_wdev_notify
get_validate	netifd_handle_wdev_get_validate

查看network.wireless支持的方法和参数,如下所示。

```
DF1A:$ ubus -v list network.wireless
'network.wireless' @b01d62a5
    "up":{}
    "down":{}
    "status":{}
    "notify":{}
    "get_validate":{}
```

● iface_object

object name为"network.interface",对应的方法如表13-22所示。

表13-22　network.interface ubus 方法

方法名	对应源代码处理函数
up	netifd_handle_up
down	netifd_handle_down
status	netifd_handle_status
prepare	netifd_handle_iface_prepare
dump	netifd_handle_dump
add_device	netifd_iface_handle_device
remove_device	netifd_iface_handle_device
notify_proto	netifd_iface_notify_proto
remove	netifd_iface_remove
set_data	netifd_handle_set_data

查看network.interface支持的方法和参数,如下所示。

```
DF1A:$ ubus -v list network.interface
'network.interface' @3bdbc845
    "up":{}
    "down":{}
    "status":{}
    "prepare":{}
    "dump":{}
    "add_device":{"name":"String","link-ext":"Boolean"}
    "remove_device":{"name":"String","link-ext":"Boolean"}
    "notify_proto":{}
    "remove":{}
    "set_data":{}
```

创建proto handler

Netifd注册完ubus对象后,会进行所支持协议的解析,创建proto handler。这个过程会把协议脚本及协议配置参数转换为Netifd内部管理的proto handler结构,最终Netifd会把所有协议都生成协议对应的proto handler结构,并把节点插入一棵AVL协议树中,以协议名作为索引。Network配置文件中的协议名(proto)与Netifd中已存在的proto handler的名字一致,该名字用于AVL树搜索对应的proto handler节点。同时proto handler还提供attach函数指针,用于和接口绑定。等AVL协议树生成后,就可以和某个接口(Interface)进行绑定。

proto handler相关处理函数调用关系如图13-33所示。

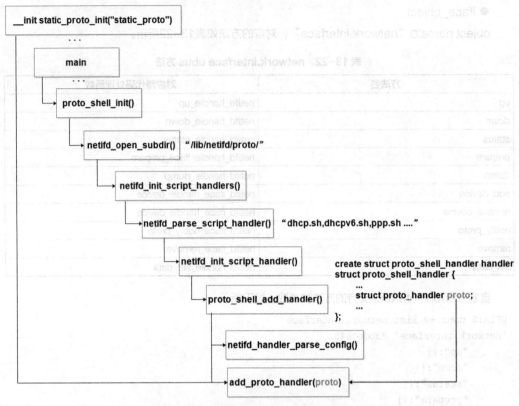

图13-33 proto handler相关处理函数调用关系

主要执行过程如下。

1. 在main函数执行之前，会先执行被标记为__init的函数（GCC特性），_init static_proto_init()函数是"static"协议的初始化函数，会创建proto handler结构，并插入AVL协议树中（AVL树的根为handlers）。所以"static"协议不是通过接下来要介绍的协议脚本来实现，而是写在Netifd代码中。插入后的结果如图13-34所示。

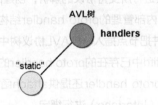

图13-34 Proto Handler协议树

2. 进入"/lib/netifd/proto"目录，遍历目录下的*.sh脚本，如dhcp.sh、dhcpv6.sh、l2tp.sh、ppp.sh。获得脚本的名称，作为协议（proto）的名称。例如dhcp.sh脚本的协议名称为"dhcp"。

3. 按照"./XXX.sh '' dump"命令格式，依次执行*.sh脚本。执行脚本后会输出JSON格式的字符串，这些字符串将作为协议的参数被Netifd保存起来。这些脚本的实现依赖于Netifd的/lib/netifd/netifd-proto.sh。

以dhcp.sh为例，执行./dhcp.sh '' dump，输出的JSON内容如下。

```
DF1A:~$ ./dhcp.sh '' dump
#输出的JSON内容如下
{ "name": "dhcp", "config": [ [ "ipaddr:ipaddr", 3 ], [ "hostname:hostname",
3 ], [ "clientid", 3 ], [ "vendorid", 3 ], [ "broadcast:bool", 7 ], [
"reqopts:list(string)", 3 ], [ "iface6rd", 3 ], [ "sendopts", 3 ], [ "delegate", 7
], [ "zone6rd", 3 ], [ "zone", 3 ], [ "mtu6rd", 3 ], [ "customroutes", 3 ] ], "no-
device": false, "no-proto-task": false, "available": false, "renew-handler": true,
"lasterror": false }
```

4. 接下来Netifd会根据输出的JSON内容，为每个协议创建对应的struct proto_handler结构体。在解析脚本的过程中，会创建struct proto_shell_handler结构，用于处理协议脚本，但它最终是为proto_handler服务的，proto_handler才是最终和Interface绑定的结构。

输出的JSON内容包括以下4项。

（1）name节点：协议的名称，如"dhcp"。

（2）config节点：协议配置参数内容。每个协议脚本输出不同，包含具体协议的参数信息和参数类型，这部分信息会放到proto_shell_handler的config结构中。后续介绍中会进行协议参数的解析。

（3）no-device、available、renew-handler、lasterror节点：用于设定struct proto_handler结构中的flags标志，分别对应宏PROTO_FLAG_NODEV、PROTO_FLAG_INIT_AVAILABLE、PROTO_FLAG_RENEW_AVAILABLE、PROTO_FLAG_LASTERROR，用于约定协议对一些事件的处理方式。

（4）no-proto-task节点：用于设定struct proto_shell_handler结构体中no_proto_task成员，用于任务结束的判定，不需要关注。

5. 解析完参数后，会把struct_proto_handler结构体中avl成员，添加到全局avl_tree handlers平衡二叉树中。

6. proto_handler结构中有一个重要的函数指针attach，这个函数会返回struct interface_proto_state结构，与Interface结构的proto数据成员进行关联，从而实现与Interface的绑定。

7. 按照上面的步骤，循环把所有协议都解析完毕，形成一棵AVL协议树。

以DHCP协议为例，从脚本到结构的构建如图13-35所示。

Wireless 初始化

Netifd无线网络部分与内核无线网络框架紧密相连，在了解无线网络前，可以先了解一下内核与无线网络相关的内容。内核无线网络框架采用无线网络子系统mac80211，使用cfg80211对无线设备进行配置和管理。图13-36所示是Linux的整体无线网络架构。

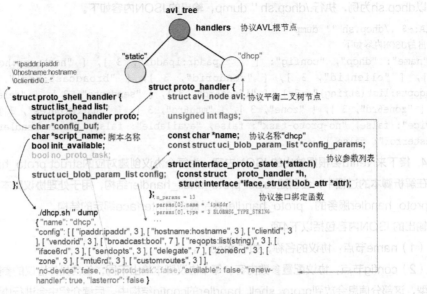

图13-35 DHCP协议从脚本到结构的构建示意图

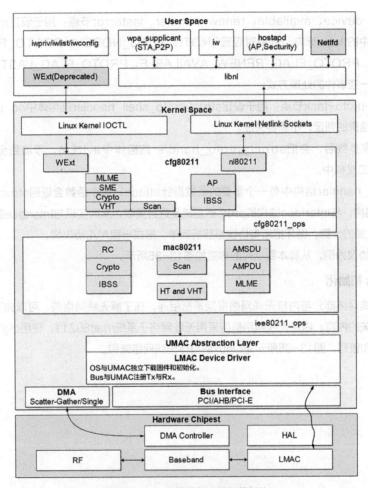

图13-36 Linux的整体无线网络架构

图中涉及的部分概念如下。

● mac80211：Linux内核无线网络子系统，是SoftMAC无线设备的驱动框架，包含mac80211.ko模块，用于与下层的MAC和驱动程序交互。

● cfg80211：用于对无线设备进行配置管理，与FullMAC、mac80211和nl80211一起工作。它包含cfg80211.ko模块，处理所有的配置，和用户空间交互。

● nl80211：用于对无线设备进行配置和管理，使用Netlink协议。

● MLME：MAC (Media Access Control) Layer Management Entity，用于管理物理层MAC状态机。

● SoftMAC：MLME由软件实现，mac80211为SoftMAC实现提供了一个API。SoftMAC设备可以对硬件进行更好的控制，用软件实现对802.11的帧管理，包括解析和产生802.11无线帧。

● FullMAC：MLME由硬件管理，使用FullMAC时，不需要使用mac80211。目前大多数802.11设备采用SoftMAC，FullMAC则较少使用。

● 用户空间采用WExt和libnl与内核无线子系统进行交互。WExt采用低效的IOCTL，目前已经被废弃，逐渐被libnl库替掉。

Netifd在进行无线接口和设备配置时，采用的方式和proto handler的处理方式类似。

在无线网络的UCI配置文件/etc/config/wireless中，描述了无线网络的设备配置（config wifi-device）和接口配置（config wifi-iface）。二者均由config wifi-iface中的device（无线设备）进行关联和绑定，配置实例如图13-37所示。

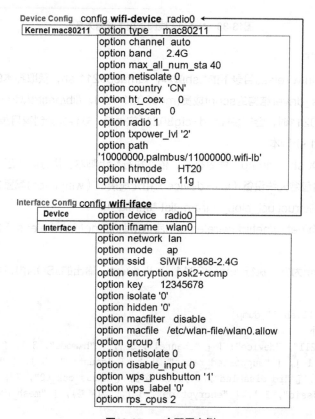

图13-37　一个配置实例

Netifd对无线网络的管理通过struct wireless_driver、struct wireless_device、struct wireless_interface结构体实现，其中wireless_device、wireless_interface分别对应/etc/config/wireless中的设备和接口配置。

Netifd源代码中对无线网络进行初始化的相关处理函数调用关系如图13-38所示。

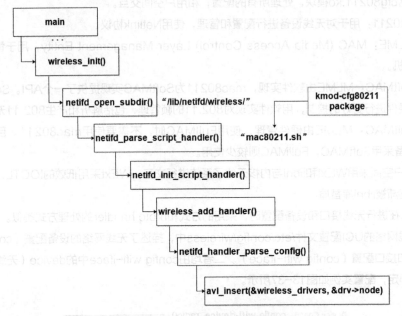

图13-38　Netifd无线脚本相关调用关系

主要执行过程如下。

1. 遍历/lib/netifd/wireless目录下的*.sh脚本，如mac80211.sh，获取脚本的名称，用于后续设置struct wireless_driver结构的script成员(mac80211.sh)。/lib/netifd/wireless目录本来是不存在的，使用mac80211时，会安装kmod-cfg80211软件包，软件包会创建目录并提供/lib/netifd/wireless/mac80211.sh脚本。

2. 按照"./XXX.sh '' dump"命令格式，依次执行*.sh脚本。执行脚本后会输出JSON格式的字符串，字符串内包含无线设备（wifi-device）和无线接口（wifi-iface)配置参数，分别保存到device和interface的struct uci_blob_param_list *config中。

Netifd提供了/lib/netifd/netifd-wireless.sh库来帮助/lib/netifd/wireless/下的脚本生成Netifd识别的JSON结构。

以mac80211.sh为例，执行./mac80211.sh '' dump，输出的JSON内容如下（省略部分内容）。

```
DF1A:~$ ./mac80211.sh '' dump
#输出的JSON内容如下
{ "name": "mac80211", "device": [ [ "channel", 3 ], [ "hwmode", 3 ], [ "htmode", 3 ],
[ "basic_rate", 1 ], [ "supported_rates", 1 ], [ "country", 3 ], [ "country_ie", 7
], [ "doth", 7 ], [ "rd_disabled", 7 ] ...(省略) [ "dsss_cck_40", 7 ] ],"iface": [ [
"mode", 3 ], [ "ssid", 3 ], [ "encryption", 3 ],...(省略), [ "mesh_power_mode", 3 ]
] }
```

3. 接下来Netifd会根据输出的JSON内容，创建struct wireless_driver结构体。输出的JSON内容包括以下3项。

（1）name节点：协议的名称，如"mac80211"。

（2）device节点：无线设备的参数配置信息，相当于wireless配置文件中config wifi-device配置支持的参数列表。

（3）iface节点：无线接口的参数配置信息，相当于wireless配置文件中config wifi-iface配置支持的参数列表。

4. 解析完参数后，会把struct wireless_driver结构体中的struct avl_node node成员添加到无线网络全局avl_tree wireless_drivers平衡二叉树中。

5. 按照上面的步骤，循环把所有无线协议都解析完毕，形成一棵无线网络AVL树，如图13-39所示。

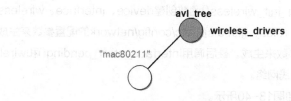

图13-39　无线网络AVL树

创建和监听NetLink消息

Netifd使用libnl与内核进行通信，接收Netlink消息。Netifd中监听了NETLINK_ROUTE和NETLINK_KOBJECT_UEVENT消息，注册两个回调函数cb_rtnl_event()和handle_hotplug_event()分别处理这两个消息。

● NETLINK_ROUTE

NETLINK_ROUTE消息分为多个消息簇：LINK（网络接口）、ADDR（网络地址）、ROUTE（路由选择消息）、NEIGH（邻接子系统消息）、RULE（策略路由规则）、QDISC（排队规则）、TCLASS（流量类）、ACTION（数据包操作）、NEIGHTBL（邻接表）、ADDRLABEL（地址标记）。每个消息簇都分为3类：RTM_NEWROUTE、RTM_DELROUTE、RTM_GETROUTE，分别用于创建消息、删除消息、检索信息。对于LINK簇，还包含修改链路消息类型的RTM_SETLINK。回调函数cb_rtnl_event()会根据内核传递过来的Netlink消息，执行下面的操作。

（1）取出消息传递的device名称ifname，从Netifd维护的device AVL树中，取出device，并向device注册的使用者（device_user和alias）广播DEV_EVENT_ADD 或DEV_EVENT_REMOVE消息。

（2）根据ifname，查看/sys/class/net/<ifname>/carrier。数值为1表示物理电缆与网卡插槽已经连接，数值为0表示物理电缆与网卡插槽断开连接。例如：

```
#表示WAN口已经插入网线
DF1A:~$ cat /sys/class/net/eth0.2/carrier
1
```

（3）如果上面第二步判断的物理连接值为1，并且device的link_active可用，则向所有device

的使用者（device_user和alias）广播 DEV_EVENT_LINK_UP 或DEV_EVENT_LINK_DOWN消息。

● NETLINK_KOBJECT_UEVENT

当内核检测到系统中出现新的设备后，会通过Netlink套接字发送uevent。Netifd只判断SUBSYSTEM的值是否为net，即是否只匹配网络子系统。根据interface传递的设备名称ifname，从Netifd维护的device AVL树中，取出device，向device注册的使用者广播消息。

初始化所有配置

初始化所有配置会把有线、无线网络的配置、接口、设备、协议等所有内容都结合起来，如创建设备、接口绑定设备、接口绑定协议等。这是整个Netifd中最为复杂和关键的内容。这部分主要集中在config_init_all()函数中，config_init_all()函数会依次调用config_init_package()、config_init_devices()、config_init_interfaces()、config_init_routes()、config_init_rules()、config_init_globals()、config_init_wireless()函数创建device、interface、wireless等结构。其中device、interface、route、rules、globals由/etc/config/network的配置参数来生成，wireless由/etc/config/wireles的配置参数来生成。最后调用interface_start_pending()和wireless_start_pending()函数来启动接口和无线网络。

初始化相关函数调用关系如图13-40所示。

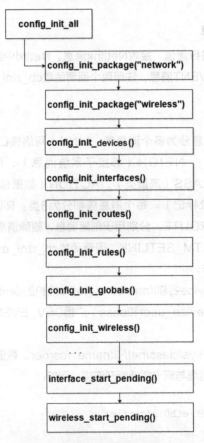

图13-40 初始化相关函数调用关系

接下来对主要过程进行说明。

1. 获得配置文件上下文

通过libuci API库封装函数config_init_package("network")和config_init_package("wireless")，获得network和wireless配置文件上下文，分别保存在变量uci_network和uci_wireless中，供后续过程解析。

2. 根据配置文件创建指定类型设备

调用config_init_devices()函数，根据network UCI配置文件中device节点，创建相应的device对象。函数会遍历uci_network缓存中的所有节点，查看是否含有"device"名称的节点，如果有这个节点，并且还包含type选项（设备类型），则根据指定的设备类型创建设备。设备类型可以分为以下几种。

● 8021ad：VLANDEV设备，支持的属性参数有type、ifname、vid。

● 8021q：VLANDEV设备，支持的属性参数有type、ifname、vid。

● bridge：桥设备，支持的属性参数有 ifname、stp、forward_delay、priority、ageing_time、hello_time、max_age、igmp_snooping、bridge_empty、multicast_querier、hash_max。

● macvlan：MACVLAN设备，支持的属性参数有ifname、macaddr、mode。

● tunnel：IP Tunnel设备，支持的属性参数有mode、local、remote、mtu、df、ttl、tos、6rd-prefix、6rd-relay-prefix、link、fmrs、info。

还有一种称为"Simple device"的设备类型，即Network设备，并不是直接由配置文件来指定的。以DF1A开发板为例，network配置文件中并没有device节点。一般情况下network配置文件中也很少出现device节点，所以这个函数并没有创建任何设备。

3. 初始化接口

调用config_init_interface()函数，根据network UCI配置文件中的interface节点信息创建对应的interface对象。

（1）函数会遍历uci_network缓存中所有interface节点，判断节点中是否存在option disabled选项，如果存在disabled选项并且其值为1，则直接退出本次遍历，否则调用interface_alloc()创建interface。判断如果存在option type bridge，则使用config_parse_bridge_interface()创建桥类型设备。

（2）函数会遍历uci_network缓存中所有alias节点，判断节点中是否存在option disabled选项，如果存在disabled选项并且其值为1，则直接退出本次遍历，否则调用interface_alloc()创建interface。

（3）调用interface_alloc()函数创建interface。接口的配置参数由uci_network提供。

（4）如果是alias类型接口，则调用interface_add_alias()函数，否则调用interface_add()函数，把新创建的interface添加到接口AVL树中。

4. 初始化route

调用config_init_routes()函数，根据network UCI配置文件中route节点配置，创建对应的route对象，并将route对象添加到接口结构成员的config_ip.route路由树中。

（1）函数会遍历uci_network缓存中所有route和route6节点，按照路由参数列表interface、

target、netmask、gateway、metric、mtu、table、valid、source、onlink、type，取出节点下面对应的选项值。参数列表如下。

```
#模板
config route或route6
        option interface 字符串类型
        option target 字符串类型
        option netmask 字符串类型
        option gateway 字符串类型
        option metric 整型
        option mtu 整型
        option table 字符串类型
        option valid 整型
        option onlink 布尔类型
        option type 字符串类型
```

配置示例如下。

```
config route
        option interface 'wan'
        option target '192.168.2.101'
        option netmask '255.255.255.0'
        option gateway '192.168.2.1'
        option metric '2'
        option mtu '1500'
```

（2）调用interface_ip_add_route()创建struct device_route实例，取出network UCI配置文件中设定的route选项参数，对device_route实例进行赋值。

（3）如果interface_ip_add_route()函数传递的接口iface为NULL，则把从UCI配置文件中取出的interface选项的值作为key，在接口AVL树interfaces中查找接口实例。找到后创建struct device_route对象，按照UCI配置值，设定 device_route信息，并把device_route添加到接口成员的struct interface_ip_settings config_ip中的route树中。

（4）更新/tmp/resolv.conf.auto文件。

5. 初始化rule

调用config_parse_rule()函数，根据network UCI配置文件中rule节点配置创建相应的iprule对象。

（1）函数会遍历uci_network缓存中所有rule和rule6节点，按照路由参数列表in、out、invert、src、dest、priority、tos、mark、lookup、action、goto，取出节点下面对应的选项值。参数列表如下。

```
#模板
config rule或rule6
        option in 字符串类型
        option out 字符串类型
        option invert 布尔类型
        option src 字符串类型
        option dest 整型
        option priority 整型
        option tos 整型
```

```
        option mark    字符串类型
        option lookup  字符串类型
        option action  字符串类型
        option goto    整型
```
配置示例如下。
```
config rule
        option mark    '0xFF'
        option in      'lan'
        option dest    '172.16.0.0/16'
        option lookup  '100'
config rule6
        option in      'vpn'
        option dest    'fdca:1234::/64'
        option action  'prohibit'
```
（2）调用iprule_add()，创建struct iprule实例，取出network UCI配置文件中设定的rule选项参数，对iprule实例进行赋值。

（3）分别取出UCI配置文件中的"in"和"out"选项的值作为key，在接口AVL树interfaces中查找接口实例。找到后取出接口实例中L3层设备的设备名称，分别赋值给iprule的in_dev和out_dev成员，同时设置iprule的flags |= IPRULE_IN 和flags |= IPRULE_OUT。

（4）把新创建的iprule节点添加到以iprules为根的AVL树中。

6. 初始化globals

配置文件中config globals节点包含独立于接口的选项，这些选项通常会影响网络的配置。这部分的解析是通过调用config_init_globals()函数来实现的，根据network UCI配置文件中globals节点的配置（只有一个），取出ula_prefix和default_ps选项。如果存在ula_prefix选项，则创建device_prefix实例，设定参数，添加到接口proto_ip.prefix AVL List Tree中。如果存在default_ps选项且值为1，则遍历设备AVL树devices，根据设备结构中成员settings.flags的标志DEV_OPT_RPS或DEV_OPT_XPS，来设置XPS和RPS优化参数。

XPS的全称是Transmit Packet Steering（发送数据包控制），RPS的全称是Receive Packet Steering（接收数据包控制）。

7. 初始化无线网络

无线网络配置的示例如下。
```
config wifi-device   radio0
        option type              mac80211
        option channel           auto
        option band              2.4G
        option max_all_num_sta   40
        option netisolate        0
        option country           'CN'
        option ht_coex           0
        option noscan            0
        option radio             1
        option txpower_lvl       '2'
        option path              '10000000.palmbus/11000000.wifi-lb'
        option htmode            HT20
```

```
        option hwmode      11g
config wifi-iface
        option device      radio0
        option ifname      wlan0
        option network     lan
        option mode        ap
        option ssid        SiWiFi-a124-2.4G
        option encryption  psk2+ccmp
        option key         12345678
        option isolate     '0'
        option hidden      '0'
        option macfilter   disable
        option macfile     /etc/wlan-file/wlan0.allow
        option group       1
        option netisolate  0
        option disable_input 0
        option wps_pushbutton '1'
        option wps_label   '0'
        option rps_cpus    2
```

（1）函数会遍历uci_wireless缓存中所有wifi-device节点，取出"type"选项值，例如mac80211。调用wireless_device_create()函数创建无线设备。

（2）在前面"无线网络初始化"中 ./mac80211.sh " dump输出的JSON内容的"device"和"iface"被保存在struct wireless_drivers结构实例的device和interface中。

```
struct wireless_driver {
        struct avl_node node;
        const char *name;
        const char *script;
        struct {
                char *buf;
                struct uci_blob_param_list *config;
        } device, interface;
};
```

（3）在AVL树wireless_drivers中，查找指定的节点，找到后取出struct wireless_driver结构中的device.config。

（4）创建的无线设备struct wireless_device结构实例把脚本解析得到的wireless_driver结构中的device.config赋值给struct wireless_device中的config成员。初始化其他wireless_device结构成员。

（5）将新创建的无线设备实例添加到全局无线AVL设备树wireless_devices中。

（6）函数会遍历uci_wireless缓存中所有wifi-iface节点，取出"device"选项值作为key，在无线AVL设备树wireless_devices中查找struct wireless_device，调用wireless_interface_create()函数创建无线接口。

（7）创建的无线设备struct wireless_interface结构实例把脚本解析得到的interface.config赋值给struct wireless_interface结构实例的config。

（8）将新创建的struct wireless_interface实例添加到struct wireless_device的struct vlist_ tree interfaces树中。

8. 启动接口

整个配置初始化结束后，开始启动所有接口。调用interface_start_pending()函数，遍历接口树interfaces中的所有接口实例，如果每个接口实例中available与autostart成员同时为true，则调用接口启动函数interface_set_up()，根据接口状态（interface_state）启动接口。

9. 启动无线网络

调用wireless_start_pending()函数启动无线网络。

（1）遍历无线设备AVL树wireless_devices，如果取出的无线设备实例的autostart成员为true，则设置状态为IFS_SETUP，调用__wireless_device_set_up()函数启动无线设备。

（2）调用wireless_device_run_handler()函数，执行具体的启动动作。启动无线设备，具体动作是通过构造命令参数，执行cfg80211配置脚本实现。以mac80211为例，构造参数如下。

无线设备结构体->drv->script 无线设备结构体->drv->name setup 无线设备结构体->name <配置>

构造好参数后，调用netifd_start_process()函数fork新进程执行构造的脚本。

以下是由无线网络启动执行脚本自动构造的脚本示例。

```
{mac80211.sh} /bin/sh ./mac80211.sh mac80211 setup radio0 {"config":{"channel":"auto","band":"2.4G","max_all_num_sta":40,"country":"CN","ht_coex":false,"noscan":false,"txpower_lvl":2,"path":"10000000.palmbus\/11000000.wifi-lb","htmode":"HT20","hwmode":"11g"},"data":{"phy":"phy0"},"interfaces":{"wlan0":{"bridge":"br-lan","config":{"ifname":"wlan0","mode":"ap","ssid":"SiWiFi-a124-2.4G","encryption":"psk2+ccmp","key":"12345678","isolate":false,"hidden":false,"macfilter":"disable","macfile":"\/etc\/wlan-file\/wlan0.allow","wps_pushbutton":true,"wps_label":false,"ifname":"wlan0","network":["lan"],"mode":"ap","isolate":false,"group":1,"netisolate":false,"disable_input":false,"rps_cpus":2}}}}
```

Netifd包含的内容众多，由于篇幅有限，本书没有对一些数据结构和细节进行详细讨论，也忽略了一些函数调用，希望读者通过本节能对Netifd和OpenWrt下的网络管理有一定的认识，这样本节的目的就达到了。至于更多的细节，请通过阅读源代码加深了解，这样能够体会更深刻。

Netifd是OpenWrt最核心的守护进程，调度着网络设备的运行。它把网络相关概念和具体设备抽象出来，和OpenWrt特有的ubus总线、UCI配置完美地结合起来，把UCI配置、接口、设备、协议等巧妙地绑定在一起，提供协议、无线设备控制的Shell扩展，灵活且优雅，是OpenWrt下最精彩的进程。

13.4 Hotplug原理

Hotplug就是我们通常所说的"热插拔"。当内核检测到系统增加或删除设备时，内核通过Netlink发送uevent，即产生一个"热插拔事件"。

Hotplug在OpenWrt下应用非常广泛，如常见网络接头的插拔、按键的按下与抬起、USB接头的插拔等事件。了解Hotplug的整个处理过程，对于OpenWrt系统的开发非常有帮助。

13.4.1 Hotplug架构

Hotplug架构整体可以分为以下两个部分。

（1）内核空间处理：内核通过设备模型（kobject）监控设备上报事件，通过Netlink通信机制,将NETLINK_KOBJECT_UEVENT事件从内核空间传递给用户空间。

(2)用户空间处理:在OpenWrt中,用户空间处理程序由procd来实现。procd创建了PF_NETLINK类型的Netlink套接字,监听内核NETLINK_KOBJECT_UEVENT消息,当接收到消息后,调用消息处理函数,进行消息预处理,然后使用预先设定的JSON匹配规则(/etc/hotplug.json文件)进行规则匹配,根据匹配结果分发给其他脚本处理。procd实现了Hotplug机制,而策略由其他脚本来实现。

Hotplug整体架构如图13-41所示。

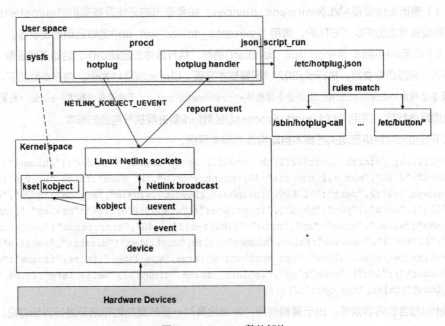

图13-41 Hotplug整体架构

13.4.2 Hotplug内核空间处理

Hotplug在内核中是通过kobject、uevent和Netlink实现的。

kobject

kobject是Linux内核设备模型的基础。kobject嵌入描述设备的结构体中,可以看作所有设备对象的基类。其内核源代码位于lib/kobject.c。

uevent

uevent是当一个kobject的状态变化时,广播出的一个对应的事件,如注册、注销等。

uevent来源于kobject的变化,kobject中预定义了一些kobject动作(源代码位于include/linux/kobject.h),用于描述kobject及基于kobject的上层结构的状态变化。定义如下:

```
enum kobject_action {
    KOBJ_ADD,
    KOBJ_REMOVE,
    KOBJ_CHANGE,
```

```
        KOBJ_MOVE,
        KOBJ_ONLINE,
        KOBJ_OFFLINE,
        KOBJ_MAX
};
```

这些动作在发送到用户空间时，通过字符串来表示，如"add""remove"，其源代码位于lib/kobject_uevent.c，对应关系如下。

```
/* the strings here must match the enum in include/linux/kobject.h */
static const char *kobject_actions[] = {
        [KOBJ_ADD] =            "add",
        [KOBJ_REMOVE] =         "remove",
        [KOBJ_CHANGE] =         "change",
        [KOBJ_MOVE] =           "move",
        [KOBJ_ONLINE] =         "online",
        [KOBJ_OFFLINE] =        "offline",
};
```

这些消息类型的含义如下。

- KOBJ_ADD：kobject（包含kobject结构的上层结构）上的添加事件。
- KOBJ_REMOVE：kobject（包含kobject结构的上层结构）上的删除事件。
- KOBJ_CHANGE：kobject（包含kobject结构的上层结构）上的内容或状态发生改变的事件。也包含设备自定义事件。
- KOBJ_MOVE：kobject（包含kobject结构的上层结构）上的更改名称或者更改Parent（如更改sysfs中目录结构）的事件。
- KOBJ_ONLINE：kobject（包含kobject结构的上层结构）的上线事件。
- KOBJ_OFFLINE：kobject（包含kobject结构的上层结构）的下线事件。

kset

kset是一组kobject的集合，相当于容器类，用于统一管理具有类似属性的kobject。只要把它们加入一个集合中即可，当一个事件发生时，可以同时通知集合中的所有kobject。

在Linux内核中预定义一些kset，则每个kset会具体实现自己的uevent预处理函数，由struct kset_uevent_ops 结构表示，这些kset_uevent_ops会在内核启动时在start_kernel()中注册。Linux-3.18.29内核中预定义的kset_uevent_ops有：dlm_uevent_ops、gfs2_uevent_ops、slab_uevent_ops、device_uevent_ops、bus_uevent_ops、module_uevent_ops。这些kset_uevent_ops会执行uevent filter、设置uevent name和添加kset特有的uevent环境变量等操作。

sysfs

内核空间与用户空间实现双向交互，可以通过sysfs文件系统来实现。sysfs本质上由kobject和kset实现。内核通过show()/store()方法来实现对用户空间数据的访问。用户空间只需要指明/sys中的某个目录，就可以通过cat/echo等命令实现对内核和驱动数据的读写，例如释放GPIO和控制GPIO端口的电平等。

sysfs下的很多kobject下都有uevent文件，这些uevent属性文件一般是可写的。例如/sys/

devices/*下的设备节点都有uevent文件，这些uevent文件都支持写入，当前支持写入的参数有"add""remove""change""move""online""offline"，与kobject定义的action对应。

```
DF1A:$ find /sys/ -type f -name uevent
...
/sys/devices/virtual/ppp/ppp/uevent
/sys/devices/virtual/tty/tty/uevent
/sys/devices/virtual/tty/ptmx/uevent
/sys/devices/virtual/tty/console/uevent
/sys/devices/virtual/misc/fuse/uevent
/sys/devices/virtual/misc/cpu_dma_latency/uevent
/sys/devices/virtual/misc/network_latency/uevent
/sys/devices/virtual/misc/memory_bandwidth/uevent
/sys/devices/virtual/misc/network_throughput/uevent
/sys/devices/virtual/workqueue/writeback/uevent
/sys/devices/virtual/workqueue/uevent
/sys/devices/platform/reg-dummy/regulator/regulator.0/uevent
/sys/devices/platform/reg-dummy/uevent
/sys/devices/platform/alarmtimer/uevent
/sys/devices/platform/cpufreq-dt/uevent
/sys/devices/platform/uevent
...
```

uevent内核实现

当kobject状态变化时，会把uevent消息通过Netlink的NETLINK_KOBJECT_UEVENT协议类型广播到用户空间程序中。uevent在内核中的实现代码为/lib/kobject_uevent.c。实现过程如下。

● 内核通过netlink_kernel_create()函数创建一个Netlink套接字，协议类型为NETLINK_KOBJECT_UEVENT。

● 当内核有事件通知时，会调用kobject_uevent()函数，内部会调用kobject_uevent_env()函数完成所有操作。

```
int kobject_uevent(struct kobject *kobj, enum kobject_action action)
int kobject_uevent_env(struct kobject *kobj, enum kobject_action action,char *envp_ext[])
```

● kobject_uevent_env()函数会执行下列操作。

（1）添加uevent的环境变量。如"ACTION""DEVPATH""SUBSYSTEM""SEQNUM"。

（2）查看事件所属kobject的kset是否需要处理uevent。取出kobject所属的kset，查看kset中是否注册了uevent处理函数，如果注册了就会执行对应处理。例如，对于设备类，会执行设备类(kset)的device_uevent_ops，额外添加"MAJOR""MINOR""DEVNAME"等环境变量。

（3）最后会调用netlink_broadcast_filtered()函数，把消息发送到用户空间程序。

13.4.3 Hotplug 用户空间处理

用户空间程序只要通过Netlink套接字注册NETLINK_KOBJECT_UEVENT协议类型消息，就可监听uevent消息。例如OpenWrt中Netifd和procd程序都对uevent消息进行了监听。

在OpenWrt中，Hotplug由用户空间程序procd来实现。作为系统的init进程，procd在整个系统启动过程中有两个阶段来处理内核uevent消息。

（1）peinit阶段：这个阶段是系统启动早期，通过"-h"参数来启用Hotplug机制，格式为"procd -h /etc/hotplug-preinit.json"，使用/etc/hotplug-preinit.json规则脚本来进行分发处理。这个阶段主要处理早期的系统通知，例如无线模块加载固件、failsafe模式等待按键输入等。这部分内容在13.1节中已经介绍过，这里就不再讨论。

（2）early阶段：procd作为系统init进程第二次被启动，这个阶段启动Hotplug机制不使用"-h"参数，而是直接调用hotplug("/etc/hotplug.json")，使用/etc/hotplug.json规则脚本来进行规则判断和分发处理。系统正常启动后，会一直保持这种方式等待内核uevent事件，可以认为接下来提到的Hotplug事件处理都是从这个阶段开始的。

procd对Hotplug的处理分为几个过程：注册、解析与执行、脚本分发、具体脚本执行。处理过程中会使用libubox库的json_script相关API，处理JSON和blogmsg数据。

注册

这部分主要的实现函数为hotplug_run()，函数内部执行流程如下。

（1）创建Netlink NETLINK_KOBJECT_UEVENT协议套接字，绑定套接字和设定套接字接收缓冲区大小。返回socket描述符为hotplug_fd。

（2）初始化uevent消息处理handle结构。定义struct json_script_ctx结构实例jctx，用于处理uevent消息。jctx注册4个规则处理handle函数，其中rule_handle_command是消息处理的核心函数。

```
static struct json_script_ctx jctx = {
#主要针对uevent变量DEVPATH进行处理
        .handle_var = rule_handle_var,
#处理过程中出现错误时,把uevent blogmsg格式转换为JSON,主要用于调试
        .handle_error = rule_handle_error,
#根据规则处理分发请求
        .handle_command = rule_handle_command,
#加载/etc/hotplug.json脚本
        .handle_file = rule_handle_file,
};
```

（3）设置hotplug_fd处理的回调函数hotplug_handler()，并把hotplug_fd添加到系统uloop监控中，等待接收uevent消息。uloop内部通过epoll对fd进行监控。

解析与执行

当内核广播uevent消息时，hotplug_fd上返回可读信息，uloop会调用fd上注册的回调函数hotplug_handler()，在函数内处理内核uevent消息。回调函数内部执行流程如下。

（1）消息转换：接收到uevent消息，将其转换为blogmsg消息，主要是为了方便后续的处理。例如"ACTION=XXX"消息，转换时会把"="替换为"\0"，然后把ACTION和XXX分别作为key和value按照blogmsg格式存放。

（2）规则解析与匹配：加载/etc/hotplug.json规则脚本文件，使用libubox库的JSON script API对JSON文件进行解析，通过预先定义的json_handler规则，对JSON文件进行语法规则匹配，支持的JSON语法命令解析规则有if、case、return、include，条件规则有eq、regex、 has、and、or、not。这些规则与/etc/hotplug.json中的描述语法规则匹配。uevent与hotplug.json匹配过程如图13-42所示。

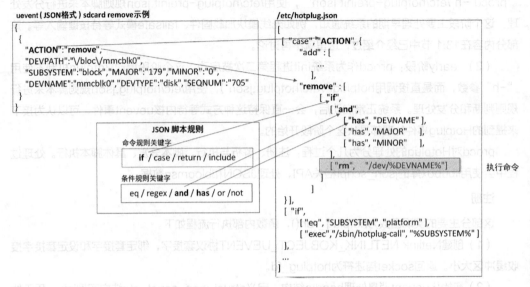

图13-42 uevent与hotplug.json匹配过程

（3）执行命令。根据语法规则条件，找到最终要执行的命令字符串，如"rm"，执行jctx对象注册的rule_handle_var()函数对参数做一些处理，然后调用jctx对象的rule_handle_command()函数执行具体的命令处理。rule_handle_command预先定义了命令处理结构体struct cmd_handler，用于处理注册的命令。结构体如下。

```
static struct cmd_handler {
    char *name;
    int atomic;
    void (*handler)(struct blob_attr *msg, struct blob_attr *data);
    void (*start)(struct blob_attr *msg, struct blob_attr *data);
    void (*complete)(struct blob_attr *msg, struct blob_attr *data, int ret);
}
```

● name：命令名称，与/etc/hotplug.json中的命令名称匹配。

● atomic：是否需要队列运行命令。1是不需要队列运行，0是需要队列运行。队列运行是指通过fork子进程来执行命令。

● handler函数指针：命令对应的执行函数。

● start函数指针：可选函数，表示命令的开始，需要队列运行时使用，用于定时任务。

● complete函数指针：可选函数，表示命令的完成，需要队列运行时使用，用于定时任务。

procd中预先定义了5种命令，分别为"makedev""rm""exec""button""load-firmware"。这5种命令的说明如表13-23所示。根据JSON解析的命令名称，与注册的cmd_handler中的名字进行匹配，如果匹配成功，则执行对应的handler。

表 13-23 Hotplug 命令说明

命令名称	ACTION	uevent 环境变量	命令说明
makedev	add、remove	ACTION、MINOR、MAJOR、DEVNAME、SUBSYSTEM、DEVPATH等	创建/dev/%DEVNAME%设备节点
rm	remove	ACTION，如果是设备节点，需要DEVNAME、MAJOR、MINOR等	删除文件
exec		ACTION、BUTTON、SUBSYTEM、DEVNAME、DEVPATH、INTERFACE等	执行外部程序
button	released、pressed	ACTION、BUTTON等	用于按键程序。ACTION包含按键pressed或released。procd添加SEEN变量，表示按键被按下的时间（秒）
load-firmware		ACTION、FIRMWARE、DEVPATH等	加载固件。先向/sys/%DEVPATH%/loading写入1，再向/sys/%DEVPATH%/data中写入数据，然后向/sys/%DEVPATH%/loading写入0

如果已知uevent对应的kobject节点在sysfs中对应的目录，可以通过cat /sys/*/uevent查看uevent环境变量。

脚本分发

在/etc/hotplug.json中匹配到对应命令后，分成3种方式处理：

（1）直接调用对应的cmd handler命令处理函数，例如"makedev""rm""Load-firmware"。

（2）对于按键，Hotplug脚本在/etc/rc.button/目录下，根据uevent传递的BUTTON变量执行对应的按键脚本，例如reset按键对应的/etc/rc.button/reset。

（3）调用/sbin/hotplug-call来执行外部脚本。/sbin/hotplug-call脚本内容如下。

```
#!/bin/sh
# Copyright (C) 2006-2010 OpenWrt.org
export HOTPLUG_TYPE="$1"
. /lib/functions.sh
PATH=/usr/sbin:/usr/bin:/sbin:/bin
LOGNAME=root
USER=root
export PATH LOGNAME USER
export DEVICENAME="${DEVPATH##*/}"

[ \! -z "$1" -a -d /etc/hotplug.d/$1 ] && {
    for script in $(ls /etc/hotplug.d/$1/* 2>&-); do (
        [ -f $script ] && . $script
    ); done
}
```

这个脚本会根据传递的参数调用/etc/hotplug.d子目录中的脚本，判断/etc/hotplug.d/中是否存在对应的子目录，如果存在则依次执行/etc/hotplug.d/子目录下所有的脚本。支持的目录有：net、input、usb、usbmisc、ieee1394、block、atm、zaptel、tty、button、platfor。注意"/etc/hotplug.d/iface"下的脚本并不是由procd执行，而是由Netifd来执行。

具体脚本的执行

根据功能编写对应的脚本，放在指定的目录中即可，Hotplug会自动传递参数和执行脚本。软件包如果需要实现Hotplug脚本，只需要在/etc/rc.button或/etc/hotplug.d对应的文件或目录中创建具体功能的Hotplug脚本即可。例如block-mount软件包提供的自动挂载磁盘的脚本/etc/hotplug.d/block/10-mount。

13.4.4 Hotplug 测试方法

procd默认的打印级为LEVEL 1，小于LEVEL 3级别的日志都不会输出到日志文件中，如果要让procd输出Hotplug信息，需要把日志打印级别调整为大于LEVEL 3。可通过设置环境变量DBGLVL来调整。

procd默认作为init进程已经启动，日志打印级别按照默认LEVEL 1设置，不方便修改。为了调试uevent信息，可以再启动一个procd作为Hotplug监控程序。

打开新终端启动procd。

```
DF1A:$ export DBGLVL=4;procd -h /etc/hotplug.json
```

当有事件发生时，可以通过demsg或logread查看日志。

```
DF1A:$ dmesg
DF1A:$ logread
```

模拟uevent事件

前面说到sysfs文件系统目录下，很多设备或总线都有uevent文件，可以通过写入uevent文件对应的方法来模拟uevent消息，当有uevent消息时，内核会将其广播到用户空间中监听Netlink消息的进程。

下面以向/sys/devices/pinctrl/uevent写入"add"事件为例来讲解。

1. 新建一个终端启动procd，对Hotplug事件进行监控。

```
DF1A:$ export DBGLVL=4;procd -h /etc/hotplug.json
```

2. 创建/etc/hotplug.d/platform/00-demo文件。因为pinctrl的SUBSYSTEM=platorm，所以会执行/etc/hotplug.d/platform/下的脚本。

```
DF1A:$ mkdir -p /etc/hotplug.d/platform
DF1A:$ touch 00-demo
```

脚本内容如下。

```
#!/bin/sh
. /lib/functions.sh
logger "[Hotplug Platorm ] $0 $1 $ACTION"
echo -e "\n[Hotplug Platorm ] $0 $1 $ACTION" > /dev/console
return 0
```

3. 新建终端，执行"add"操作。非串口终端不能显示输出到/dev/console的信息。

```
DF1A:$ echo "add" > /sys/devices/pinctrl/uevent
```

4. 查看日志。

控制台输出（仅串口终端输出）如下。

```
DF1A:$ echo "add" > /sys/devices/pinctrl/uevent
```

```
[Hotplug Platorm ] /sbin/hotplug-call platform add
[Hotplug Platorm ] /sbin/hotplug-call platform add
```

输出两次是因为作为init进程的procd也收到了Netlink消息，也会调用这个脚本。

在demsg中可以查看procd的完整调试信息。

```
DF1A:$ dmesg
...
[ 7339.941240] procd: {{"ACTION":"add","DEVPATH":"\/devices\/pinctrl","SUBSYST
EM":"platform","DRIVER":"pinctrl-sfax8","OF_NAME":"pinctrl","OF_FULLNAME":"\/
pinctrl","OF_COMPATIBLE_0":"siflower,sfax8-pinctrl","OF_COMPATIBLE_N":"1","MODALIAS
":"of:NpinctrlT<NULL>Csiflower,sfax8-pinctrl","SEQNUM":"620"}}
[ 7339.943901] procd: Command: execprocd: /sbin/hotplug-callprocd: platformprocd:
[ 7339.944336] procd: Message:procd: ACTION=addprocd: DEVPATH=/devices/
pinctrlprocd: SUBSYSTEM=platformprocd: DRIVER=pinctrl-sfax8procd: OF_
NAME=pinctrlprocd: OF_FULLNAME=/pinctrlprocd: OF_COMPATIBLE_0=siflower,sfax8-
pinctrlprocd: OF_COMPATIBLE_N=1procd: MODALIAS=of:NpinctrlT<NULL>Csiflower,sfax8-
pinctrlprocd: SEQNUM=620procd:
[ 7339.946232] procd: Launched hotplug exec instance, pid=2997
[ 7339.979204] procd: Finished hotplug exec instance, pid=2997
```

用logread查看输出信息。

```
DF1A:$ logread
...
Fri Dec  6 13:09:05 2019 user.debug kernel: [ 7339.941240] procd:
{{"ACTION":"add","DEVPATH":"\/devices\/pinctrl","SUBSYSTEM":"platform","D
RIVER":"pinctrl-sfax8","OF_NAME":"pinctrl","OF_FULLNAME":"\/pinctrl","OF_
COMPATIBLE_0":"siflower,sfax8-pinctrl","OF_COMPATIBLE_N":"1","MODALIAS":"of:Npinc
trlT<NULL>Csiflower,sfax8-pinctrl","SEQNUM":"620"}}
Fri Dec  6 13:09:05 2019 user.debug kernel: [ 7339.943901] procd: Command:
execprocd: /sbin/hotplug-callprocd: platformprocd:
Fri Dec  6 13:09:05 2019 user.debug kernel: [ 7339.944336] procd: Message:procd:
ACTION=addprocd: DEVPATH=/devices/pinctrlprocd: SUBSYSTEM=platformprocd:
DRIVER=pinctrl-sfax8procd: OF_NAME=pinctrlprocd: OF_FULLNAME=/pinctrlprocd: OF_
COMPATIBLE_0=siflower,sfax8-pinctrlprocd: OF_COMPATIBLE_N=1procd: MODALIAS=of:Npi
nctrlT<NULL>Csiflower,sfax8-pinctrlprocd: SEQNUM=620procd:
Fri Dec  6 13:09:05 2019 user.debug kernel: [ 7339.946232] procd: Launched hotplug
exec instance, pid=2997
Fri Dec  6 13:09:05 2019 user.notice admin: [Hotplug Platorm ] /sbin/hotplug-call
platform add
Fri Dec  6 13:09:05 2019 user.notice admin: [Hotplug Platorm ] /sbin/hotplug-call
platform add
Fri Dec  6 13:09:05 2019 user.debug kernel: [ 7339.979204] procd: Finished hotplug
exec instance, pid=2997
```

13.4.5 Hotplug 实例

USB设备是比较常用的外部设备，通常用于扩展系统存储。接下来介绍如何利用Hotplug实现U盘、USB移动硬盘等外部设备自动挂载的原理和方法。

约束与说明

（1）本文中涉及的实例需要根据实际情况作相应的适配，本实例仅作演示。

（2）如果已经安装了block-mount软件包，这个软件包会包含/etc/hotplug.d/block/10-mount脚本，该脚本会执行自动挂载等操作。为了避免冲突，我们暂时把该脚本移动到上一级目录/etc/hotplug.d中，本实例演示完毕后，再把文件移回/etc/hotplug.d/block/目录。

（3）本实例中USB设备为U盘，内部仅有一个分区。

Hotplug开发

当USB设备插入/拔出时，内核USB总线会匹配USB设备和驱动程序，内核会产生uevent事件并通过Netlink进行广播，应用层Hotplug收到这个内核广播事件后，会根据uevent事件信息内容，在hotplug.json中进行匹配。

1. 查看uevent信息

参考前面章节中的测试方法，当开发板中插入U盘后，得到的uevent信息和procd执行的命令如下所示（具体输出不同，下列内容只作为过程参考）。

```
{{"ACTION":"add","DEVPATH":"\/devices\/10000000.palmbus\/17000000.usb\/usb1\/1-1","SUBSYSTEM":"usb","MAJOR":"189","MINOR":"39","DEVNAME":"bus\/usb\/001\/040","DEVTYPE":"usb_device","PRODUCT":"14cd\/8168\/201","TYPE":"0\/0\/0","BUSNUM":"001","DEVNUM":"040","SEQNUM":"1014"}}
procd: Command: makedev  /dev/bus/usb/001/040 0644
procd: Command: exec /sbin/hotplug-call usb
{{"ACTION":"add","DEVPATH":"\/devices\/10000000.palmbus\/17000000.usb\/usb1\/1-1\/1-1:1.0","SUBSYSTEM":"usb","DEVTYPE":"usb_interface","PRODUCT":"14cd\/8168\/201","TYPE":"0\/0\/0","INTERFACE":"8\/6\/80","MODALIAS":"usb:v14CDp8168d0201dc00dsc00dp00ic08isc06ip50in00","SEQNUM":"1015"}}
procd: Command: exec /sbin/hotplug-call usb
{{"ACTION":"add","DEVPATH":"\/devices\/10000000.palmbus\/17000000.usb\/usb1\/1-1\/1-1:1.0\/host18","SUBSYSTEM":"scsi","DEVTYPE":"scsi_host","SEQNUM":"1016"}}
{{"ACTION":"add","DEVPATH":"\/devices\/10000000.palmbus\/17000000.usb\/usb1\/1-1\/1-1:1.0\/host18\/scsi_host\/host18","SUBSYSTEM":"scsi_host","SEQNUM":"1017"}}
{{"ACTION":"add","DEVPATH":"\/devices\/10000000.palmbus\/17000000.usb\/usb1\/1-1\/1-1:1.0\/host18\/target18:0:0","SUBSYSTEM":"scsi","DEVTYPE":"scsi_target","SEQNUM":"1018"}}
{{"ACTION":"add","DEVPATH":"\/devices\/10000000.palmbus\/17000000.usb\/usb1\/1-1\/1-1:1.0\/host18\/target18:0:0\/18:0:0:0","SUBSYSTEM":"scsi","DEVTYPE":"scsi_device","MODALIAS":"scsi:t-0x00","SEQNUM":"1019"}}
{{"ACTION":"add","DEVPATH":"\/devices\/10000000.palmbus\/17000000.usb\/usb1\/1-1\/1-1:1.0\/host18\/target18:0:0\/18:0:0:0\/scsi_disk\/18:0:0:0","SUBSYSTEM":"scsi_disk","SEQNUM":"1020"}}
{{"ACTION":"add","DEVPATH":"\/devices\/10000000.palmbus\/17000000.usb\/usb1\/1-1\/1-1:1.0\/host18\/target18:0:0\/18:0:0:0\/scsi_device\/18:0:0:0","SUBSYSTEM":"scsi_device","SEQNUM":"1021"}}
{{"ACTION":"add","DEVPATH":"\/devices\/10000000.palmbus\/17000000.usb\/usb1\/1-1\/1-1:1.0\/host18\/target18:0:0\/18:0:0:0\/bsg\/18:0:0:0","SUBSYSTEM":"bsg","MAJOR":"250","MINOR":"0","DEVNAME":"bsg\/18:0:0:0","SEQNUM":"1022"}}
procd: Command: makedev /dev/bsg/18:0:0:0 0644
{{"ACTION":"add","DEVPATH":"\/block\/sdb","SUBSYSTEM":"block","MAJOR":"8","MINOR":"16","DEVNAME":"sdb","DEVTYPE":"disk","SEQNUM":"1024"}}
procd: Command: makedev /dev/sdb 0644
```

```
procd: Command: exec /sbin/hotplug-call block
{{"ACTION":"add","DEVPATH":"\/block\/sdb\/sdb1","SUBSYSTEM":"block","MAJOR":"8","M
INOR":"17","DEVNAME":"sdb1","DEVTYPE":"partition","SEQNUM":"1025"}}
procd: Command: makedev/dev/sdb1 0644
procd: Command: exec/sbin/hotplug-call block
```

当U盘被拔出时，得到的uevent信息和procd执行的命令如下所示（具体输出不同，下列内容只作为过程参考）。

```
{{"ACTION":"remove","DEVPATH":"\/devices\/10000000.palmbus\/17000000.usb\/usb1\/1-
1\/1-1:1.0\/host18\/target18:0:0\/18:0:0:0\/bsg\/18:0:0:0","SUBSYSTEM":"bsg","MAJO
R":"250","MINOR":"0","DEVNAME":"bsg\/18:0:0:0","SEQNUM":"1026"}}
procd: Command: rm /dev/bsg/18:0:0:0
{{"ACTION":"remove","DEVPATH":"\/devices\/10000000.palmbus\/17000000.usb\/usb1\/1-
1\/1-1:1.0\/host18\/target18:0:0\/18:0:0:0\/scsi_device\/18:0:0:0","SUBSYSTEM":"sc
si_device","SEQNUM":"1027"}}
{{"ACTION":"remove","DEVPATH":"\/devices\/10000000.palmbus\/17000000.usb\/usb1\/1-
1\/1-1:1.0\/host18\/target18:0:0\/18:0:0:0\/scsi_disk\/18:0:0:0","SUBSYSTEM":"sc
si_disk","SEQNUM":"1028"}}
{{"ACTION":"remove","DEVPATH":"\/block\/sdb\/sdb1","SUBSYSTEM":"block","MAJOR":"8"
,"MINOR":"17","DEVNAME":"sdb1","DEVTYPE":"partition","SEQNUM":"1029"}}
procd: Command: rmp /dev/sdb1
procd: Command: exec /sbin/hotplug-call block
{{"ACTION":"remove","DEVPATH":"\/devices\/virtual\/bdi\/8:16","SUBSYSTEM":"bdi","S
EQNUM":"1030"}}
{{"ACTION":"remove","DEVPATH":"\/block\/sdb","SUBSYSTEM":"block","MAJOR":"8","MINO
R":"16","DEVNAME":"sdb","DEVTYPE":"disk","SEQNUM":"1031"}}
procd: Command: rm /dev/sdb
procd: Command: exec /sbin/hotplug-call block
{{"ACTION":"remove","DEVPATH":"\/devices\/10000000.palmbus\/17000000.usb\/usb1\/1-
1\/1-1:1.0\/host18\/target18:0:0\/18:0:0:0","SUBSYSTEM":"scsi","DEVTYPE":"scsi_
device","MODALIAS":"scsi:t-0x00","SEQNUM":"1032"}}
{{"ACTION":"remove","DEVPATH":"\/devices\/10000000.palmbus\/17000000.usb\/
usb1\/1-1\/1-1:1.0\/host18\/target18:0:0","SUBSYSTEM":"scsi","DEVTYPE":"scsi_
target","SEQNUM":"1033"}}
{{"ACTION":"remove","DEVPATH":"\/devices\/10000000.palmbus\/17000000.usb\/usb1\/1-
1\/1-1:1.0\/host18\/scsi_host\/host18","SUBSYSTEM":"scsi_host","SEQNUM":"1034"}}
{{"ACTION":"remove","DEVPATH":"\/devices\/10000000.palmbus\/17000000.usb\/usb1\/1-
1\/1-1:1.0\/host18","SUBSYSTEM":"scsi","DEVTYPE":"scsi_host","SEQNUM":"1035"}}
{{"ACTION":"remove","DEVPATH":"\/devices\/10000000.palmbus\/17000000.usb\/usb1\/1-
1\/1-1:1.0","SUBSYSTEM":"usb","DEVTYPE":"usb_interface","PRODUCT":"14cd\/8168\/201
","TYPE":"0\/0\/0","INTERFACE":"8\/6\/80","MODALIAS":"usb:v14CDp8168d0201dc00dsc00
dp00ic08isc06ip50in00","SEQNUM":"1036"}}
procd: Command: exec /sbin/hotplug-call usb
{{"ACTION":"remove","DEVPATH":"\/devices\/10000000.palmbus\/17000000.usb\/
usb1\/1-1","SUBSYSTEM":"usb","MAJOR":"189","MINOR":"39","DEVNAME":"bus\/
usb\/001\/040","DEVTYPE":"usb_device","PRODUCT":"14cd\/8168\/201","TYPE":"0\/0\/0"
,"BUSNUM":"001","DEVNUM":"040","SEQNUM":"1037"}}
procd: Command: rm /dev/bus/usb/001/040
procd: Command: exec /sbin/hotplug-call usb
```

2. 匹配和执行命令

根据uevent的信息，会在hotplug.json中找到3条匹配命令。

（1）匹配makedev命令。该命令仅在ACTION="add"时执行，该命令会创建对应的设备节点。例如插入USB设备时，会创建"/dev/bus/usb/001/040"" /dev/bsg/18:0:0:0""/dev/sdb""/dev/sdb1"设备节点，掩码为0644。

（2）匹配rm命令。该命令仅在ACTION="remove"时执行，命令会删除对应的设备节点。例如USB设备被拔出时，会删除"/dev/bsg/18:0:0:0""/dev/sdb""/dev/sdb1"" /dev/bus/usb/001/040"设备节点。

```
[
    [ "case", "ACTION", {
        "add": [
            [ "if",
                [ "and",
                    [ "has", "MAJOR" ],
                    [ "has", "MINOR" ],
                ],
                [
                    ...
                    [ "if",
                        [ "has", "DEVNAME" ],
                        [ "makedev", "/dev/%DEVNAME%", "0644" ]
                    ],
                ],
            ],
            ...
        ],
        "remove" : [
            [ "if",
                [ "and",
                    [ "has", "DEVNAME" ],
                    [ "has", "MAJOR" ],
                    [ "has", "MINOR" ],
                ],
                [ "rm", "/dev/%DEVNAME%" ]
            ]
        ]
    } ],
```

（3）根据uevent中的"SUBSYSTEM"信息，匹配"block"和"usb"，执行exec命令，hotplug-call会执行/etc/hotplug.d/block和/etc/hotplug.d/usb目录下的所有脚本，传递ACTION、DEVPATH、SUBSYSTEM、DEVNAME、DEVTYPE等变量。

```
[
    ...
    [ "if",
        [ "eq", "SUBSYSTEM",
            [ "net", "input", "usb", "usbmisc", "ieee1394", "block",
"atm", "zaptel", "tty", "button" ]
```

```
            ],
            [ "exec", "/sbin/hotplug-call", "%SUBSYSTEM%" ]
        ],
...
]
```

3. 编写Hotplug脚本

由于在本示例中用到的是USB存储设备,所以可以在block目录放置自动挂载/卸载处理脚本。

编写/etc/hotplug.d/block/01-mount-hotplug(如果相同目录下有10-mount,可以先将其移除),内容如下。

```
#!/bin/sh

logger -t block-hotplug $DEVPATH $ACTION $DEVNAME $DEVTYPE

case "$ACTION" in
        add)
                [ "$DEVTYPE" = "partition" ] && {
                        echo "$DEVNAME" | grep 'sd[a-z][1-9]' || exit 0
                        test -d /mnt/$DEVNAME || mkdir /mnt/$DEVNAME
                        mount -o iocharset=utf8,rw /dev/$DEVNAME /mnt/$DEVNAME ||
mount -o rw /dev/$DEVNAME /mnt/$DEVNAME
                }
                ;;
        remove)
                [ "$DEVTYPE" = "partition" ] && {
                        sync
                        echo "$DEVNAME" | grep 'sd[a-z][1-9]' || exit 0
                        umount /mnt/$DEVNAME && rmdir /mnt/$DEVNAME
                }
                ;;
esac
```

设置脚本权限。

```
DF1A:$ chmod +x /etc/hotplug.d/block/01-mount-hotplug
```

4. 插入U盘进行测试

系统中自动创建了设备节点。

```
DF1A:$ ls -l /dev/sda*
brw-r--r--    1 admin    root        8,   0 Sep 11 18:56 /dev/sda
brw-r--r--    1 admin    root        8,   1 Sep 11 18:56 /dev/sda1
```

用logread读取日志,查看传递参数,ACTION=add。

```
Wed Sep 11 19:01:10 2019 user.notice block-hotplug: /block/sda add sda disk
Wed Sep 11 19:01:10 2019 user.notice block-hotplug: /block/sda/sda1 add sda1
partition
```

仅/dev/sda1分区会自动挂载。

```
DF1A:$ mount|grep sda1
/dev/sda1 on /mnt/sda1 type fuseblk (rw,nosuid,nodev,relatime,user_id=0,group_
id=0,allow_other,blksize=4096)
```

5. 拔掉U盘进行测试

挂载的分区会被卸载。

```
DF1A:$ mount|grep sda1
```

用logread读取日志，查看传递参数，ACTION=remove。

```
Wed Sep 11 19:01:07 2019 user.notice block-hotplug: /block/sda/sda1 remove sda1 partition
Wed Sep 11 19:01:07 2019 user.notice block-hotplug: /block/sda remove sda disk
```

设备节点被删除。

```
DF1A:$ ls /dev/sda*
ls: /dev/sda*: No such file or directory
```

14 扩展与实战

14.1 PHP/Python开发环境

14.1.1 PHP开发环境

在x86的服务器上，很多应用系统或是网站都是基于LAMP(Linux + Apache + MySQL + PHP)开发的。在本开发板中也可以搭建出一套类似的网站或Web应用的开发平台，本章介绍采用Lighttpd+PHP+SQLite来完成这一系统搭建。

PHP的中文名字为"超文本预处理器"(Hypertext Preprocessor)，它拥有大量的开源库，其特点主要有：运行快、学习简单、库资源丰富。一般情况下，具备任何一门编程语言背景的朋友只要花费1~2个小时便能入门，在一周时间内基本可以掌握开发技巧。在OpenWrt系统下，我们可以采用FastCGI技术让PHP和Lighttpd配合工作，其性能和内存资源占用率都非常均衡。

Lighttpd是一个德国人发起的开源Web服务器软件，其特点是高性能、安全、快速、兼容性好、内存开销低、CPU占用率低等。Lighttpd是所有轻量级Web技术中最优秀的服务器软件之一。它支持FastCGI、CGI、Auth、Compress、Rewrite、Alias等必备的Web服务器技术。PHP采用FastCGI模式与Lighttpd连接，采用在内存中驻留一定数量的进程等待请求的方式，在消耗一定内存的情况下，降低了CPU占用率，提升了性能。

SQLite是一款轻量级的关系数据库，它的设计目标就是为嵌入式而生，占用系统资源非常少，甚至连进程都没有，且数据结构的敏感程度也非常低，但是它的语法与其他流行的关系数据库（比如MySQL）是非常接近的，因此很适合用在OpenWrt系统中。由繁入简，不论采用何种数据库技术，最终归根结底都是要对I/O进行操作，当一种技术能最直接、最有效地操作I/O时，才是最需要的技术，才最适合嵌入式应用，SQLite就是这样的技术。

关于SQLite的常见误区如下。

（1）SQLite也有服务：SQLite没有服务，SQLite总是嵌入各种编程语言当中，不需要连接到所谓的数据库服务。

（2）SQLite的数据库创建：确实可以在SQLite中创建数据库，但事实上那只是个文件，SQLite都是直接针对文件操作的。

（3）SQLite速度非常慢：这是一个典型的误区，SQLite采用I/O直接操作，因此是迄今性能最高的数据库之一,SQLite的缺点在于访问的同时性,因为I/O操作存在文件锁的问题。

(4) SQLite文件锁:在SQLite中可以通过特定操作解决这个问题,就是把多个表分离成不同的库(文件)。

(5) SQLite的数据稳定性:SQLite在数据稳定性方面要比传统数据库更强,尤其是具有断电异常时的保护机制。

PHP的安装

查看已编译的PHP模块情况。

```
DF1A:$ opkg update
DF1A:$ opkg find php5-*
```

安装PHP,可以根据需要安装模块。

```
DF1A:$ opkg install php5-cli php5-cgi php5-fastcgi php5-mod-session php5-mod-sqlite3
DF1A:$ opkg install php5-mod-ctype php5-mod-gd php5-mod-sockets php5-mod-mcrypt
DF1A:$ opkg install php5-mod-mbstring php5-mod-curl
```

测试PHP是否已安装。

```
DF1A:$ php-cli --version
PHP 5.6.17 (cli) (built: Nov 29 2019 09:31:31)
Copyright (c) 1997-2015 The PHP Group
Zend Engine v2.6.0, Copyright (c) 1998-2015 Zend Technologies
```

PHP在OpenWrt下的主要解析器有以下3个。

- php-cli:在命令行模式下执行PHP程序的解析器。
- php-cgi:在CGI模式下执行PHP程序的解析器。
- php-fcgi:在FastCGI模式下执行PHP程序的解析器。

Lighttpd安装

查看已编译的Lighttpd模块情况。

```
DF1A:$ opkg update
DF1A:$ opkg list lighttpd-*
```

安装Lighttpd常见软件包。

```
DF1A:$ opkg install lighttpd lighttpd-mod-fastcgi lighttpd-mod-access lighttpd-mod-alias
DF1A:$ opkg install lighttpd-mod-redirect lighttpd-mod-rewrite lighttpd-mod-evasive
```

配置Lighttpd+PHP

准备Web文件存放路径(/www已经被uhttpd占用,因此使用/www2):

```
#如果当前设备启用了TF卡作存储器,也可以相应地使用TF卡的目录作为存放位置
DF1A:$ mkdir -p /www2
```

/etc/lighttpd/lighttpd.conf配置文件参数说明如表14-1所示。

表 14-1　/etc/lighttpd/lighttpd.conf 配置文件参数说明

参数	说明	可选值及说明
server.modules	启用哪些Lighttpd模块	已安装模块名称
server.document-root	设置www主目录	目录路径
server.errorlog	设置错误信息文件路径，如果不设置这个参数就表示禁用	文件名路径
server.port	设置Web端口，默认为80	端口，数值
index-file.names	设置索引主文件名称	字符串
static-file.execute-extesions	设置可执行文件的名称	字符串
server.pid-file	设置Lighttpd服务的pid文件存放名称	文件名
server.upload-dirs	设置临时Web上传文件位置	目录路径
evasive.max-conns-per-ip	设置同时允许多少个请求一起发给Lighttpd，这个参数是mod_evasive模块提供的	字符串

在FastCGI模式下，在lighttpd.conf中可直接配置PHP的FastCGI模式参数，如表14-2所示。

表 14-2　配置 FastCGI 模式的参数说明

参数	说明	可选值及说明
bin-path	FastCGI解析器，就是/usr/bin/php-fcgi	字符串
bin-environment	Lighttpd输出的环境变量，在PHP下可以接收。 PHP_FCGI_CHILDREN：产生进程数量 PHP_FCGI_MAX_REQUESTS：最大请求	字符串
min-procs	最小进程数量	数值
max-procs	最大进程数量	数值
idle-timeout	空闲超时时间	数值，单位为秒

将/etc/lighttpd/lighttpd.conf修改为下列内容（端口使用8080，避免与uhttpd冲突）。

```
server.modules = (
        "mod_access",
        "mod_alias",
        "mod_redirect",
        "mod_rewrite",
        "mod_fastcgi",
        "mod_evasive"
)
server.document-root = "/www2"
server.errorlog = "/var/log/lighttpd/error.log"
server.port = 8080
index-file.names = ( "index.php", "index.html", "index.htm" )
static-file.exclude-extensions = ( ".php" )
server.pid-file = "/var/run/lighttpd.pid"
server.upload-dirs = ( "/tmp" )
evasive.max-conns-per-ip = 18
fastcgi.server = ( ".php" =>
  ((
      "socket" => "/tmp/php-fastcgi.socket",
      "bin-path" => "/usr/bin/php-fcgi",
      "bin-environment" => (
          "PHP_FCGI_CHILDREN" => "2",
```

```
        "PHP_FCGI_MAX_REQUESTS" => "100"
    ),
    "min-procs" => 1,
    "max-procs" => 2,
    "idle-timeout" => 20
))
)
```

编辑/etc/php.ini文件。

```
#将doc_root修改为如下表达方式,禁用该参数,防止PHP程序无法被访问
;doc_root = "/www"
```

启动lighttpd服务。

```
#设置启动服务
DF1A:$ /etc/init.d/lighttpd enable
#手动启动
DF1A:$ /etc/init.d/lighttpd start
```

示例：环境检测

创建/www2/info.php文件。

```
<?php
phpinfo();
?>
```

通过浏览器访问该文件，效果如图14-1所示。

PHP Version 5.6.17

System	Linux SiWiFi8867 3.18.29 #24 SMP PREEMPT Wed Nov 20 09:26:51 CST 2019 mips
Build Date	Nov 29 2019 09:31:01
Configure Command	'./configure' '--target=mipsel-openwrt-linux' '--host=mipsel-openwrt-linux' '--build=x86_64-linux-gnu' '--program-prefix=' '--program-suffix=' '--prefix=/usr' '--exec-prefix=/usr' '--bindir=/usr/bin' '--sbindir=/usr/sbin' '--libexecdir=/usr/lib' '--sysconfdir=/etc' '--datadir=/usr/share' '--localstatedir=/var' '--mandir=/usr/man' '--infodir=/usr/info' '--disable-nls' '--enable-cli' '--enable-cgi' '--enable-fpm' '--enable-shared' '--disable-static' '--disable-rpath' '--disable-debug' '--without-pear' '--with-config-file-path=/etc' '--with-config-file-scan-dir=/etc/php5' '--disable-short-tags' '--with-zlib=/home/hoowa/sf16a18-sdk-4.2.10/chaos_calmer_15_05_1/staging_dir/target-mipsel_mips-interAptiv_uClibc-0.9.33.2/usr' '--with-zlib-dir=/home/hoowa/sf16a18-sdk-4.2.10/chaos_calmer_15_05_1/staging_dir/target-mipsel_mips-interAptiv_uClibc-0.9.33.2/usr' '--with-pcre-regex=/home/hoowa/sf16a18-sdk-4.2.10/chaos_calmer_15_05_1/staging_dir/target-mipsel_mips-interAptiv_uClibc-0.9.33.2/usr' '--disable-phar' '--enable-calendar=shared' '--enable-ctype=shared' '--with-curl=shared,/home/hoowa/sf16a18-sdk-4.2.10/chaos_calmer_15_05_1/staging_dir/target-mipsel_mips-interAptiv_uClibc-0.9.33.2/usr' '--enable-fileinfo=shared' '--with-gettext=shared,/home/hoowa/sf16a18-sdk-4.2.10/chaos_calmer_15_05_1/staging_dir/target-mipsel_mips-interAptiv_uClibc-0.9.33.2/usr/lib/libintl-full' '--enable-dom=shared' '--enable-exif=shared' '--enable-ftp=shared' '--with-gd=shared' '--without-freetype-dir' '--with-jpeg-dir=/home/hoowa/sf16a18-sdk-4.2.10/chaos_calmer_15_05_1/staging_dir/target-mipsel_mips-interAptiv_uClibc-0.9.33.2/usr' '--with-png-dir=/home/hoowa/sf16a18-sdk-4.2.10/chaos_calmer_15_05_1/staging_dir/target-mipsel_mips-interAptiv_uClibc-0.9.33.2/usr' '--without-xpm-dir' '--without-t1lib' '--enable-gd-native-ttf' '--disable-gd-jis-conv' '--with-gmp=shared,/home/hoowa/sf16a18-sdk-

图14-1 用浏览器访问info.php文件的效果

示例：UCI信息读取

创建/www2/get_wireless.php文件。

```
<?php
$wireless_radio0 = 'uci show wireless.radio0';
$wireless_radio0_iface0 = 'uci show wireless.@wifi-iface[0]';
```

```
?>
<html>
 <head>
  <title> get_2.4g </title>
  <meta http-equiv="Content-Type" content="text/html; charset=utf-8" />
 </head>
 <body>
<H3>2.4GHz设备配置</H3>
<pre>
<?=$wireless_radio0?>
</pre>
<H3>2.4GHz网络配置</H3>
<pre>
<?=$wireless_radio0_iface0?>
</pre>
 </body>
</html>
```

用浏览器访问get_wireless.php文件的效果如图14-2所示。

2.4GHz设备配置

```
wireless.radio0=wifi-device
wireless.radio0.type='mac80211'
wireless.radio0.channel='auto'
wireless.radio0.band='2.4G'
wireless.radio0.max_all_num_sta='40'
wireless.radio0.netisolate='0'
wireless.radio0.country='CN'
wireless.radio0.ht_coex='0'
wireless.radio0.noscan='0'
wireless.radio0.radio='1'
wireless.radio0.txpower_1v1='2'
wireless.radio0.path='10000000.palmbus/11000000.wifi-1b'
wireless.radio0.htmode='HT20'
wireless.radio0.hwmode='11g'
```

2.4GHz网络配置

```
wireless.cfg033579=wifi-iface
wireless.cfg033579.device='radio0'
wireless.cfg033579.ifname='wlan0'
wireless.cfg033579.network='lan'
wireless.cfg033579.mode='ap'
wireless.cfg033579.ssid='SiWiFi-8868-2.4G'
wireless.cfg033579.encryption='psk2+ccmp'
wireless.cfg033579.key='12345678'
wireless.cfg033579.isolate='0'
wireless.cfg033579.hidden='0'
wireless.cfg033579.macfilter='disable'
wireless.cfg033579.macfile='/etc/wlan-file/wlan0.allow'
wireless.cfg033579.group='1'
wireless.cfg033579.netisolate='0'
wireless.cfg033579.disable_input='0'
wireless.cfg033579.wps_pushbutton='1'
wireless.cfg033579.wps_label='0'
wireless.cfg033579.rps_cpus='2'
```

图14-2 用浏览器访问get_wireless.php文件的效果

示例：用PHP操作SQLite数据库

在本章开头已经介绍过不需要单独安装SQLite服务，各种语言中已经包含了相关的库，直接调用就好了，它是嵌入式的。接下来介绍SQLite的命令，其作用是在命令行下查看或管理数据库文件。

安装sqlite3-cli软件包。

```
DF1A:$ opkg update
DF1A:$ opkg install sqlite3-cli
```

sqlite3命令的语法格式如下。

```
sqlite3 [库文件名]
```

用sqlite3命令创建数据库及生成数据表。

```
DF1A:$ sqlite3 /www2/test.db
SQLite version 3.8.11.1 2015-07-29 20:00:57
Enter ".help" for usage hints.
#创建表格
sqlite> create table test(
   ...> name varchar(60),
   ...> age int(3)
   ...> );
#插入测试数据
sqlite> insert into test(name,age) values('hoowa',18);
sqlite> insert into test(name,age) values('zhengqiwen',50);
#查看插入效果
sqlite> select * from test;
hoowa|18
zhengqiwen|50
#退出sqlite3-cli
sqlite> .exit
```

查看数据库文件。

```
DF1A:$ ls -l /www2/test.db
-rw-r--r--    1 admin    root          2048 Nov 29 10:29 /www2/test.db
```

创建/www2/db.php文件。

```php
<?php
//以读写形式打开数据库文件。如果文件不存在就尝试创建
$db = new SQLite3("/www2/test.db",SQLITE3_OPEN_READWRITE | SQLITE3_OPEN_CREATE);
//设置sqlite存储编码
$db->query('PRAGMA encoding="UTF-8"');
//设置sqlite同步缓存方式
$db->query('PRAGMA synchronous = NORMAL');

//查询
$result=@$db->query("select * from test");
if ($db->lastErrorCode() !== 0) {
    //如果查询有错则显示错误信息
    echo "Error: ".$db->lastErrorMsg();
    exit;
}
//准备输出
$output="";
while ($row = $result->fetchArray(SQLITE3_ASSOC)) {
    $output.="<p><b>name: </b> ".$row['name'];
    $output.=" <b>age: </b> ".$row['age']."</p>";
```

```
}
$result->finalize(); //输出结束
//关闭数据库
$db->close();
?>
<html>
 <head>
  <title> DB </title>
  <meta http-equiv="Content-Type" content="text/html; charset=utf-8" />
 </head>
 <body style="background-color: #f3f3f3">
<?=$output;?>
 </body>
</html>
```

用浏览器访问db.php文件的效果如图14-3所示。

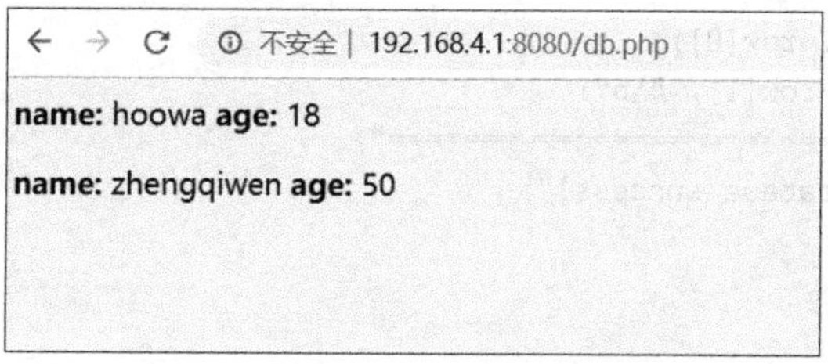

图14-3　用浏览器访问db.php文件的效果

14.1.2 Python开发环境

Python是一种面向对象的解释型计算机程序设计语言，由Guido van Rossum于1989年底发明，第一个公开发行版发行于1991年。Python语法简洁而清晰，具有丰富和强大的类库。它常被称作"胶水语言"，能够把用其他语言（尤其是C/C++）制作的各种模块很轻松地联结在一起。

Python需要的存储器空间较大，请参考前述章节将Overlay分区挂载到TF卡中以实现扩展，否则接下来的实例可能无法测试成功。如果当前Overlay分区已挂载到TF卡中，则不需要这一步骤。

Python安装

查看已编译的Python模块情况。

```
DF1A:$ opkg update
DF1A:$ opkg find python3-*
```

安装Python软件包。

```
DF1A:$ opkg install python3
```

测试Python是否已安装。

```
DF1A:$ python3 --version
Python 3.4.3
```

示例：用Python完成sqlite3操作

这一实例演示用Python通过终端方式访问sqlite3数据库，请保留test.db文件，如已删除请参考PHP部分重新创建该数据库文件。

创建db.py文件。

```python
import sqlite3
#打开数据库
sql = sqlite3.connect('/www2/test.db')
c = sql.cursor()
print("Open Sqilte3 /www2/test.db success!")
#向数据库插入tony的记录
c.execute("insert into test(name,age) values('tony',33)")
#提交记录,让sqlite3写入文件
sql.commit()
print("Insert success!");
#向sqlite3查询所有数据
print("====================================")
cursor = c.execute("select name,age from test")
for row in cursor:
    print("NAME = ", row[0])
    print("AGE = ", row[1], "\n")
print("====================================")
print("Operation database success!")
#关闭数据库
sql.close()
```

运行程序进行测试。

```
DF1A:tmp$ python3 db.py
Open Sqilte3 /www2/test.db success!
Traceback (most recent call last):
  File "db.py", line 11, in <module>
    conn.commit()
NameError: name 'conn' is not defined
DF1A:tmp$ python3 db.py
Open Sqilte3 /www2/test.db success!
Insert success!
====================================
NAME =  hoowa
AGE =  18
NAME =  zhengqiwen
AGE =  50
NAME =  tony
AGE =  33
====================================
Operation database success!
```

14.1.3 其他编程语言

本书所附带的源代码中除以上两种编程语言,还拥有Lua(系统已安装,另外有大量扩展库)、Java、Node.js、Perl、Ruby等语言的扩展包。愿意动手的读者可以参考相关章节自行编译这些软件包,通过WinSCP(Windows系统下)或scp命令(Linux系统下)将软件包复制到开发板中实现这些编程语言环境的搭建。

14.2 GPIO灯与按键控制

14.2.1 GPIO介绍

GPIO（General Purpose Input Output，通用输入/输出端口）是采用复杂工业总线简化出来的I/O扩展总线技术，市面上大部分的处理器已支持该总线。

简单来说，GPIO就是一条从CPU延伸出来的线，如果将其设置为输入，则可以连接到按键上，实现按键动作检测的操作，系统将接收到二进制数据1或0（高低电平）；如果将其设置为输出，则可以连接到LED上，通过发送二进制数据1或0（高低电平）控制LED发光或熄灭。

在以前的章节中介绍过开发板上预留的GPIO接口（PG1/PG2/PG3组），实际上那些GPIO都存在共用关系，在教学固件中通过DTS定义已经部分定义为特定功能。本章节中，我们将使用GPIO25和开发板上的SYS_LED介绍LED的控制功能，使用GPIO33（PG3组排针）介绍按键功能，方便读者了解定义按键的全过程。

接下来以GPIO34为例演示如何对GPIO进行控制，GPIO34在开发板的固件中处于空闲状态。在这一测试过程中，需要读者准备万用表及杜邦线。

从内核导出GPIO34的文件符号。

```
DF1A:$ cd /sys/class/gpio/
#注意,每次启动后都需要导出一次,或在自己的程序中完成这一步骤
DF1A:gpio$ echo 34 > export
#查看符号目录是否创建
DF1A:gpio$ ls
export     gpio34      gpiochip0    unexport
```

以输出模式测试GPIO34。

```
DF1A:gpio$ cd gpio34
DF1A:gpio34$ echo out > direction
#设置GPIO34为低电平
DF1A:gpio34$ echo 0 > value
#此时万用表选择为直流电压20V挡,正表笔连接PG3组的IO34,负表笔连接PG3组的GND,测出电压为0V
#设置GPIO34为高电平
DF1A:gpio34$ echo 1 > value
#此时万用表选择为直流电压20V挡,正表笔连接PG3组的IO34,负表笔连接PG3组的GND,测出电压为3.3V左右
```

用内核查看GPIO状态的方法如下（注意gpio-34的显示）。

```
DF1A:gpio34$ cat /sys/kernel/debug/gpio
GPIOs 0-70, platform/pinctrl, gpio:
GPIOs 0-70, platform/pinctrl, gpio:
 gpio-5    (?                   ) out hi
 gpio-6    (?                   ) out hi
 gpio-24   (?                   ) out hi
 gpio-25   (sys_led              ) out hi
 gpio-33   (hoowa                ) in  lo
 gpio-34   (sysfs                ) out hi
 gpio-36   (gpio-leds            ) out hi
 gpio-55   (eth_led0             ) in  lo
 gpio-56   (eth_led1             ) in  lo
```

```
gpio-61  (gpio-leds           ) out hi
```

以输入模式测试GPIO34。

```
#SF16A18芯片手册中说明GPIO34默认为下拉,在开发板中表现为低电平
#设定GPIO34为输入模式
DF1A:gpio34$ echo in > direction
#使用杜邦线一头接入PG3组的IO34,另外一头接入PG3组的GND
DF1A:gpio34$ cat value
0
#IO34接入GND都是低电平状态,无电流导通,值为0表示低电平
#使用杜邦线一头接入PG3组的IO34,另外一头接入PG4组的VCC3V3(请勿接错,防止烧板)
DF1A:gpio34$ cat value
1
#IO34接入VCC3V3,电流从VCC3V3导入IO34,值为1表示高电平
```

高低电平有效的作用如下。

```
#当active_low为0的时候,value=0表示低电平,value=1表示高电平
#当active_low为1的时候,value=1表示低电平,value=0表示高电平
DF1A:gpio34$ cat active_low
0
#可以设定active_low翻转value
DF1A:gpio34$ echo 1 > active_low
```

释放GPIO34。

```
DF1A:gpio34$ cd ..
DF1A:gpio$ echo 34 > unexport
```

14.2.2 按键功能的实现

根据以上内容,可以大致了解GPIO的基本特性。接下来以按键为例,介绍SF16A18芯片如何通过GPIO实现按键功能。本节内容从DTS对按键定义开始,展现从无到有的按键注册过程。开发板固件使用了SF16A18芯片的官方软件包led-button实现按键动作的处理。

测试前请准备一条杜邦线,用于通过短接PG组排针接口来模拟按键被按下动作,也可以制作或准备一个轻触按键连接到相应接口。接下来以GPIO33为例进行介绍。

在固件源代码中增加GPIO33的按键定义(在开发板中已实现,以下为展示)

1. 在DTS中注册按键。

```
#编辑源代码下的DTS文件
#linux-3.18.29-dev/linux-3.18.29/arch/mips/boot/dts/sf16a18_fullmask_df1a.dts
#目前开发板已经定义了GPIO33,以下为参考
&gpio_keys {
        status = "okay";
        //hoowa@33为标签,可以任意编写。如果要增加一个按键,需要增加标签数量
        hoowa@33 {
                label = "hoowa";    //标签名称,可以根据需要编写
                /*linux,input-type = <0x5>;*/
                linux,code = <0x102>;  //对应的按键编码
                gpios = <&gpio 33 1>;  //设定GPIO编号为GPIO33,1为GPIO_ACTIVE_HIGH模式
                poll-interval = <10>;  //内部延时,建议保持
```

```
            debounce-interval = <20>; //按键去抖时长,建议保持
    };
};
```

按键编码参考源代码:linux-3.18.29-dev/linux-3.18.29/include/dt-bindings/input/input.h

2. 在led-button软件包的plat.h中增加对DF1A开发板的支持。

```
#编辑led-button源代码文件
chaos_calmer_15_05_1/package/siflower/bin/led-button/src/plat.h
#增加一个新的BOARD枚举定义,例如BOARD_9
```

plat.h对开发板的支持如图14-4所示。

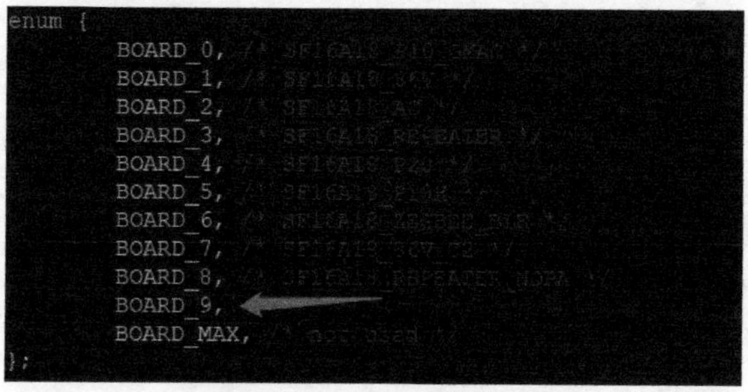

图14-4 plat.h对开发板的支持

3. 在led-button软件包的plat.c中,将BOARD_9的枚举定义与sf16a18-df1a字符串对应。

```
#编辑led-button源代码文件
chaos_calmer_15_05_1/package/siflower/bin/led-button/src/plat.c
```

plat.c对开发板的支持如图14-5所示。

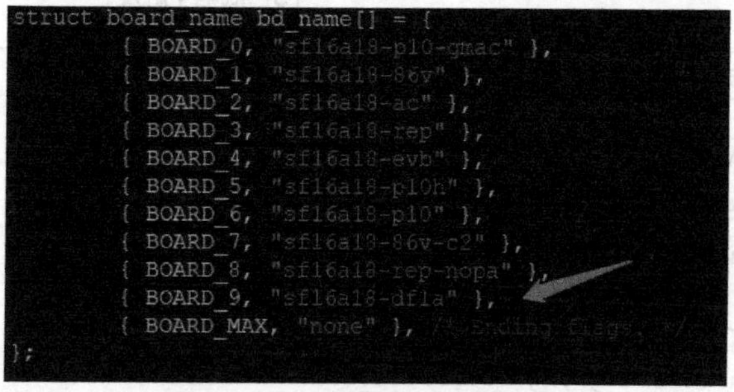

图14-5 plat.c对开发板的支持

4. 编译固件。

```
#需要删除掉已编译的.dtb文件再编译系统
HOST:sf16a18-sdk-4.2.10$ cd linux-3.18.29-dev/linux-3.18.29/arch/mips/boot/dts/
HOST:dts$ rm -f sf16a18_fullmask_df1a.dtb*
HOST:dts$ cd ../../../../../../
HOST:sf16a18-sdk-4.2.10$ cd chaos_calmer_15_05_1/
HOST:chaos_calmer_15_05_1$ make V=99
#编译完成后烧入新固件实现底层支持
```

UCI按键触发配置

查看标签为hoowa@33的按键,它已经进入devicetree。

```
DF1A:$ ls /sys/firmware/devicetree/base/palmbus@10000000/gpio-keys/
#address-cells    compatible      name            status
#size-cells       hoowa@33        poll-interval
```

UCI配置btn_ctrl_cfg内容如下。

```
DF1A:$ uci show btn_ctrl_cfg
btn_ctrl_cfg.@btn_config[0]=btn_config
btn_ctrl_cfg.@btn_config[0].btn_code='0x198'
btn_ctrl_cfg.@btn_config[0].btn_action='1'
btn_ctrl_cfg.@btn_config[0].btn_cmd='sh /bin/sf_reset.sh'
btn_ctrl_cfg.@btn_config[1]=btn_config
btn_ctrl_cfg.@btn_config[1].btn_code='0x198'
btn_ctrl_cfg.@btn_config[1].btn_action='3'
btn_ctrl_cfg.@btn_config[1].btn_cmd='sh /bin/sf_reset.sh'
btn_ctrl_cfg.@btn_config[2]=btn_config
btn_ctrl_cfg.@btn_config[2].btn_code='0x211'
btn_ctrl_cfg.@btn_config[2].btn_action='2'
btn_ctrl_cfg.@btn_config[2].btn_cmd='sh /usr/share/led-button/wps.sh'
```

在btn_ctrl_cfg中增加匿名节点btn_config,可用参数如表14-3所示。

表14-3 匿名节点 btn_config 可用参数

参数	说明	可选值
btn_code	在上述DTS中按键定义的linux,code值	使用十六进制0x方式表达,以免识别错误
btn_action	按键触发的机制	1:长按下,大于等于4秒,小于12秒。 2:短按下,小于4秒。 3:超长按下,大于12秒。 4:高电平触发。 5:低电平触发。 6:边沿触发
btn_cmd	触发后执行的命令	系统命令或脚本

在UCI中增加hoowa@33按键的触发测试。

```
#增加btn_config匿名节点,请使用该匿名节点号添加
DF1A:$ uci add btn_ctrl_cfg btn_config
cfg080e5e
#设定的btn_code要与DTS中按键的linux,code相同
DF1A:$ uci set btn_ctrl_cfg.cfg080e5e.btn_code='0x102'
#我们采用短按
DF1A:$ uci set btn_ctrl_cfg.cfg080e5e.btn_action='2'
#我们使用logger命令将这次按键被按下记录下来
DF1A:$ uci set btn_ctrl_cfg.cfg080e5e.btn_cmd='logger "hoowa key press detected"'
DF1A:$ uci commit btn_ctrl_cfg
```

使最新设置生效。

```
DF1A:$ /etc/init.d/led_button_init stop
DF1A:$ /etc/init.d/led_button_init start
```

测试按键效果

在前台启动logread -f等待日志事件。

```
#可通过按组合键Ctrl+C停止查看
DF1A:$ logread -f
```

将杜邦线一头接入GPIO33，另一头插入PG5组的VCC3V3电源（通过插入后拔下来模拟按键短按），或将自己焊接的真实轻触按键连接上去。按下按键后，logger获取到的日志如图14-6所示。

```
Thu Dec 12 08:48:50 2019 user.debug syslog: led-button evdev_handler: signum is 22 vs SIGIO is 22
Thu Dec 12 08:48:50 2019 user.debug syslog: led-button get board: sf16a18-df1a, len = 12
Thu Dec 12 08:48:50 2019 user.debug syslog: led-button first press!
Thu Dec 12 08:48:50 2019 user.debug syslog: led-button cmd = sh /bin/check_btn_cfg 102  6
Thu Dec 12 08:48:51 2019 user.debug syslog: led-button plat_btn_handler: 18 >> system ret = 1
Thu Dec 12 08:48:51 2019 user.debug syslog: led-button cmd = sh /bin/check_btn_cfg 102  5
Thu Dec 12 08:48:51 2019 user.debug syslog: led-button plat_btn_handler: 18 >> system ret = 1
Thu Dec 12 08:48:51 2019 user.debug syslog: led-button EV_SYN recieved!
Thu Dec 12 08:48:51 2019 user.debug syslog: led-button time is 1576111731, 327369 vs 1576111730, 967334
Thu Dec 12 08:48:51 2019 user.debug syslog: led-button type is 1, code is 258, value is 1
Thu Dec 12 08:48:51 2019 user.debug syslog: led-button cmd = sh /bin/check_btn_cfg 102  2
Thu Dec 12 08:48:51 2019 user.notice admin: "hoowa key press detected"
Thu Dec 12 08:48:51 2019 user.debug syslog: led-button plat_btn_handler: 18 >> system ret = 0
Thu Dec 12 08:48:51 2019 user.debug syslog: led-button EV_SYN recieved!
```

图14-6　按下按键后，logger获取到的日志

14.2.3 点亮LED

SYS_LED是开发板中已焊接好的LED，使用的是GPIO25，该GPIO与I2C2_DAT、PCM1_SIN、UART2_RTS共用（开发板DTS已默认释放掉以上使用）。电路原理为：3.3V高电平连接到LED的正极，GPIO25连接负极。当GPIO25为高电平时，电路不导通，LED不亮起；当程序将GPIO25改为低电平时，电流将流过LED，将其点亮。SYS_LED的电路原理如图14-7所示。

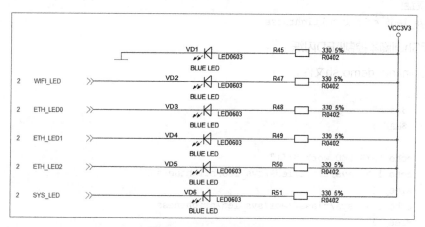

图14-7　SYS_LED的原理图

增加GPIO25驱动（开发板已实现，以下为展示）

linux-3.18.29-dev/linux-3.18.29/arch/mips/boot/dts/sf16a18_fullmask_df1a.dts的内容如下。

```
&leds {
    status = "okay"; //开启LED定义
```

```
        sys_led {  //这个是LED的标签名称,不能重复
                label = "sys_led"; //LED的标签名称,不能重复
                gpios = <&gpio 25 1>; //设置GPIO编号,1为GPIO_ACTIVE_HIGH模式
        };
};
```

编译固件（开发板已实现，以下为展示）

```
#需要删除掉.dtb文件再编译系统
HOST:sf16a18-sdk-4.2.10$ cd linux-3.18.29-dev/linux-3.18.29/arch/mips/boot/dts/
HOST:dts$ rm -f sf16a18_fullmask_df1a.dtb*
HOST:dts$ cd ../../../../../../
HOST:sf16a18-sdk-4.2.10$ cd chaos_calmer_15_05_1/
HOST:chaos_calmer_15_05_1$ make V=99
#编译完成后烧入新固件实现底层支持
```

测试LED是否有效

```
#查看是否有sys_led
DF1A:$ cd /sys/class/leds/
DF1A:leds$ ls
eth_led0            siwifi-phy0::rx   siwifi-phy1::rx   sys_led
eth_led1            siwifi-phy0::tx   siwifi-phy1::tx
DF1A:leds$ cd sys_led/
#查看当前LED是否点亮
DF1A:sys_led$ cat brightness
0
#brightness为0表示LED没有点亮,为1表示点亮
#设定点亮LED
DF1A:sys_led$ echo 1 > brightness
#设定熄灭LED
DF1A:sys_led$ echo 0 > brightness
```

使用Shell脚本控制LED闪烁

创建/bin/led_demo.sh文件。

```
#!/bin/sh
MAX="1 2 3 4 5"
for I in $MAX
do
        echo "SYS_LED work ${I}"
        echo 1 > /sys/class/leds/sys_led/brightness
        sleep 1
        echo 0 > /sys/class/leds/sys_led/brightness
        sleep 1
done
```

执行脚本。

```
#设置可执行
DF1A:$ chmod +x /bin/led_demo.sh
#执行看效果
DF1A:$ /bin/led_demo.sh
```

OpenWrt触发器

以上对LED的测试能够取得成功。此外，SF16A18芯片的OpenWrt系统提供了一些触发器，可以根据某些硬件状态实现LED的闪烁。

LED的触发器在UCI的system配置中实现，以led节点方式表示。

```
#查看全部支持的触发器
DF1A:sys_led$ cat /sys/class/leds/sys_led/trigger
[none] mmc0 timer default-on netdev transient phy0rx phy0tx phy0assoc phy0radio
phy1rx phy1tx phy1assoc phy1radio gpio heartbeat morse oneshot
```

触发器的全部支持情况如表14-4～表14-7所示（部分触发器在SF16A18芯片上尚无法工作，下面仅列出已测试能够使用的触发器）。

表14-4 触发器default-on，默认是否亮起

项	说明	可选值
default	LED触发前状态	0表示关闭，1表示开启，可不设置
sysfs	LED设备名称	参考DTS设定标签名称，如sys_led
trigger	触发器名称	default-on

表14-5 触发器heartbeat，根据CPU的负载情况闪烁

项	说明	可选值
default	LED触发前状态	0表示关闭，1表示开启，可不设置
sysfs	LED设备名称	参考DTS设定标签名称，如sys_led
trigger	触发器名称	heartbeat

表14-6 触发器timer，定时闪烁

项	说明	可选值
default	LED触发前状态	0表示关闭，1表示开启，可不设置
delayon	点亮时间	毫秒级
delayoff	熄灭时间	毫秒级
sysfs	LED设备名称	参考DTS设定标签名称，如sys_led
trigger	触发器名称	timer

表14-7 触发器netdev，网络触发器

项	说明	可选值
default	LED触发前状态	0表示关闭，1表示开启，可不设置
dev	以太网设备	设备名称，可用ifconfig查看
mode	触发方法	可选link、rx、tx或全部触发
sysfs	LED设备名称	参考DTS设定标签名称，如sys_led
trigger	触发器名称	netdev

下面以heartbeat触发器示例进行展示。

```
#设置触发器
DF1A:$ uci set system.sys_led=led
DF1A:$ uci set system.sys_led.sysfs=sys_led
DF1A:$ uci set system.sys_led.trigger=heartbeat
DF1A:$ uci commit system
```

```
#查看
DF1A:$ uci show system.sys_led
system.sys_led=led
system.sys_led.sysfs='sys_led'
system.sys_led.trigger='heartbeat'
#重启LED触发器进程使之生效
DF1A:$ /etc/init.d/led restart
setting up led sys_led
#此时LED会根据CPU的负载情况闪烁
```

下面以eth0.2（开发板中该网卡连接外网）的收发作为触发条件进行展示。

```
#设置触发器
DF1A:$ uci set system.sys_led=led
DF1A:$ uci set system.sys_led.dev='eth0.2'
DF1A:$ uci set system.sys_led.mode='link rx tx'
DF1A:$ uci set system.sys_led.sysfs='sys_led'
DF1A:$ uci set system.sys_led.trigger='netdev'
DF1A:$ uci commit system
#查看
DF1A:$ uci show system.sys_led
system.sys_led=led
system.sys_led.sysfs='sys_led'
system.sys_led.dev='eth0.2'
system.sys_led.mode='link rx tx'
system.sys_led.trigger='netdev'
#重启LED触发器进程使之生效
DF1A:$ /etc/init.d/led restart
setting up led sys_led
#此时通过WAN口收发数据包,LED会闪烁
```

14.3 UART-TTL串口

14.3.1 UART介绍

通用异步接收发送设备（Universal Asynchronous Receiver/Transmitter），通常简称作UART。UART是一种串行异步收发协议，应用十分广泛，在物联网领域用于各类外围设备的连接与控制、管理等。

作为异步串口通信协议的一种，UART的工作原理是将传输数据的每个字符一位接一位地传输。在UART通信协议中，信号线上的状态位高电平代表"1"，低电平代表"0"。协议基本参数如下。

● 起始位：先发出一个逻辑"0"的信号，表示传输字符的开始。

● 数据位：紧接在起始位之后。数据位的个数可以是4、5、6、7、8等，构成一个字符。通常采用ASCII码，从最低位开始传送，靠时钟定位。

● 校验位：数据位加上这一位后，使得"1"的位数应为偶数（偶校验）或奇数（奇校验），以此来校验数据传送的正确性。

● 停止位：它是一个字符数据的结束标志，可以是1位、1.5位、2位的高电平。

- 空闲位：处于逻辑"1"状态，表示当前线路上没有数据传送。
- 波特率：它是衡量数据传送速率的指标，表示每秒钟传送的符号数（symbol）。一个符号代表的信息量（比特数）与符号的阶数有关。例如传输使用256阶符号，每8bit代表一个符号，数据传送速率为120字符/秒，则波特率就是120波特，比特率是120×8=960bit/s。这两者的概念很容易搞错。在双方传输设备中，如有一方时钟出现偏差超过一定允许值的问题，则接收到的数据很可能被读取为乱码。

UART的硬件接口如下。

- TTL：表示采用晶体管-晶体管逻辑（Transistor-Transistor Logic,TTL）电平的串口。对于输出电路，电压大于等于（≥）2.4V为逻辑1；电压小于等于（≤）0.4V为逻辑0；对于输入电路，电压大于等于（≥）2.0V为逻辑1，电压小于等于（≤）0.8V为逻辑0。
- RS-232：美国电子工业协会于1962年发布的串行通信接口标准。该标准对串行通信的物理接口及逻辑电平都做了规定，其输出的电平称为 RS-232 电平。开发板中无RS-232接口，不过可以通过电平转换模块将TTL转换为RS-232。
- RS-485：RS-232接口可以实现点对点的通信方式，但这种方式不能实现联网功能。于是，为了解决这个问题，一个新的标准RS-485产生了。RS-485的数据信号采用差分传输方式，也称作平衡传输，它使用一对双绞线，将其中一线定义为A，另一线定义为B。开发板中的J3凤凰端子连接器将SF16A18内部的UART3转换为RS-485协议的接口。

使用TTL电平的UART接口连接方法如表14-8所示，DF1A开发板中的UART接口如表14-9所示。

表14-8 使用 TTL 电平的 UART 接口连接方法

CPUA	CPUB	说明
TTL-RX	TTL-TX	CPUB向CPUA发送信号
TTL-TX	TTL-RX	CPUA向CPUB发送信号
GND	GND	共用地

表14-9 DF1A 开发板中的 UART 接口（DTS 中已定义）

接口类型	功能	说明
UART0	UART0_TX（IO18） UART0_RX（IO19）	用于系统调试接口，OpenWrt下设备为/dev/ttyS0
UART1	UART1_TX（IO26） UART1_RX（IO27） UART1_RTS（IO29） UART1_CTS（IO30）	关闭，与IIS2、PCM0复用。默认DTS定义为IIS2接口，需要修改DTS改变用途
UART2	UART2_TX（IO22） UART2_RX（IO23） UART2_CTS（IO24） UART2_RTS（IO25）	TX与RX为PG1组引出UART2模式。CTS、RTS被其他设备占用。 OpenWrt下设备为/dev/ttyS1
UART3	UART3_RX（IO31） UART3_TX（IO32） UART3_RTS（IO34） UART3_CTS（IO35）	TX与RX被开发板转换为RS-485接口。CTS、RTS被其他设备占用。 OpenWrt下设备为/dev/ttyS2

14.3.2 UART-TTL的使用

我们以PG1组的UART2（系统下为/dev/ttyS1）举例说明UART-TTL的使用及测试方法。

在计算机中进行以下准备。

1. 将一个USB-TTL的转换电路板接入计算机，并且查找到其生成的COM口编号。在本例中，计算机内生成的COM口编号为COM5。

2. 连接信号线。

（1）使用杜邦线将USB-TTL的RX连接到PG1组的IO22（UART2_TX）。

（2）使用杜邦线将USB-TTL的TX连接到PG1组的IO23（UART2_RX）。

3. 在计算机中下载"串口调试工具"，设置计算机的波特率并且连接。设置串口为COM5（以实际情况为准）、波特率为115 200波特、数据位8、校验位None、停止位1、流控None。

4. 单击启动按钮，进入准备状态，如图14-8所示。

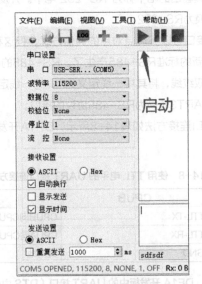

图14-8 串口调试工具

进行收发测试。

```
#microcom是DF1A开发板固件附带的一个串口测试工具,其他参数已经默认设置,只需要设置波特率即可
DF1A:$ microcom -s 115200 /dev/ttyS1
#microcom启动后将进入等待状态,这时键盘输入的内容将原封不动地发送到计算机上。计算机上发送的内容也
将返回开发板中
#microcom使用组合键Ctrl+x表示退出
```

计算机和DF1A通过UART2进行通信测试的界面如图14-9所示。

图14-9 计算机和DF1A通过UART2进行通信测试的界面

图14-9中"hello df1a"是由计算机发送给DF1A开发板的信息。由DF1A开发板发送给计算机的信息是"hello pc"（由于microcom中输入内容并不回显，所以看不到键盘输入的"hello pc"）。

14.4 ZigBee物联网通信

14.4.1 ZigBee 简介

ZigBee，中文名称为蜂舞协议，是一种短距离低速的无线通信协议标准，底层采用IEEE 802.15.4标准规定的媒体访问层和物理层。

ZigBee最早由美国Honeywell（霍尼韦尔）公司提出，于1998年开始构建和发展，推出一种能够自我组网的无线点对点（ad-hoc）网络标准，后期成立的ZigBee Alliance商业组织，于2001年向电气电子工程师学会（IEEE）提出的提案被纳入IEEE 802.15.4标准之中，并于2005年正式发布为ZigBee 1.0（又称ZigBee2004）工业规范。自此ZigBee渐渐成为各业界共通的低速短距无线通信技术之一。ZigBee发展到1.2版本时，其应用范围已经延伸到家庭娱乐与控制、无线感测网络（WSN）、工业控制、嵌入式感测、医疗数据搜集、烟幕与擅闯警示、建筑自动化等领域，并有各式各样的应用层标准。为了满足物联网（IoT）需求，ZigBee联盟推出2.0版，主打智慧能源规范；2015年正式推出了ZigBee 3.0，统一了ZigBee标准，让应用之间具备高度互通性、传输安全，同时具有低功耗的优势，为物联网提供了统一的产业标准。

ZigBee具有如下特点。

- 低功耗。在低功耗模式下，根据电池容量大小，ZigBee设备可支持半年到几年。
- 低成本。
- 低速率。工作在20~250kbit/s速率，提供20kbit/s(868MHz)、40kbit/s(915MHz)、250kbit/s(2.4GHz)数据吞吐，满足各种应用需求。
- 近距离。相邻节点间传输距离范围介于10~100m；增加发射功率，传输距离可达1~3km；如果通过路由和节点实现扩展，传输距离可以更远。
- 低时延。响应速度快，一般从睡眠到工作态只需要15ms，节点接入网络需要30ms，进一步节能。
- 高容量。采用星形、网状、树状网络结构，由一个主节点管理若干子个节点，最多的主节点可以管理256个子节点，同时主节点还有上一层节点管理，每个网络最高可支持65 000个节点。
- 高安全性。安全可靠，支持3级安全模式，包括无安全设定、访问控制清单（ACL），采用128bit AES对称加密。
- 自组网。只要信道和PAN ID（网络标识符）相同，就可以自行组网，支持碰撞避免、重试和应答确认。
- 使用免执照频段。使用工业科学医疗（ISM）频段、915MHz频段（美国）、868MHz频段（欧洲）、2.4GHz频段（全球）。

ZigBee网络中的节点

ZigBee标准将网络节点按照功能划分为：协调器（Coordinator）、路由器（Router）和终端设备（End Device）。

- 协调器

每个ZigBee网络中只允许有一个协调器，协调器会为每个设备分配一个唯一的网络地址。为整个网络选择一个唯一的16位的PAN ID和信道。采用PAN ID，网络中的设备就可以通过网络地址进行通信。协调器负责初始化、终止、转发网络中的消息。

- 路由器

路由器是一种支持关联的设备，用于扩展网络覆盖的物理范围和数据包路由的功能，实现其他节点的消息转发功能。

- 终端设备

ZigBee终端设备是具体执行数据采集并将数据传输出去的设备，它不能转发其他节点的消息，只能负责网络数据的采集和传输。

拓扑结构

得益于其拓扑结构，ZigBee具有强大的组网能力。ZigBee支持3种拓扑结构，分别为星形（Star）、树状（Mesh-Tree）和网状（Mesh）拓扑，如图14-10所示。每种拓扑结构都有其特点和优势。以下对这3种拓扑结构分别进行介绍。

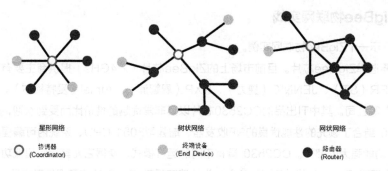

图14-10　ZigBee网络拓扑

- 星形网络：是最简单的一种网络拓扑形式。由一个协调器节点、多个终端设备节点和路由器节点组成。每一个终端设备节点只能和协调器节点进行通信。如果需要在两个终端设备节点之间进行通信，必须通过协调器节点进行信息的转发。这种拓扑形式的缺点是节点之间的数据路由只有唯一的一条路径，协调器是网络的中心。
- 树状网络：包括一个协调器、多个路由器和终端设备节点。协调器连接多个路由器和终端设备节点，路由器节点可以连接多个路由器节点和终端设备节点，这一模式可以重复多个层级。
- 网状网络：包含一个协调器、多个路由器和终端设备节点。网状拓扑具有更加灵活的信息路由规则，在可能的情况下，路由器节点之间可以直接通信。这种路由机制使得信息通信变得更有效率，而且即使有一条路由路径出现了问题，信息也可以自动沿着其他路由路径进行传输。

ZigBee协议栈

ZigBee协议分为两部分，IEEE 802.15.4定义了PHY（物理层）和MAC（介质访问控制层）技术规范。ZigBee联盟定义了NWK（网络层）、APS（应用支持子层）、APL（应用层）技术规范。ZigBee协议栈就是将各个层定义的协议都集合在一起，以函数的形式实现，并给用户提供API（应用层），用户可以直接调用。ZigBee协议栈如图14-11所示。

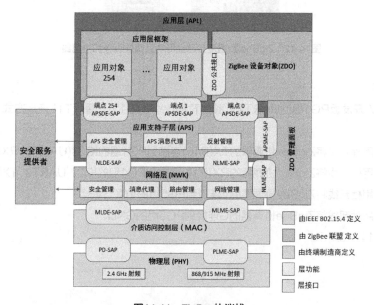

图14-11　ZigBee协议栈

14.4.2 ZigBee物联网实战

本节将展示一个ZigBee物联网实例。

首先需要选择ZigBee芯片。目前市场上的ZigBee芯片（2.4GHz）提供商主要有TI（德州仪器）、EMBER（ST）、JENNIC（捷力）、NXP（恩智浦）、Atmel（爱特梅尔）、Microchip（微芯科技）等公司。其中TI出品的CC2530芯片以其非常优越的性价比而受到欢迎，应用最为广泛。CC2530结合了领先的性能优良的RF收发器、增强型8051 CPU、系统内可编程闪存、8KB RAM和其他功能强大的部件。CC2530具有不同的运行模式，使得它尤其适应超低功耗要求的系统。TI还提供了ZigBee协议栈（Z-Stack），为快速实现ZigBee解决方案提供了方便。

以下的示例选用TI的CC2530芯片方案，采购成熟的ZigBee模块。实验采用一个ZigBee协调器和一个带电池的ZigBee温/湿度传感器低功耗终端设备，选择星形网络拓扑结构。

配置如表14-10所示。

表14-10 ZigBee 模组信息

芯片类型	节点类型	连接方式	PAN ID	信道	备注
TI CC2530	协调器	与DF1A开发板PG1组UART2采用TTL进行连接。波特率为38 400波特、数据位8、校验位None、停止位1、流控None	6688	15	
TI CC2530	带电池的低功耗温/湿度传感器终端	与协调器通过ZigBee网络进行通信	6688	15	采用CR2032电池供电，默认每10s采集一次数据，可待机150天左右

ZigBee温/湿度传感器低功耗终端设备和协调器的实物如图14-12所示。

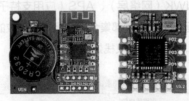

图14-12 ZigBee温/湿度传感器低功耗终端设备和协调器

硬件准备

使用DF1A开发板PG1组的UART2（系统下为/dev/ttyS1），采用TTL连接方式与ZigBee协调器相连接。

（1）使用杜邦线将ZigBee协调器的TX(P16)连接到PG1组的IO23(UART2_RX)。

（2）使用杜邦线将ZigBee协调器的RX(P17)连接到PG1组的IO22(UART2_TX)。

（3）使用杜邦线将ZigBee协调器的V连接到PG5组的VCC3V3。

（4）使用杜邦线将ZigBee协调器的G连接到PG5组的GND。

连接如图14-13所示。

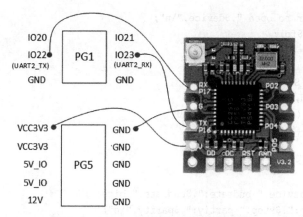

图14-13 ZigBee协调器与开发板连接方式

终端设备节点在得到电池供电后默认自动启动。终端设备节点采用CR2032型电池供电,每隔10s上报温/湿度采集值,也可以更改上报间隔时间。可待机时间约150天,上报数据的格式默认为字符串形式,也可以通过模块厂商提供的工具转换为十六进制协议格式。终端设备中如果设置了与协调器相同的PAN ID和信道,就会被协调器自动分配节点地址后加入协调器所在的网络。

ZigBee模组可通过淘宝购买。

安装依赖软件包

本示例中,采用PHP语言在DF1A开发板中编写程序,实现对协调器数据的读写。当终端设备有数据上报时,协调器会收到上报数据,并把数据传输到DF1A开发板串口侧,这时只要对DF1A开发板进行串口监听,读取相关数据即可。

安装PHP语言支持及其模块。

```
DF1A:$ opkg update
DF1A:$ opkg install php5-cli php5-mod-dio
```

编写测试程序

创建文件名为serial的PHP程序。

```php
#!/usr/bin/php-cli
<?php

if (count($argv) != 6) {
    echo "syntax: serial_read [device]";
    echo " [budrate] [bits] [stop] [parity]\n";
    exit;
}
$device = $argv[1];
$budrate = $argv[2];
$bits = $argv[3];
$stop = $argv[4];
$parity = $argv[5];
$fd = dio_open($device, O_RDWR | O_NOCTTY | O_NONBLOCK);
if ($fd == FALSE) {
```

```php
        echo "error to open ".$device."\n";
        dio_close($fd);
        exit;
}
dio_fcntl($fd, F_SETFL, O_SYNC);
dio_tcsetattr($fd, array(
        'baud'   => (int)$budrate,//38400
        'bits'   => (int)$bits,//8
        'stop'   => (int)$stop,//1,
        'parity' => (int)$parity//0
));
echo "device ".$device." budrate:".$budrate." bits:";
echo $bits." stop:".$stop." partiy:".$parity."\n";
while(1)
{
        $data = dio_read($fd, 128);
        if ($data) {
                echo $data;
        }
}
dio_close($fd);
?>
```

将程序设置为可执行。

```
DF1A:$ chmod +x serial
```

执行程序进行测试。

```
DF1A:$ ./serial /dev/ttyS1 38400 8 1 0
```

终端设备通过ZigBee输出温/湿度信息，实际效果如图14-14所示。

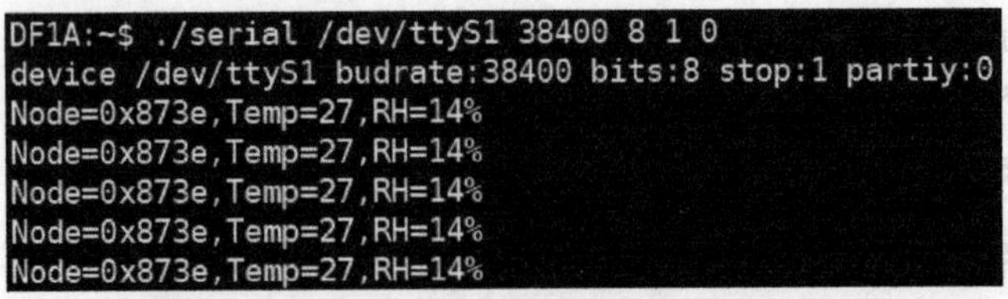

```
DF1A:~$ ./serial /dev/ttyS1 38400 8 1 0
device /dev/ttyS1 budrate:38400 bits:8 stop:1 partiy:0
Node=0x873e,Temp=27,RH=14%
Node=0x873e,Temp=27,RH=14%
Node=0x873e,Temp=27,RH=14%
Node=0x873e,Temp=27,RH=14%
Node=0x873e,Temp=27,RH=14%
```

图14-14 终端设备通过ZigBee输出温/湿度信息

14.5 工业物联网网关

通过工业物联网的网关采集设备的运行状态、生产过程、故障状况等数据，实现IT（信息技术）和OT（操作运营技术）的融合，就能建立工厂的数字孪生状态，在IT系统中反映实体装备的全生命周期。

14.5.1 工业物联网

工业物联网是将具有感知、监控能力的各类传感器或控制器，以及移动通信、智能分析等技术不断融入工业生产过程各个环节，从而大幅提高制造效率，改善产品质量，降低产品成本和资源消耗，最终实现将传统工业提升为智能化工业的新技术。从应用形式上，工业物联网的应用具有实时性、自动化、嵌入式（软件）、安全性和信息互通互联性等特点。

工业物联网网关的概念应运而生，它能够在OT和IT之间搭建起一座真正的桥梁。网关一般都具备下述的功能。

- 联网方式：通过WAN、Wi-Fi、4G、5G等接入互联网。
- 设备接入：通过串口或网口连接设备的传感器或控制器。
- 数据采集：根据接入设备的协议，采集设备的各种状态数据。
- 数据处理：根据业务规则，过滤、加工采集到的设备状态数据。
- 数据上报：将处理过的数据通过物联网协议上报到服务器进行存储。

目前，国内提供工业物联网服务的企业越来越多，大部分企业有自己的物联网网关产品，实现的基本功能大同小异。读者们读完本章节，也可以自行搭建一个工业物联网的实验环境。

图14-15是一例智能工厂实施项目的整体架构图，其中采用了羽智科技的物联网网关。网关连接工厂设备采集数据，通过OPC UA对工厂设备进行信息化建模，为ERP、MES、SCADA等上位系统提供统一的设备信息访问接口。

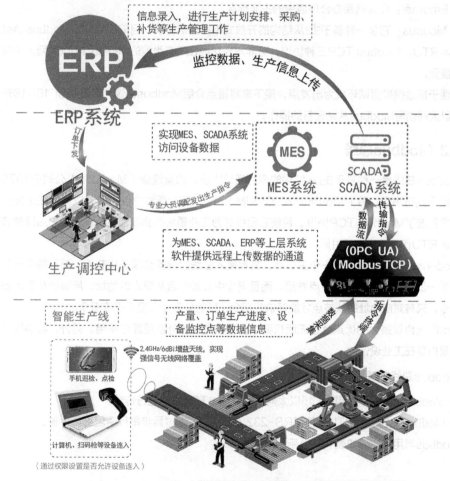

图14-15 智能工厂示意图

14.5.2 工业总线介绍

工业总线是在自动化发展过程中为了解决不同厂家的仪器仪表、控制器、执行机构之间互联互通的需求而建立的。目前主要厂家有欧系的西门子、倍福、ABB、施耐德等，美系的罗克韦尔、艾默生等，日系的三菱电机、欧姆龙、松下等。随着国内工业自动化的发展，目前国内厂家（如汇川、禾川、台达）的市场份额和产品性能也在大幅提升。

常见的工业总线如下。

● Profibus：它作为一种快速总线，被广泛应用于分布式外围组件(PROFIBUS-DP)。

● EtherCAT：全称是Ethernet for Control Automation Technology，它是一种用于工业自动化的实时以太网解决方案，性能优越，使用简便。

● CANopen：通过有效利用总线带宽，CANopen在相对较低的数据传输速率时也能实现较短的系统响应时间。它秉承了CAN的传统优点，例如数据安全性高且具备多主站能力。

● ControlNet：它是一种开放式标准现场总线系统。该总线协议允许循环数据和非循环数据通过总线同时进行交换，而且两者之间互不影响。

● Ethernet：以太网是办公环境中的主流标准。

● Modbus：它是一种基于主/从结构的开放式串行通信协议，目前主要分为Modbus AsCII、Modbus RTU、Modbus TCP三种协议，可非常轻松地在所有类型的串行接口和以太网上实现，已被广泛接受。

以便于搭建模拟测试环境为出发点，接下来将重点介绍Modbus协议，以及在SF16A18开发板上通过以太网读写Modbus从站设备的操作。

14.5.3 Modbus总线

Modbus总线是一种基于主/从结构的串行通信协议，由莫迪康（Modicon）公司于1979年发布，初期仅支持串口，分为Modbus ASCII和Modbus RTU两种协议。随着以太网在工业领域的普及，它新增了Modbus TCP协议。目前它已经成为工业领域的标准通信协议。本书后续章节将以Modbus RTU作为重点进行讲解。

Modbus总线中有且只有一个主节点，其他接入总线的设备都属于从节点，而且每个从节点都有一个唯一的地址。指令由主节点发起，而且指令中必须包含从节点的地址，所有的从节点都能够收到指令，只有对应地址的从节点才能够响应指令，其他从节点必须无视该指令。

Modbus协议最大的优点就是无版权要求，无须支付额外费用即可使用。另外，其协议简单，非常容易内置在工业设备中。

Modbus总线物理层

● Modbus TCP：使用RJ-45以太网接口，通过TCP协议进行传输。

● Modbus ASCII和RTU：使用RS-232/485接口，通过异步串行通信进行传输。

Modbus常用功能码如表14-11所示。

表 14-11 Modbus 常用功能码

功能码	名称
01	读线圈寄存器
02	读离散输入寄存器
03	读保持寄存器
04	读输入寄存器
05	写单个线圈寄存器
06	写单个保持寄存器
15	写多个线圈寄存器
16	写多个保持寄存器

Modbus RTU协议的格式

Modbus RTU协议发送指令如表14-12所示。

表 14-12 发送指令

内容	长度
从站地址	1字节
功能码	1字节
起始地址	2字节
读取数量	2字节
校验码（CRC16）	2字节

发送指令的示例如下。

```
#读取05号从节点的第100号寄存器的数值
0x05 0x03 0x00 0x64 0x00 0x01 0xC4 0x51
```

Modbus RTU协议接收响应如表14-13所示。

表 14-13 接收响应

内容	长度
从站地址	1字节
功能码	1字节
返回长度	1字节
返回数据	2×N字节
校验码（CRC16）	2字节

接收响应的示例如下。

```
#返回的数据为0x00 0x10,转换为十进制的结果为16
0x05 0x02 0x00 0x10 0xA0 0xE4
```

14.5.4 MQTT服务器环境搭建

目前，3种最常见的物联网通信协议是MQTT、AMQP和COAP。这些协议特性各有优势和劣势，读者可以根据自己的业务场景选择合适的协议。

● MQTT：以前被称为"SCADA协议"，它是一种易于实现、轻量级、经ISO批准的消息传递协议，特别适用于远程通信和带宽受限的情况。MQTT的发布订阅、低功耗、小尺寸和通过最小化数据包高效分发数据等特点，使其成为工业物联网部署和移动应用的好选择。本章节后续内容主要以MQTT为例进行讲解。

● AMQP：这是一种开放标准的、功能丰富的消息队列协议，它提供可靠和安全的消息队列、路由和定向。AMQP提供了高互操作性，允许各种各样的通信模式和消息传递应用。

● CoAP：专门用于连接资源受限（如电源有限或内存小）的设备。对CoAP的新扩展允许将几个CoAP资源作为一个组来定义和处理，并减少了传输时间。

MQTT是基于TCP的发布订阅机制协议，由IBM公司于1999年提出，现在主流的版本是2014年发布的3.1.1，最新版本是2019年发布的5.0。

MQTT协议的基本原理

MQTT协议的基本原理如图14-16所示。MQTT协议提供一对多的消息发布订阅能力，可以解除工业物联网的耦合问题。MQTT协议需要客户端和服务器端两种角色同时存在，协议中主要有3种身份：发布方（Publisher，客户端）、消息代理（Broker，服务器端）、订阅方（Subscriber，客户端）。消息发布方也可以同时是订阅方，实现生产者和消费者的解耦。

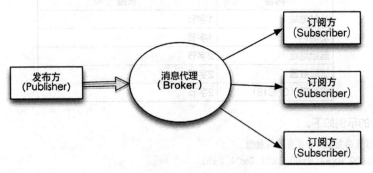

图14-16 MQTT协议的基本原理

在HOST中编译、运行Mosquitto

Mosquitto是实现MQTT协议的开源消息代理软件。接下来介绍Mosquitto的编译、运行。

下载源代码。

```
HOST:~$ wget https://mosquitto官网地址/files/source/mosquitto-1.4.15.tar.gz
```

解压缩。

```
HOST:~$ tar xvf mosquitto-1.4.15.tar.gz
```

进入源代码目录。

```
HOST:~$ cd mosquitto-1.4.15/
```

安装必要组件。

```
HOST:mosquitto-1.4.15$ sudo apt install libssl-dev libc-ares-dev uuid-dev
```

编译源代码并安装。

```
HOST:mosquitto-1.4.15$ make
HOST:mosquitto-1.4.15$ sudo make install
```

建立软连接，将/usr/local/lib下安装的.so文件映射到/usr/lib下。

```
HOST:mosquitto-1.4.15$ sudo ln -s /usr/local/lib/libmosquitto.so.1 /usr/lib/
libmosquitto.so.1
```

启动mosquitto消息代理。

```
HOST:mosquitto-1.4.15$ mosquitto
1577067287: mosquitto version 1.4.15 (build date 2019-12-23 10:13:04+0800) starting
1577067287: Using default config.
1577067287: Opening ipv4 listen socket on port 1883.
1577067287: Opening ipv6 listen socket on port 1883.
```

从日志输出的信息可知,mosquitto使用了默认的配置文件,默认的端口为1883,后续如要进行修改端口、用户名、密码验证等操作,可以通过/etc/mosquitto.conf.example生成mosquitto.conf实现修改。

新开一个A终端,订阅freeiris.hello主题。

```
HOST:~$ mosquitto_sub -h 127.0.0.1 -t "freeiris.hello"
```

再开一个B终端,发布freeiris.hello主题。

```
HOST:~$ mosquitto_pub -h 127.0.0.1 -t "freeiris.hello" -m "hello freeiris"
```

执行上述操作后,A终端将收到"hello freeiris"。

```
hello freeiris
```

至此,Mosquitto服务器环境搭建完成,可以关闭掉终端B,保留终端A用于开发板测试。

DF1A开发板与Mosquitto通信

安装软件包。

```
DF1A:$ opkg update
DF1A:$ opkg install mosquitto-client-nossl
```

通过mosquitto_pub命令,发布freeiris.hello主题(假设IP地址为172.16.10.121)。

```
DF1A:$ mosquitto_pub -h 172.16.10.121 -t "freeiris.hello" -m "hello freeiris"
```

执行上述操作后,A终端将收到"hello freeiris"。

```
HOST:siflower$ mosquitto_sub -h 127.0.0.1 -t "freeiris.hello"
hello freeiris
```

14.5.5 实战工业物联网

通过以上章节,读者应该对Modbus、MQTT通信技术有了初步了解,并且已经搭建了一个MQTT消息代理服务器。接下来的内容将介绍怎样建立一个完整的工业物联网网关。

为了便于搭建模拟环境,我们通过Modbus Slave模拟支持Modbus TCP的PLC,可以新建一个包,通过Modbus读取PLC数据,并通过MQTT协议,将数据上报到服务器。另外,服务器也可以通过MQTT协议将控制指令下发给开发板,开发板再将控制指令转发给PLC,完成一个支持双向通信的网关。

准备模拟PLC环境

考虑到PLC设备较为昂贵,为了以较低的成本进行学习,可以在Windows计算机上安装Modbus Slave软件,模拟实现PLC功能。该安装文件可以在互联网上自行搜索下载,安装完成之后,启动Modbus Slave软件,按照下面的步骤,模拟PLC。

● 新建连接，单击"Connection"下的"Connect"，如图14-17所示。

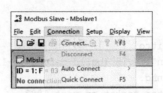

图14-17 Connection菜单

● 在Connection Setup界面，选择"Connection"为"Modbus TCP/IP"，"Port"（端口）为800，单击"OK"保存，如图14-18所示。

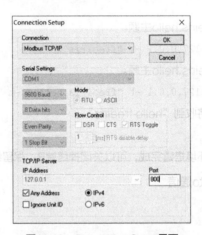

图14-18 Connection Setup界面

● 选择"Setup"→"Setup Definition"进行从站设置。在Slave Definition界面，将"Slave ID"设为1，将"Function"设为"03 Holding Register（4X）"（03保持寄存器），如图14-19所示。

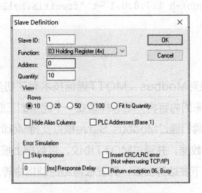

图14-19 Slave Definition界面

单击"OK"确定之后，就能在计算机上实现模拟PLC。地址从40001开始，总共有10个点位的数据，每个点位占两个字节。

· 434 ·

在任意点位单击鼠标右键,将跳出点位详细设定菜单,如图14-20所示。

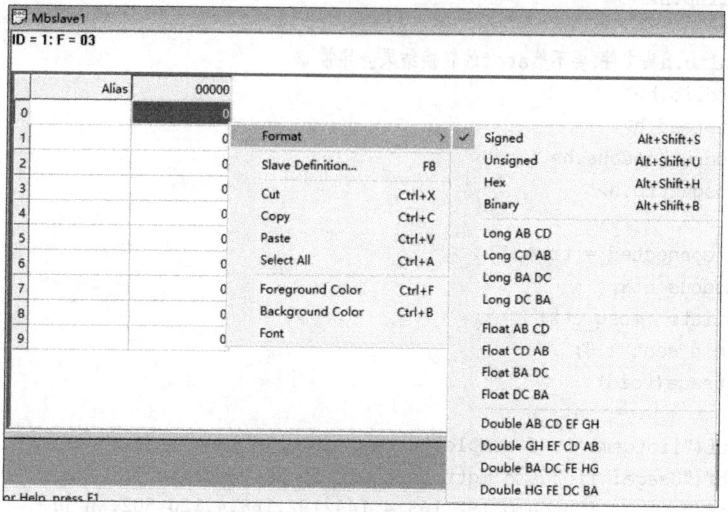

图14-20 在任意点位单击鼠标右键

● 对于图14-20中的菜单,需要重点介绍一下Format(数据格式)。

Signed:有符号16位数,数值范围为-32768~32767。

Unsigned:无符号16位数,数值范围为0~65535。

Hex:十六进制数,数值范围为0x0000~0xFFFF。

Binary:二进制数,数值范围为0000 0000 0000 0000~1111 1111 1111 1111;

由于大小端模式存在区别,不同的CPU处理Long、Float、Double这几个超过2个字节的数据类型时,高低位是不一样的。

■ 大端模式,是指数据的高字节保存在内存的低地址中,而数据的低字节保存在内存的高地址中,这样的存储模式类似于把数据当作字符串顺序处理:地址由小向大增加,而数据从高位往低位放;这和我们的阅读习惯一致。

■ 小端模式,是指数据的高字节保存在内存的高地址中,而数据的低字节保存在内存的低地址中,这种存储模式将地址的高低和数据位权有效地结合起来,高地址部分权值高,低地址部分权值低。

以unsigned int value = 0x1234为例,在内存中大小端模式的存储顺序是不一样的。表14-14展示出大小端模式存储顺序的区别。

表14-14 大小端模式存储顺序

内存地址	小端模式存放内容	大端模式存放内容
0x8000	0x34	0x12
0x8001	0x12	0x34

网关软件环境

为了实现完整演示,我们在接下来的示例中创建一个软件包iiot-demo。该软件包位于源代码chaos_calmer_15_05_1/package/utils/iiot-demo/下。

iiot-demo的主程序main.c的内容如下。

```c
#include <errno.h>
#include <stdio.h>
//必须引入stdlib.h头文件,要不然atof的转换结果会异常
#include <stdlib.h>
#include <pthread.h>
#include <modbus/modbus.h>
#include <mosquitto.h>

static bool connected = true;
modbus_t *modbus_ctx;
struct mosquitto *mosq_ctx;
static int mid_sent = 0;
void print_usage(void)
{
    printf("iiotdemo is a simple gateway demo.\n");
    printf("Usage: iiotdemo mqtt_host plc_ip plc_port.\n");
    printf("       iiotdemo 192.168.4.144 192.168.4.120 502.\n");
}
void my_message_callback(struct mosquitto *mosq, void *obj, const struct mosquitto_message *message)
{
    //接收服务器下发修改设备PLC的控制温度指令
    if(message->payloadlen != 0)
    {
        uint16_t temp_reg[2];
        float *temp = (float *)temp_reg;
        //将字符串转换为float写入PLC
        *temp = atof(message->payload);
        int rc = modbus_write_registers(modbus_ctx, 0, 2, temp_reg);
        if(rc == -1)
        {
            printf("%s\n", modbus_strerror(errno));
        }
    }
}
void my_connect_callback(struct mosquitto *mosq, void *obj, int result)
{
    //订阅服务器下发的控制指令
    mosquitto_subscribe(mosq, NULL, "com.freeiris.iiot.control.iiotdemo-12345", 1);
}
void my_disconnect_callback(struct mosquitto *mosq, void *obj, int rc)
{
    connected = false;
}
void my_publish_callback(struct mosquitto *mosq, void *obj, int mid)
{
}
void my_log_callback(struct mosquitto *mosq, void *obj, int level, const char *str)
```

```c
{
        //如果MQTT有异常,可以去掉下面的注释
        //printf("%s\n", str);
}
void plc_thread(void*param)
{
        int i;
        int rc;
        printf("plc_thread\n");
        while(connected)
        {
                //读者要注意,直接用下面的命令读取的时候,40001相当于第0个通道
                //读取PLC当前温度
                uint16_t temp_reg[2];
                rc = modbus_read_registers(modbus_ctx,0,2,temp_reg);
                if (rc == -1) {
                        printf("%s\n", modbus_strerror(errno));
                        return -1;
                }
                float *temp = (uint32_t *)temp_reg;
                //读取PLC当前产量
                uint16_t yields_reg[2];
                rc = modbus_read_registers(modbus_ctx,2,2,yields_reg);
                if (rc == -1) {
                        printf("%s\n", modbus_strerror(errno));
                        return -1;
                }
                uint32_t *yields = (uint32_t *)yields_reg;
                char payload[256] = {0};
                //%.2f的意思是保留2位小数
                sprintf(payload,"iiotdemo-12345,%.2f,%ld",*temp,*yields);
                mosquitto_publish(mosq_ctx,&mid_sent,"com.freeiris.iiot.report",strl
en(payload),payload,0,0);
                sleep(1);
        }
}
int main(int argc, char* argv[])
{
        if(argc != 4)
        {
                print_usage();
                return 1;
        }
        //定义Modbus实例,连接到PLC
        modbus_ctx = modbus_new_tcp(argv[2], atoi(argv[3]));
        modbus_set_slave(modbus_ctx, 1);
        //连接到PLC,如果无法连通,直接报错,返回1
        if (modbus_connect(modbus_ctx) == -1) {
                printf("can't connect to PLC!\n");
                return 1;
```

```c
        }
        /*将设备名称作为MQTT的CientID,ClientID当作网关的唯一识别ID,本实例直接用12345,读者也
可以尝试用GUID来解决重复问题*/
        char client_id[256] = {0};
        strcpy(client_id,"iiotdemo-12345");
        mosq_ctx = mosquitto_new(client_id, true, NULL);
        //如果失败,报错并返回
        if(!mosq_ctx){
                switch(errno){
                        case ENOMEM:
                                printf("Error: Out of memory.\n");
                                break;
                        case EINVAL:
                                printf("Error: Invalid id.\n");
                                break;
                }
                mosquitto_lib_cleanup();
                return 1;
        }
        //设置MQTT实例的回调函数
        mosquitto_log_callback_set(mosq_ctx, my_log_callback);
        mosquitto_connect_callback_set(mosq_ctx, my_connect_callback);
        mosquitto_disconnect_callback_set(mosq_ctx, my_disconnect_callback);
        mosquitto_publish_callback_set(mosq_ctx, my_publish_callback);
        mosquitto_message_callback_set(mosq_ctx, my_message_callback);
        //连接到MQTT服务器
        int rc = mosquitto_connect(mosq_ctx, argv[1], 1883, 60);
        if(rc == 0)
        {
                pthread_t plc_thread_id = 0;
                pthread_create(&plc_thread_id,NULL,(void *) plc_thread,NULL);
                //通过mosquitto_loop_forever函数,让MQTT保持连接状态,就可以收到消息代理下发
的内容
                rc = mosquitto_loop_forever(mosq_ctx, -1, 1);
                printf("exit iiot demo..........................\n");
                mosquitto_destroy(mosq_ctx);
                mosquitto_lib_cleanup();
        }
        else
        {
                printf("can't connect to mqtt server!\n");
                return 1;
        }
        return 0;
}
```

安装软件包。

```
DF1A:$ opkg update
DF1A:$ opkg install iiot-demo
```

启动实验环境

完整的实验环境如图14-21所示。在不同设备上启动实验环境的准备步骤如下。

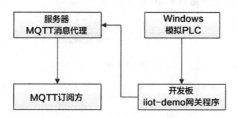

图14-21　完整的实验环境

● 计算机Windows系统：启动Modbus Slave模拟PLC设备，被开发板连接（本例IP为172.16.10.185）。

● 开发板：启动iiot-demo网关程序，连接模拟PLC与mosquitto消息服务器（本例IP为172.16.10.203）。

● MQTT服务器（消息代理）：启动mosquitto消息服务器，与开发板（本例IP为172.16.10.121）连接。

● MQTT订阅方：HOST中的订阅方程序。

在开发板中启动iiot-demo的方法如下。

```
DF1A:~$ iiot-demo 172.16.10.121 172.16.10.185 800
```

启动完成后，在HOST的MQTT服务器中可以看到连入信息。

```
HOST:mosquitto-1.4.15$ mosquitto
1577074495: mosquitto version 1.4.15 (build date 2019-12-23 10:13:04+0800) starting
1577074495: Using default config.
1577074495: Opening ipv4 listen socket on port 1883.
1577074495: Opening ipv6 listen socket on port 1883.
1577074497: New connection from 172.16.10.203 on port 1883.
1577074497: New client connected from 172.16.10.203 as iiotdemo-12345 (c1, k60).
```

在HOST中启动订阅方程序，检测网关上行消息。

```
#该程序启动后将发现每秒获取到一次上报
HOST:~$ mosquitto_sub -h 172.16.10.121 -t "com.freeiris.iiot.report"
iiotdemo-12345,0.00,0
iiotdemo-12345,0.00,0
......
```

网关与模拟PLC工作测试

改变模拟PLC数据的操作如下（修改Modbus Slave的值）。

● 在0通道值的位置单击鼠标右键，选择Format，选择Float CD AB模式，然后双击，将值修改为80.8，选择"OK"确定。

● 在2通道值的位置单击鼠标右键，选择Format，选择Long CD AB模式，然后双击，将值修改为2019，选择"OK"确定。

● Alias可以任意填写。

Modbus Slave值修改后的效果如图14-22所示。

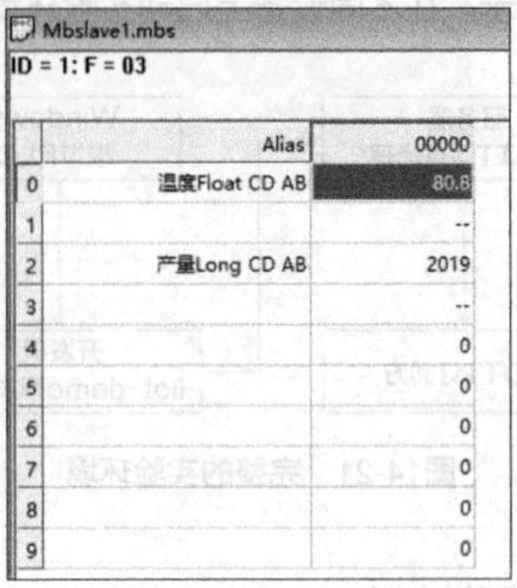

图14-22 Modbus Slave值修改后的效果

查看HOST中的订阅端程序，可以发现数据已经出现变化。

```
iiotdemo-12345,80.80,2019
iiotdemo-12345,80.80,2019
iiotdemo-12345,80.80,2019
```

接下来，在HOST中再开启一个终端控制PLC设备的温度。

```
#通过发布的方法将指令发送给iiot-demo,iiot-demo将完成控制Modbus Slave修改温度
HOST:~$ mosquitto_pub -h 172.16.10.121 -t "com.freeiris.iiot.control.iiotdemo-12345" -m "90.9"
```

通过iiot-demo修改PLC的温度所得的效果如图14-23所示。

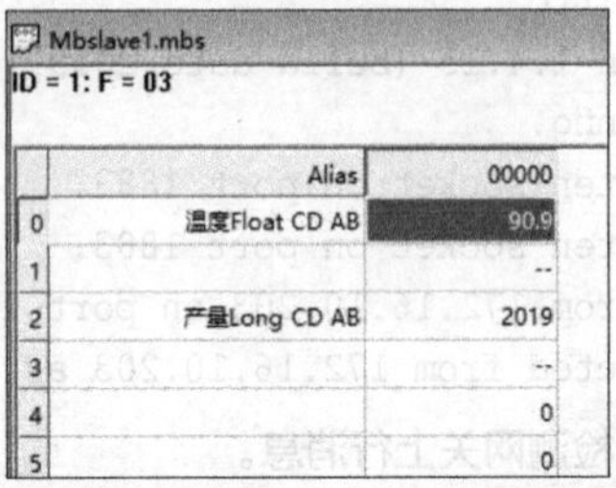

图14-23 通过iiot-demo修改PLC的温度所得的效果

至此，我们就完成了整个工业物联网网关的简单环境搭建。在网关的环境搭建中还有很多细节，以下指出其中的一部分，请有兴趣的读者进一步深入思考。

● MQTT的Client ID必须确保唯一性，例如通过硬件的SN、MAC地址或者GUID实现唯一性识别。

● PLC是通用的控制器，不同的设备参数存储在不同点位，应该通过配置文件对映射关系进行确认。

● 上位服务器通过订阅，可以将不同PLC中变化的数据存储到数据库中。

最后，提供两个开源的PLC通信SDK，读者可以尝试将它们移植到OpenWrt中：

● Snap7：西门子PLC的开源协议库。

● Libplctag：EIP开源协议库。